Applications of Interval Computations

Applied Optimization

Volume 3

*The titles published in this series are listed at the end of this volume.*

# Applications of Interval Computations

Edited by

R. Baker Kearfott
*University of Southwestern Louisiana*
and
Vladik Kreinovich
*University of Texas at El Paso*

KLUWER ACADEMIC PUBLISHERS

A C.I.P. Catalogue record for this book is available from the Library of Congress.

ISBN 0-7923-3847-2

---

Published by Kluwer Academic Publishers,
P.O. Box 17, 3300 AA Dordrecht, The Netherlands.

Kluwer Academic Publishers incorporates
the publishing programmes of
D. Reidel, Martinus Nijhoff, Dr W. Junk and MTP Press.

Sold and distributed in the U.S.A. and Canada
by Kluwer Academic Publishers,
101 Philip Drive, Norwell, MA 02061, U.S.A.

In all other countries, sold and distributed
by Kluwer Academic Publishers Group,
P.O. Box 322, 3300 AH Dordrecht, The Netherlands.

*Printed on acid-free paper*



Printed in the Netherlands

# CONTENTS

# CONTRIBUTORS


**Götz Alefeld**
Institut für Angewandte Mathematik
Universität Karlsruhe
D-76128 Karlsruhe, Germany,
email: `goetz.alefeld@mathematik.uni-karlsruhe.de`

**Wyllis Bandler**
Department of Computer Science
Florida State University
Tallahassee, Florida 32306-4019, USA

**Daniel Berleant**
Dept. of Computer Systems Engineering
University of Arkansas
Fayetteville, AR 72701, USA
email: `djb@engr.uark.edu`

**Charles L. Fefferman**
Department of Mathematics
Princeton University
Princeton NJ 08544, USA

**Sevgui Hadjihassan**
Laboratoire des Signaux et Systèmes
CNRS/ESE, Plateau de Moulon,
91192 Gif sur Yvette, France
email: `sevgui@lss.supelec.fr`

**Eero Hyvönen**
VTT Information Technology
P.O. Box 1201
02044 VTT Finland
email: `eero.hyvonen@vtt.fi`

**Max E. Jerrell**
College of Business Administration
Northern Arizona University
Flagstaff, AZ 86011-5066, USA
email: `jerrell@nauvax.ucc.nau.edu`

**R. Baker Kearfott**
Department of Mathematics
University of Southwestern Louisiana
U.S.L. Box 4-1010
Lafayette, LA 70504-1010, USA
email: `rbk@usl.edu`

**Ladislav J. Kohout**
Department of Computer Science
Florida State University
Tallahassee, Florida 32306-4019, USA
email: `kohout@cs.fsu.edu`

**Vladik Kreinovich**
Department of Computer Science
University of Texas at El Paso
El Paso, TX 79968, USA
email: `vladik@cs.utep.edu`

**Günter Mayer**
Fachbereich Mathematik
Universität Rostock
D–18051 Rostock, Germany
email: `guenter.mayer@mathematik.uni-rostock.de`

**Hung T. Nguyen**
Department of Mathematical Sciences
New Mexico State University
Las Cruces, NM 88003-8001, USA
email: hungueyn@nmsu.edu

**Stefano De Pascale**
VTT Information Technology
P.O. Box 1201
02044 VTT Finland
email: stefano.depascale@vtt.fi

**Luc Pronzato**
Laboratoire I3S, CNRS URA-1376
250 av. A. Einstein, Sophia Antipolis
66560 Valbonne, France
email: pronzato@mimosa.unice.fr

**Luis Mateus Rocha**
Department of Systems Science
and Industrial Engineering
T.J. Watson School
State University of New York
Binghamton, NY 13902, USA
email: ba05099@bingsuns.cc.binghamton.edu

**Jiří Rohn**
Faculty of Mathematics and Physics
Charles University
Malostranské nám. 25
118 00 Prague, Czech Republic
and
Institute of Computer Science
Academy of Sciences
Pod vodárenskou věží 2
182 07 Prague, Czech Republic
email: rohn@uivt.cas.cz

**Michael J. Schulte**
Department of Electrical and Computer Engineering
University of Texas at Austin
Austin, TX, 78712, USA
email: schulte@devil.ece.utexas.edu

**Luis A. Seco**
Department of Mathematics
University of Toronto
Toronto, Canada M5S 1A4
email: seco@math.toronto.edu

**Alexander L. Semenov**
Novosibirsk Division of the
Russian Research Institute of
Artificial Intelligence
pr. Lavrent'eva 6
Novosibirsk, Russia, 630090
email: semenov@iis.nsk.su

**Earl E. Swartzlander, Jr.**
Department of Electrical
and Computer Engineering
University of Texas at Austin
Austin, TX, 78712, USA

**G. William Walster**
SunSoft,
A Sun Microsystems, Inc. Business
2550 Garcia Avenue, MS UMTV12-40
Mountain View, CA 94043-1100, USA
email bill.walster@Eng.Sun.Com

**Eric Walter**
Laboratoire des Signaux et Systèmes
CNRS/ESE, Plateau de Moulon
91192 Gif sur Yvette, France
email: walter@lss.supelec.fr

# PREFACE

## Primary Audience for the Book

- Specialists in numerical computations who are interested in algorithms with automatic result verification.
- Engineers, scientists, and practitioners who desire results with automatic verification and who would therefore benefit from the experience of successful applications.
- Students in applied mathematics and computer science who want to learn these methods.

## Goal Of the Book

This book contains surveys of applications of interval computations, i.e., applications of numerical methods with automatic result verification, that were presented at an international workshop on the subject in El Paso, Texas, February 23–25, 1995. The purpose of this book is to disseminate detailed and surveyed information about existing and potential applications of this new growing field.

## Brief Description of the Papers

At the most fundamental level, interval arithmetic operations work with sets: The result of a single arithmetic operation is the set of all possible results as the operands range over the domain. For example, $[0.9, 1.1] + [2.9, 3.1] = [3.8, 4.2]$, where $[3.8, 4.2] = \{x + y | x \in [0.9, 1.1] \text{ and } y \in [3.8, 4.2]\}$. The power of interval arithmetic comes from the fact that (i) the elementary operations and standard functions can be computed for intervals with formulas and subroutines; and (ii) *directed roundings* can be used, so that the images of these operations (e.g.

$[3.8, 4.2]$ in the preceding example) ***rigorously*** contain the ***exact*** result of the computations. These facts allow

- rigorous enclosure of roundoff error, truncation error, and errors in data;
- computation of rigorous bounds on the ranges of functions.

In fact, interval arithmetic is a convenient and effective means of obtaining information used in place of Lipschitz constants. Such information, combined with bounds on the ranges of functions that interval arithmetic supplies, is widely recognized as valuable in algorithms for global optimization. In "A Review of Techniques in the Verified Solution of Constrained Global Optimization Problems," Kearfott reviews the literature, including several books on the subject, outlines some of the basic techniques, including some of his own experimental ones, and gives advice.

A related fundamental problem, the numerical analysis of linear systems of equations, is treated in the articles by Alefeld et al. and by Rohn. Interval arithmetic computations, naively arranged, sometimes implicitly assume independent variation among quantities among which there actually are relationships. This interval dependency leads to overestimation (i.e. lack of sharpness) in computed solution sets. For example, handling a symmetric linear system with uncertainty in the coefficients as an arbitrary, unsymmetric system leads to such interval dependency. In "The Shape of the Symmetric Solution Set," Alefeld, Kreinovich and Mayer describe the exact solution set to symmetric linear systems as an intersection of linear and quadratic inequalities. In "Linear Interval Equations: Computing Enclosures with Bounded Relative Or Absolute Overestimation is NP-Hard," Jiří Rohn shows that there is no polynomial time algorithm that computes bounds on the solution sets to *all* interval linear systems with overestimation less than $\delta$, for any $\delta > 0$. This result can guide algorithm developers. Fortunately, heuristic methods such as interval Gaussian elimination or the interval Gauss–Seidel method work well for many systems in practice, when the widths of the coefficient intervals are small and also in many other cases.

Hadjihassan et al. and Jerrell use these fundamental numerical analysis considerations in manufacturing design and in economic input-output models, respectively. In both works, uncertainties in the input data are encompassed by describing them as intervals. In "Quality Improvement via the Optimization of Tolerance Intervals During the Design Stage," Hadjihassan, Walter and Pronzato use a combination of interval and non-interval techniques to optimize the tolerance of manufactured items, subject to uncertainties in the inputs. In "Applications of Interval Computations to Regional Economic Input-Output Models," Jerrell effectively uses fundamental properties of interval linear systems in Leontief input/output models. Jerrell thus obtains meaningful bounds for the impact of Northern Arizona University on the economy of Coconino County, Arizona.

Related to global optimization, computational existence and uniqueness proofs are possible with interval computations. Fefferman and Seco's contribution, "Interval Arithmetic in Quantum Mechanics" is an example of a computer-assisted proof outside of general global optimization algorithms. Fefferman and Seco discuss the role of interval arithmetic in establishment of an inequality related to a precise asymptotic formula for the ground state of a non-relativistic atom.

Also related to global optimization is ***constraint propagation***, increasingly popular in recent years in fields such as robotics and computer-aided geometric design, to verify intersection or non-intersection, etc. Several authors have also applied the technique in general global optimization and nonlinear equation systems codes and software. In this technique, sometimes implemented in languages such as Prolog, relationships among intermediate quantities in arithmetic expressions in constraints are used recursively to compute ever-narrower bounds on solution variables. This volume contains two software systems that allow such computations. In "Interval Computations on the Spreadsheet," Hyvönen and De Pascale explain an extension of Microsoft Excel that allows interval computations, and constraint propagation in particular. In "Solving Optimization Problems with Help of the UniCalc Solver," Semenov describes an integrated interactive environment, available commercially, that uses subdefinite programming, related to interval constraint propagation, to solve various types of optimization problems.

Experts in statistics and in fuzzy logic have found the ability of intervals to represent uncertainties valuable. Various higher-order structures have been defined and utilized. In "Automatically Verified Arithmetic on Probability Distributions and Intervals," Berleant surveys statistical applications of interval computations, while the surveys "Nested Intervals and Sets: Concepts, Relations to Fuzzy Sets, and Applications" by Nguyen and Kreinovich, "Fuzzy Interval Inference Utilizing the Checklist Paradigm and BK-Relational Products" by Kohout and Bandler, and "Computing Uncertainty in Interval Based Sets" by Rocha, Kreinovich, and Kearfott describe applications to fuzzy logic and expert systems.

In the past, widespread application of interval computations has been limited by availability of hardware and software support. Also, experts in interval computations have urged the development of faster, hardware-oriented systems. In "Software and Hardware Techniques for Accurate, Self-Validating Arithmetic," Schulte and Swartzlander review some of the available programming language extensions, packages and problem solving environments for interval computations. They then discuss hardware design alternatives and propose a specific hardware design and software interface for variable precision interval arithmetic. In "Stimulating Hardware and Software Support for Interval Arithmetic," Walster presents his view of why many hardware and software vendors do not presently provide interval computations as integral parts of their products. He then discusses steps the community of interval computations experts should take to make such capabilities more widely available.

Finally, our introduction gives a more detailed elementary explanation of the fundamentals of interval arithmetic, as well as lengthier elementary explanations of the surveys that follow.

## Our Thanks:

- to all the authors for their excellent job;
- to the anonymous referees for their thorough and unrewarding job of reviewing the papers: refereeing was especially tough because of the strict deadlines;
- to Kluwer Academic Publishers, especially to John Martindale, who proposed the idea of this book and helped us make it happen, and to his helpful assistant Arlene Apone;
- to Professor Pardalos, who kindly agreed to publish this volume in the optimization series that he is supervising, and who encouraged us in this endeavor;
- to the sponsors whose support made the workshop possible:
  - to the *Institute of Manufacturing and Materials Management*;
  - to the *University of Texas at El Paso*;
  - to the journal of *Reliable Computing*;
- to the organizations that helped us by publishing the information about this workshop:

- to the *American Mathematical Society*;
- to the *Special Interest Group on Artificial Intelligence (SIGART)* of the *Association for Computing Machinery (ACM)*;
- to the *Computer Society (CS)* of the *Institute of Electrical and Electronic Engineers (IEEE)*;

- to the granting agencies: this work was supported in part by:
  - the *National Science Foundation (NSF)*, through grants CCR-9203730, CDA-9015006, and EEC-9322370, and
  - the *National Aeronautics and Space Administration (NASA)*, through grant No. NAG 9-757.

- and, last but not the least, thanks to you, readers of this book, because it is for you that this book has been composed.

*Thanks to all of you!*

R. Baker Kearfott and Vladik Kreinovich
August, 1995

# 1

# APPLICATIONS OF INTERVAL COMPUTATIONS: AN INTRODUCTION

**R. Baker Kearfott* and Vladik Kreinovich****

**Department of Mathematics, University of Southwestern Louisiana, U.S.L. Box 4-1010, Lafayette, LA 70504-1010, USA, email: rbk@usl.edu*

***Department of Computer Science, University of Texas at El Paso, El Paso, TX 79968, email vladik@cs.utep.edu*

## ABSTRACT

The main goal of this introduction is to make the book more accessible to readers who are not familiar with interval computations: to beginning graduate students, to researchers from related fields, etc. With this goal in mind, this introduction describes the basic ideas behind interval computations and behind the applications of interval computations that are surveyed in the book.

## 1 WHAT ARE INTERVAL COMPUTATIONS?

### *Most Real-Life Application Problems Lead to (Constrained or Unconstrained) Global Optimization*

In a typical real-life application problem, we need to find an alternative $x$ that satisfies given constraints and that optimizes the value of a given characteristic $y = f(x)$ (the optimized function $f(x)$ is usually called an *objective function*). For example:

- in *data processing*, we must find a model $x$ that best fits the known data; in this case, $f(x)$ describes how well $x$ fits;



*R. B. Kearfott and V. Kreinovich (eds.), Applications of Interval Computations*, 1–22.

- in *control*, we must find the control strategy $x$ that optimizes the given criterion $f(x)$ (e.g., smoothness of the resulting trajectory);
- in *decision making*, we must find an alternative $x$ from a given set $X$ of possible alternatives that guarantees the best outcome $f(x)$.

From the mathematical viewpoint, all these real-life application problems can be described as (constrained and unconstrained) *global optimization problems.*

## *In Some Cases, We Must Solve Systems of Equations and/or Inequalities*

In some real-life problems, there are so many constraints on the problem that even finding a single solution $x$ that satisfies all these constraints is a difficult task. This often happens in design problems; e.g., if we want to design a spacecraft that can fulfill the task of taking photos of a distant planet under restrictions on weight and cost. For these problems, there is no question of a choice between several possible solutions (we are lucky if we have one), and therefore, it makes no practical sense to choose an objective function for this choice. Therefore, such problems are formulated as the problems of satisfying given constraints. Each of these constraints is either an inequality (e.g., "total cost $\leq$ a given amount"), or an equality (e.g., "at a given moment of time, the position of the spacecraft must be exactly a given number of miles above the explored planet"). Therefore, in mathematical terms, these problems are formulated as solution of a system of equations and/or inequalities.

In other cases (e.g., in economic problems), we know the constraints well, but the objective is only informally formulated (e.g., "welfare of the nation"). In these cases, it is difficult to formulate the objective function in precise mathematical terms; therefore, it makes sense to provide the decision-makers with the entire set of the alternatives $x$ that satisfy all given constraints, so that these decision-makers will make the final decision. In such problems, we also need to solve a system of equations and/or inequalities.

## *Numerical Algorithms Usually Result in Approximate Solutions*

There exist many numerical algorithms for solving optimization problems (see, e.g., [50]). Algorithms also exist for solving systems of equations and inequalities.

The majority of these algorithms are iterative, so, when we stop after a certain number of steps, we only get an *approximation* $\tilde{x}$ to the desired solution $x$.

## *For Many Practical Cases, We Need Guaranteed Estimates of the Solution's Accuracy*

A natural question is: how good is the approximate solution $x$? That is, how close is the approximate solution to the desired solution $x$?

In many practical problems, we need a *guaranteed* solution, i.e., a solution that is guaranteed to be sufficiently close to an optimum (i.e., for which the distance $\rho(\tilde{x}, x)$ between $x$ and $\tilde{x}$ does not exceed a certain given number $\varepsilon$). To get such a guarantee, we definitely need to estimate the closeness $\rho(\tilde{x}, x)$.

## *Traditional Error Estimating Techniques of Numerical Mathematics Do Not Always Lead to Satisfactory Estimates*

In the majority of traditional numerical methods, at each step of the iteration process, we get a number (or, a sequence of numbers) that represents an increasingly better approximation to the solution. After we compute this approximate solution, we use some estimates (e.g., estimates based on Lipschitz constants) to describe how close $\tilde{x}$ is to $x$. The resulting method does not always lead to good estimates:

- For some numerical algorithms, *no methods are known* yet for estimating the distance $\rho(\tilde{x}, x)$. Designing such an estimation technique for each new algorithm is usually a very complicated problem. Heuristic algorithms are constantly appearing, and there are simply not enough researchers in the numerical analysis community to find estimates for all of them.

- Error estimates of numerical mathematics are usually ***overestimating***. One of the reasons for that is that these methods are usually based only on the final result $\tilde{x}$ of the computations. We could get better estimates if we could use more information about the behavior of the function $f(x)$. To get this information, we either need to do more numerical experiments with the function $f$, or we can use the values that have been computed during the previous computation steps. In both cases, we need ***extra computation time***, and this computation time is added to the running time of the numerical algorithm itself, which is often already large.

## An Alternative Error Estimation Approach – A Natural Idea

An ideal situation would be if we could estimate the errors of the result not ***after*** the iteration process (thus, adding time), but ***simultaneously*** with the iteration process. This is the main idea behind the ***interval computations*** methodology originally proposed by R. E. Moore [13, 14, 15].

This idea will help us solve the above-mentioned problems:

- First, it will enable us to estimate errors for the approximate solution while computing the solution itself, thus saving ***computation time***.
- Second, since the error estimation will take place throughout the entire process of computing $\tilde{x}$, we will be able to use intermediate results in this error estimation and therefore, hopefully, come up with better estimation results (i.e., results that are ***less overestimating***).
- Finally, this idea will enable us to easily find estimates for new algorithms: Indeed, it is difficult to compute an error estimate for the result of a lengthy computation. However, each lengthy computation consists of a sequence of elementary steps ($+$, $-$, $\cdot$, $/$, ...). Thus, if we trace this error step-by-step, then this complicated problem is reduced to the problems of estimating the error of a sum (product, etc.) of two previous quantities, when we already know the errors in each quantity.

## Why "Interval"?

If we follow the idea described above, then on each computation step, instead of a single (approximate) number $\tilde{r}$, we will compute $\tilde{r}$ ***and*** an upper bound

$\Delta$ for the error of this approximation. In other words, after each computation step, we will get a *pair* $(\tilde{r}, \Delta)$.

If we know $\tilde{x}$ and $\Delta$, this means that the actual value $r$ of the corresponding quantity can take any value from $\tilde{r} - \Delta$ to $\tilde{r} + \Delta$. In other words, this pair means that instead of the (ideal) single value of $r$, there is a *set* of possible values of $r$, and this set is an *interval* $\mathbf{r} = [\underline{r}, \overline{r}]$, where $\underline{r} = \tilde{r} - \Delta$ and $\overline{r} = \tilde{r} + \Delta$.

Because of this, the above idea is called *interval computations.*

## *Basic Formulas of Interval Arithmetic*

For intervals, the above-mentioned step-by-step process works as follows: at each elementary computation step, we perform one of the arithmetic operations $c := a \odot b$, where the two values $a$ and $b$ have already estimated on the previous steps of this algorithm. Since the values $a$ and $b$ have already been estimated by our algorithm, this means that we already have the *intervals* $[\underline{a}, \overline{a}]$ and $[\underline{b}, \overline{b}]$ of possible values of $a$ and $b$. Since we know that $a \in [\underline{a}, \overline{a}]$ and $b \in [\underline{b}, \overline{b}]$, we can conclude that $c$ belongs to the set of all possible values of $a \odot b$ for such $a$ and $b$, i.e., to the set

$$\{a \odot b | a \in [\underline{a}, \overline{a}], \text{and} b \in [\underline{b}, \overline{b}]\}.$$

**Example.** If $\odot = +$, and if we know that $a \in [0.9, 1.1]$ and $b \in [2.9, 3.1]$, then we can conclude that

$$a + b \in \{a + b | a \in [0.9, 1.1] \text{and} b \in [2.9, 3.1]\}.$$

One can easily see that this set is the interval $[3.8, 4.2]$.

In general, the new set (of possible values of $c$) is called the *result of applying the operation* $\odot$ *to intervals* $[\underline{a}, \overline{a}]$ *and* $[\underline{b}, \overline{b}]$, and it is denoted by

$$[\underline{a}, \overline{a}] \odot [\underline{b}, \overline{b}].$$

For basic arithmetic operations, it is easy to find explicit expressions for the resulting set:

- $[\underline{a}, \overline{a}] + [\underline{b}, \overline{b}] = [\underline{a} + \underline{b}, \overline{a} + \overline{b}]$;
- $[\underline{a}, \overline{a}] - [\underline{b}, \overline{b}] = [\underline{a} - \overline{b}, \overline{a} - \underline{b}]$;
- $[\underline{a}, \overline{a}] \cdot [\underline{b}, \overline{b}] =$ $[\min(\underline{a} \cdot \underline{b}, \underline{a} \cdot \overline{b}, \overline{a} \cdot \underline{b}, \overline{a} \cdot \overline{b}), \max(\underline{a} \cdot \underline{b}, \underline{a} \cdot \overline{b}, \overline{a} \cdot \underline{b}, \overline{a} \cdot \overline{b})]$;
- $[\underline{b}, \overline{b}]^{-1} = [1/\overline{b}, 1/\underline{b}]$ if $0 \notin [\underline{b}, \overline{b}]$;
- $[\underline{a}, \overline{a}]/[\underline{b}, \overline{b}] = [\underline{a}, \overline{a}] \cdot [\underline{b}, \overline{b}]^{-1}$.

These operations are called *interval arithmetic*. To these operations, we must add interval analogues of standard functions (exp, sin, max, etc).

For iterative algorithms, we can also take into consideration the fact that, in these algorithms, we repeatedly compute estimates for one and the same quantity. Usually, we use a previous estimate $x^{(k)}$ to compute the next (usually, better) estimate $x^{(k+1)}$. If we use interval operations, then on these two successive steps, we get *intervals* $\mathbf{x}^{(k)}$ and $\mathbf{x}^{(k+1)}$. Since both intervals contain the desired value $x$, we can conclude that $x$ belongs to the *intersection* of these intervals. Therefore, as an interval that contains $x$, on the next step of interval computations, we can take not the interval $\mathbf{x}^{(k+1)}$, but the (usually smaller) interval $\mathbf{x}^{(k+1)} \cap \mathbf{x}^{(k+1)}$.

## *These Formulas Must be Modified to Take Roundoff Errors into Consideration*

The formulas of interval arithmetic are only valid for idealized computers, in which all arithmetic operations are performed exactly. Real-life computers truncate or round the result of multiplication and division, thus adding additional *roundoff* errors. To take these errors into consideration, we must modify the above formulas by using the so-called *directed roundings*, i.e.:

- a rounding $\phi_-(r)$ that always returns a lower bound $\phi_-(r) \le r$, such that $\phi_-(r) \le r$ is a machine number; and
- a rounding $\phi_+(r)$ that always returns an upper bound $\phi_+(r) \ge r$, such that $\phi_+(r) \le r$ is a machine number.

For example, the modified formula for $[\underline{b}, \overline{b}]^{-1}$ is (for $0 \notin [\underline{b}, \overline{b}]$):

$$[\underline{b}, \overline{b}]^{-1} = [\phi_-(1/\overline{b}), \phi_+(1/\underline{b})].$$

## *This Methodology can Also Take into Consideration Uncertainty in the Input Data*

In many real-life situations, the input data come from measurements. Measurements are not 100% precise. Therefore, if we have made the measurement with an accuracy $\Delta$, and obtain the measurement result $\tilde{y}$, this means that the actual value $y$ of the measured quantity can take any value from the interval $\mathbf{y} = [\tilde{y} - \Delta, \tilde{y} + \Delta]$. The desired solution $x$ depends on the exact value of $y$ from this interval. Hence, it makes sense to produce the set of all possible solutions $x$ that correspond to all possible values $y \in \mathbf{y}$.

To generate this estimate, we can start with the intervals for input data (instead of the precise values), and then follow interval computations step-by-step.

## *Brief Summary: The Power of Interval Computations*

The power of interval arithmetic comes from the following facts:

1. The elementary operations and standard functions (such as sin, cos, and exp, found in standard Fortran) can be computed for intervals by using simple formulas and subroutines; and
2. *directed roundings* can be used, so that the images of these operations (e.g. $[3.8, 4.2]$ in the preceding example) *rigorously* contain the *exact* result of the (ideal) computations.

These facts allow

- rigorous enclosure of roundoff error, truncation error, and errors in data;
- computation of rigorous bounds on the ranges of functions.

## *Naive and Sophisticated Interval Computations*

Interval computations were first applied (in the 1960's) *naively*, that is, to existing numerical algorithms. In some cases, reasonable estimates appeared as a result. In other cases, the resulting intervals were too wide to be useful. These overestimates made many researchers and practitioners reluctant to use interval methods. Many researchers, not familiar with the field, are still under the impression that interval computations are not a useful tool.

However, in the last three decades, many ***sophisticated*** ideas and algorithm modifications have been proposed that enable replacing the excessively wide intervals of naive interval computations with reasonably narrow ones. In some cases, radically new algorithms emerged that were specifically tailored for interval computations. For a general survey of basic modern interval techniques, see, e.g., [6]; basic interval optimization techniques are described in [7].

As a result of these improvements, modern interval computations have been successfully ***applied*** to many real-life problems.

# 2 INTERNATIONAL WORKSHOP ON APPLICATIONS OF INTERVAL COMPUTATIONS: HOW THIS BOOK ORIGINATED

To promote real-life applications of interval computations, an International Workshop on Applications of Interval Computations (APIC'95) was organized in El Paso, Texas, on February 23–25, 1995. For this workshop, 80 researchers from 15 countries (Austria, Brazil, Canada, China, Czech Republic, Denmark, Finland, France, Germany, Mexico, Poland, Russia, Spain, Ukraine, and the USA) submitted papers describing numerous applications of interval computations:

- To *Engineering*:
  - to manufacturing, including:
    - quality control;
    - detection of defects in computer chips;
    - flexible manufacturing;
  - to automatic control, including:
    - control of airplane engines;
    - control of electric power plants;
  - to robotics;

  - to airplane inertial navigation;
  - to civil engineering, including traffic control.

- To *Ergonomics* and *Social Sciences*:
  - to learning curves (that describe how people learn);
  - to project management;
  - to service systems;
  - to sociology.

- To Physics:
  - to laser beams;
  - to particle accelerators;
  - to astrophysics (planet atmospheres);
  - to image processing in radioastronomy.

- To *Geology* and *Geophysics*.

- To *Chemistry*, including:
  - to spectral analysis.

- To *Computer Science and Engineering*:
  - to expert systems;
  - to communication networks, especially computer networks.

- To *Economics*:
  - to planning;
  - to banking.

Extended abstracts of these presentations appeared in [12]. The majority of these papers described specific applications. (Technical details of different specific applications will also appear in the special issue of the international journal *Reliable Computing*.)

In addition to the papers describing technical details of specific applications, we also had several general survey papers, of general interest to the public. It is these surveys that we present in this book.

*This book contains surveys of the applications of interval computations, i.e., applications of numerical methods with automatic result verification.*

# 3 GENERAL OPTIMIZATION PROBLEMS

## *A General Survey*

Interval arithmetic is a convenient and effective means of obtaining information used in the place of Lipschitz constants. Such information, combined with bounds on the ranges of functions that interval arithmetic supplies, is is widely recognized as valuable in algorithms for *global optimization*. In this book, a general survey of such algorithms, titled "A Review of Techniques in the Verified Solution of Constrained Global Optimization Problems" [10] is given by Kearfott. In this survey, Kearfott:

- reviews the literature on the subject (including several books);
- outlines the basic techniques (including some of his own experimental ones); and
- gives advice on the use of different techniques.

## *Optimization Techniques Based on Constraint Propagation*

Traditionally, before applying numerical techniques to a problem, we analyze it to formulate a compact and understandable description. Namely:

- First, we describe all the physical properties characterizing the object, and all the equations and inequalities that describe relations between these properties.
- Usually, some of these variables can be expressed in terms of the others. As a result, we can express these variables (called *dependent*) in terms of the remaining few *independent* variables. We can then reformulate all the restrictions in terms of these independent variables.

Thus, we obtain a constrained optimization problem $f(x) \to \max$ with few variables $x = (x_1, ..., x_n)$ and with (often rather complicated) objective function and constraints. (They are complicated because they come from repeated substitution of expressions into other ones, and each such substitution increases the complexity of the resulting expression.)

This reduction is useful, because the running time of existing global optimization techniques grows rapidly with the number of variables, so, the fewer variables we have, the faster we can find the optimum.

The reduction itself, however, is time-consuming. It makes sense if we have many similar problems. In this case, we do this reduction once, for a generic problem, and then use the results of this reduction to save time on all these problems.

There are fields, however, such as robotics and computer-aided geometric design, in which preliminary reduction is not possible. In problems so arising, many variables describe the system: e.g.:

- To describe the state of a robot, we need to know the coordinates and velocities of all its joints, as well as the coordinates of all the points of the geometric environment in which the robot currently operates.
- To characterize a design, we also usually need many design parameters.

Not all of these characteristics are independent: the lengths of the joints restrict their respective positions; design demands impose heavy restrictions on the design parameters so that only few of them remain independent, etc. So, in all these cases, we can apply reduction to get an optimization problem with fewer variables. However, we cannot make these reductions once and forever, because the problems are constantly changing:

- For a manufacturing robot, the problem drastically changes if a new part is added.
- For a design problem, the reduction changes every time we add or delete one of the design requirements. (Such change in design requirements is typical for a design process.)

In these cases, we should solve unreduced optimization problems, i.e., problems in which there are many different variables, but also many constraints that in effect leave only few of the variables independent. Because of the large number of variables, general optimization techniques do not behave well for such problems, so we need new methods.

The corresponding technique is called *constraint propagation*. In this technique, relationships among intermediate quantities in arithmetic expressions in constraints are used recursively to compute ever-narrower bounds on solution variables.

It turns out that constraint propagation can be ***successfully applied*** not only to the above-mentioned problems like robotics, in which symbolic reduction is not practical, but also to ***general*** optimization and equation-solving problems. Often, even if there is a way to reduce the number of constraints, it is actually advantageous to put in more constraints (but make these constraints simpler) and then apply constraint propagation techniques. Thus, constraint propagation is a new powerful general optimization tool.

Several authors have applied the technique in global optimization and nonlinear equation systems codes and software. It turns out that the

In the survey "Solving Optimization Problems with Help of the UniCalc Solver" [21], presented in this book, Semenov describes an integrated interactive environment, available commercially, that uses so-called ***sub-definite programming*** (a generalization of interval computations to sets more general than intervals) to solve various types of optimization problems.

## *Two Surveys of Specific Techniques*

In addition to the general surveys of different interval optimization techniques, the book also contains two surveys of interval optimization techniques tailored to problems in two specific fields: manufacturing (Hadjihassan et al [5]) and quantum mechanics (Fefferman and Seco [39]).

## *Applications to Manufacturing*

In "Quality Improvement via The Optimization Of Tolerance Intervals During The Design Stage" [5], Hadjihassan et al use the fundamental numerical analysis considerations described above in manufacturing design. Uncertainties in the input data are encompassed by describing them as intervals. Namely, in [5], Hadjihassan, Walter and Pronzato use a combination of interval and non-interval techniques to solve the following real-life optimization problem: In mass manufacturing, we can usually only reproduce the parameters of the manufactured objects with a certain accuracy $\delta$ (typically, 5% or 10%). As a result, when we fix the nominal values $x_i$ of these parameters, we can only guarantee that the actual values of the parameters of the manufactured objects are inside the interval $[x_i - \delta \cdot x_i, x_i + \delta \cdot x_i]$. Hence, even if we choose nominal design parameters $x = (x_1, ..., x_n)$ that guarantee the desired value $y^*$ of the performance characteristic $y = \eta(x_1, ..., x_n)$, the actual values of $y$ may differ from $y^*$. The quality $f(x)$ of a design can be characterized, crudely speaking, by the worst case deviation of $y$ from $y^*$. In Hadjihassan et al [5], interval computations help to solve the problem of finding the optimal (best quality) design, i.e., a design $x$ for which $f(x)$ is minimum.

## *Applications to Quantum Mechanics*

Many optimization problems occur in fundamental physics. Actually, fundamental physical theories are usually formulated not in terms of differential equations (as in Newton's time), but in terms of an appropriate ***variational principle***: The characteristics $x$ of the particles and fields must be chosen to optimize the value of an objective function $S(x)$ (called ***action***). This use of optimization is not simply a mathematical trick: it has a fundamental justification in quantum mechanics (see, e.g., [4]).

Because of the optimization formulation of fundamental physical theories, numerical optimization techniques, especially interval-type techniques that provide automatic verification, appear to be of great use to physicists.

The majority of unsolved fundamental problems in physics, however, are mainly related to areas such as quantum field theory, (quantum) cosmology, etc, for which the equations are not yet confidently known. Since we are not sure about the equations anyway, there is not much need for a ***guaranteed*** solution. Therefore, physicists use heuristic techniques, and additional computational efforts needed to estimate the accuracy do not seem to be justified.

There is, however, a class of fundamental problems where the corresponding optimization criterion (and related equations) have been known for over 70 years, and the main problems are computational: these are the problems of non-relativistic quantum mechanics. The fundamental equation was described by Schroedinger as early as 1924; this equation describes all the properties of the simplest atoms (hydrogen and helium) with such a good accuracy that physicists were convinced that what was previously known as chemistry would be replaced by fundamental quantum physics. This did not happen because, to describe an atom with $n$ electrons, we need to find a function of $3n$ variables (by solving an optimization problem or an equivalent differential equation). Even if we take 2 possible values of each of these coordinates, we still need $2^{3n}$ values only to describe function. For realistic $n$, this problem is clearly computationally intractable.

To solve this problem, several approximate equations and optimization formulations have been proposed:

- Some of them are computationally easy, but lead to very crude (and thus practically useless) estimates.
- Other approximate equations lead to better estimates, but are still relatively computationally complicated. One such approximation, called *Thomas–Fermi* theory, is considered in the survey "Interval Arithmetic in Quantum Mechanics" [39] by Fefferman and Seco, that is presented in the book.

We are interested in the characteristics of the atoms with large $n$ (because for small $n$, we can simply solve the exact equations). It is known that for large very $n$, these characteristics tend to behave in a regular fashion, so, to get a good estimate, we can expand the values of the characteristics into an asymptotic series in $1/n$, and take only a few terms in this expansion. Hence, we need to find ***asymptotic formulas*** for $n \to \infty$.

Several heuristic optimization methods have been proposed to find the asymptotics of the optimum in this problem. These methods, however, are known to produce good results for some physical problems, but poor estimates for others. Therefore, there is a real need to apply techniques with automatic result verification, and thus get guaranteed estimates on the asymptotics.

Such estimates are obtained in a survey [39] by Fefferman and Seco. Their contribution is thus an example of a computer-assisted proof outside of general global optimization algorithms. Fefferman and Seco explicitly discuss the role of interval arithmetic in establishment of an inequality related to a precise asymptotic formula for the ground state of a non-relativistic atom.

# 4 GENERAL SYSTEMS OF EQUATIONS AND INEQUALITIES

In this book, a survey of ***constraint propagation*** techniques for solving such systems is presented. In this technique, sometimes implemented in languages such as Prolog, relationships among intermediate quantities in arithmetic expressions in constraints are used recursively to compute ever-narrower bounds on solution variables. As we have already mentioned, constraint propagation is increasingly popular in recent years in fields such as robotics and computer-aided geometric design. Systems of equations that it allows us to solve can tell, e.g., whether there is an intersection or non-intersection, e.g.:

- whether after a planned movement the robot will hit the obstacle, or move safely;
- whether a cutting instrument of a manufacturing robot will actually start cutting the desired part;
- whether in a computer design all components of the car's design will actually fit into the car frame (intersection in this case will mean that they won't), etc.

Several authors have applied this technique in general global optimization and nonlinear equation systems codes and software.

This volume contains descriptions of two software systems that allow such computations. We have already mentioned a survey by Semenov [21]. In "Interval Computations on the Spreadsheet" [7], Hyvönen and De Pascale explain an extension of Microsoft Excel that allows interval computations, and constraint propagation in particular.

# 5 LINEAR INTERVAL PROBLEMS

## *Why Linear Problems?*

In many real-life situations, the general problem (outlined in the beginning of this introduction) can be simplified. A typical situation in which simplification is possible is when we know that the components $x_1$, .., $x_n$ of the desired solution $x$ are relatively small. Then, we can expand all the terms in all the equations that determine $x$ into power series in $x_i$, and neglect quadratic and higher order terms.

As a result, we get a problem with:

- a linear objective function $y = f(x) = \sum c_j x_j$, and
- linear constraints: i.e.,
  - linear equalities $\sum a_{ij} x_j = b_j$, and/or
  - linear inequalities $\sum a_{ij} x_j \geq b_i$.

This problem is called a *linearization* of the original non-linear problem.

In other words, in many real-life situations, we must solve linear problems. If all the coefficients $a_{ij}$, $b_i$, and $c_j$ of the corresponding linear functions are precisely known, then we have a *linear programming* problem for which many efficient algorithms are known.

If we do not know the objective functions, then, as we have mentioned, we must describe the set $X$ of all the vectors $x$ that satisfy the given linear constraints. This description can also be given by linear programming.

## Interval Linear Problems

In many real-life problems, the only source of the values of the coefficients $a_{ij}$ and $b_i$ is measurement. Since measurements are never absolutely precise, as a result of these measurements, we only get *intervals* $\mathbf{a}_{ij}$ and $\mathbf{b}_j$ that contain the (unknown) values of these coefficients. In this case, we are interested in describing the set $X$ of all the vectors $x$ that satisfy the given linear constraints, for *some* $a_{ij} \in \mathbf{a}_{ij}$ and $b_j \in \mathbf{b}_j$.

## How to Describe the Solution of an Interval Linear Problem

For symmetric and non-symmetric $a_{ij}$, the set $X$ described above is characterized in the paper "The Shape of the Symmetric Solution Set" [1] by Alefeld, Kreinovich, and Mayer, presented in this book. Namely, this set is an an intersection of half-spaces (sets described by linear inequalities) and quadrics (i.e., sets defined by quadratic inequalities).

## General Interval Linear Problems are Hard to Solve

The description of a solution of an interval linear problem provided in Alefeld et al [1] is computationally complicated. A natural question is: is it really a hard problem and no easy algorithm is possible, or there is a simple algorithm, but we simply have not found it yet?

The answer to this question is given in a paper by Rohn "Linear Interval Equations: Computing Enclosures with Bounded Relative Or Absolute Overestimation is NP-Hard" [19], presented in the book. Namely, Rohn shows that, crudely speaking, there is no feasible (polynomial time) algorithm that computes bounds on the solution sets $X$ to *all* interval linear systems with overestimation less than $\delta$, for any $\delta > 0$.

## *Existing Algorithms Can Often Solve Interval Linear Problems Efficiently*

This general negative result means that, for every algorithm for solving interval linear problems, there are hard cases in which this algorithm either runs too long, or results in a drastic overestimation of $X$.

In practice, several (heuristic) methods such as interval Gaussian elimination or the interval Gauss–Seidel method work well for many real-life systems. In particular, they work well when the widths of the coefficient intervals are small, and also in many other cases.

## *Applications to Economics*

One case where linear models naturally appear is economics. Namely, the dependency $y = f(x_1, ..., x_n)$ of the amount $y$ of goods produced in an economic sector on the amounts of goods $x_i$ used in production in this production can be described (with a reasonably good accuracy) by a linear function. The corresponding linear model was first proposed by the Nobel laureate economist W. Leontief, and is therefore called a *Leontief model.* A typical problem here is to determine the desired amounts of production for each sector so that all consumer demands are satisfied, and there is no unnecessary production (i.e., all goods are either directly consumed, or used in production of other goods in the chain that leads to consumption). In his pioneering research, Leontief assumed that the coefficients are known precisely. In this case, we have a linear programming problem, that is (in principle) feasible.

In real life, however, the only way to determine the coefficients is from the empirical data. Empirical economic data is not precise. Therefore, instead of precise values for the coefficients, we have *intervals* of possible values. Thus, in real life, Leontief's problem leads to an interval linear system.

Luckily, in spite of the general computational complexity result, specific interval linear systems stemming from these economic problems are feasible. In this book, a survey of corresponding methods and applications is given by Jerrell in "Applications of Interval Computations to Regional Economic Input-Output Models" [9]. As a case study, Jerrell obtains meaningful bounds for the impact of Northern Arizona University on the economy of Coconino County, Arizona, and on the state of Arizona in general.

## 6 INTERVAL COMPUTATIONS CAN ALSO HANDLE POSSIBLE ADDITIONAL INFORMATION ABOUT THE INPUT DATA

We have already mentioned that interval computations can naturally handle situations in which, instead of the exact values of the input parameters $p$, intervals $\mathbf{p}$ of possible values of these parameters are known.

In many real-life cases, in addition to these intervals, we have additional information about $p$:

- In some cases, we know not only the *interval* of possible values of $p$, but we also have some information about the *probabilities* of different values. If we know all the probabilities for all possible values of all parameters, then we can apply traditional statistical techniques. However, if we only know some probabilities for some of the parameters, and only interval information for the others, we need a special technique for combining these two types of information. In this book, a survey of such techniques and their applications is given by Berleant "Automatically Verified Arithmetic on Probability Distributions and Intervals" [4].
- In some cases, in addition to the interval in which the value of $p$ is guaranteed to lie, we have experts who produce narrower intervals to which, they claim, the actual values "most certainly" belong. In this case, in addition to the interval of "definitely possible" values of $x$ (produced by interval computation techniques), it is desirable to supply the user with

narrower intervals that result from the experts' narrower input intervals. (These new intervals will be correct enclosures if the experts did not err in their estimates.) In this book, this problem is described and analyzed in a survey "Nested Intervals and Sets: Concepts, Relations to Fuzzy Sets, and Applications" [16] by Nguyen and Kreinovich. It turns out that the resulting formalism is very close to the formulas proposed in so-called *fuzzy logic* (often without rigorous justification). Namely, to get these formulas, as a "fuzzy truth value" $\mu(A)$ of a statement $A$, we must take, crudely speaking, the ratio $\mu(A) = N(A)/N$ of those experts ($N(A)$) who believe $A$ to be true to the total number $N$ of the experts asked. Thus, this version of interval computation leads to a new justification of the formulas of fuzzy logic.

Two other surveys from this book describe further modifications of this idea:

- The first modification is described by Kohout and Bandler in "Fuzzy Interval Inference Utilizing the Checklist Paradigm and BK-Relational Products" [11]. This modification is based on the fact that we can only ask so many questions to the experts. In particular, if we have, say, $n$ statements $A_1, ..., A_n$ of interest, then we can ask an expert's opinion about these $n$ statements, maybe about some of their logical combinations (of the type $A_1 \& A_2$), but *not* about all $> 2^n$ possible logical combinations of these statements. As a result, there may be some statements $A_i$ and $A_j$ for which we only know the truth values $\mu(A_i)$ and $\mu(A_j)$, but not $\mu(A_i \& A_j)$. Suppose that we are interested in knowing the degree of belief in $A_i \& A_j$. If, e.g., $\mu(A_i) = \mu(A_j) = 0.5$. Then there are two possibilities:
  - It could be that the same experts believe in $A_i$ as in $A_j$. In this case, $\mu(A_i \& A_j) = \mu(A_i) = 0.5$.
  - It can also be that the experts who believe in $A_i$ are exactly those who *do not* believe in $A_j$. In this case, $\mu(A_i \& A_j) = 0$.

  In general, for a given $\mu(A_i)$ and $\mu(A_j)$, we have an *interval* of possible values of $\mu(A_i \& A_j)$. If we already have intervals of possible values for $\mu(A)$ and $\mu(B)$, then we need an interval operation to compute the interval of possible values of $\mu(A \& B)$. Here, interval computations come in handy. As a case study, Kohout and Bandler describes an implementation of this idea in an expert medical system *Clinaid*.

- Another modification is described by Rocha et al in "Computing Uncertainty in Interval Based Sets" [18]. The underlying idea is that experts are often inconsistent. Therefore, instead of a single small interval of possible values, we get several (often non-intersecting) intervals that supposedly describe the same quantity. Some intervals are supported by more experts, some by less. To describe the degree of support for each interval $\mathbf{x}$, we can use a ratio $N(\mathbf{x})/\mathbf{N}$ similar to that described above. Since each expert selects some interval, these ratios add up to 1 and can thus be mathematically interpreted as (subjective) probabilities. Therefore, to describe the expert's opinions, we must have a ***probability distribution on the set of all intervals.*** An interval is a particular case of this distribution, when only one probability is different from 0. In Rocha et al [18], standard operations of interval arithmetic are extended to these new objects. Applications to knowledge representation in expert systems are described.

# 7 SOFTWARE AND HARDWARE SUPPORT FOR INTERVAL COMPUTATIONS

If interval methods are so good, why aren't they widely applied? One of the main reasons is lack of easily available software. Now, people rarely program "from scratch": they try to use existing software as much as possible. Therefore, to make interval computations usable, the computations must be supported by ***software*** packages.

This brings us to the next issue: a software package will not be used if it is not sufficiently fast. Everyone is interested in buying a faster computer, and "a faster computer" means that computer on which typical "benchmarking" programs run faster. The bulk of existing software does not use intervals at all; therefore, when designing new computers, computer engineers try to speed up the operations with numbers that are most frequently used in the usual programs. As a result, operations with numbers are fast, while elementary operations with intervals (that form the basis for interval computations) are much slower. Because of that, experts in interval computations have urged the development of faster, ***hardware***-oriented systems for interval computations.

In their survey "Software and Hardware Techniques for Accurate, Self-Validating Arithmetic" [20], Schulte and Swartzlander review some of the available programming language extensions, packages and problem solving environments for interval computations. They then discuss hardware design alternatives, and propose a specific hardware design and software interface for variable precision interval arithmetic.

Finally, in "Stimulating Hardware and Software Support for Interval Arithmetic" [22], Walster presents his view of why many hardware and software vendors do not presently provide interval computations as integral parts of their products. He then discusses steps the community of interval computations experts should take to make such capabilities more widely available.

# Acknowledgements

This work was supported in part by the *National Science Foundation (NSF)*, through grants CCR-9203730, CDA-9015006, and EEC-9322370, and by the *National Aeronautics and Space Administration (NASA)*, through grant No. NAG 9-757.

# REFERENCES

[1] G. Alefeld, V. Kreinovich, and G. Mayer, "The Shape of the Symmetric Solution Set," *This Volume.*

[2] D. Berleant, "Automatically Verified Arithmetic on Probability Distributions and Intervals", *This Volume.*

[3] C. L. Fefferman and L. A. Seco, "Interval Arithmetic in Quantum Mechanics", *This Volume.*

[4] R. P. Feynman, R. B. Leighton, and M. L. Sands, *The Feynman Lectures On Physics*, Addison-Wesley, Redwood City, CA, 1989.

[5] S. Hadjihassan, E. Walter, and L. Pronzato, "Quality Improvement via The Optimization Of Tolerance Intervals During The Design Stage," *This Volume.*

[6] R. Hammer, M. Hocks, U. Kulisch, D. Ratz, *Numerical toolbox for verified computing. I. Basic numerical problems*, Springer Verlag, Heidelberg, N.Y., 1993.

[7] E. R. Hansen, *Global optimization using interval analysis*, Marcel Dekker, N.Y., 1992.

[8] E. Hyvönen and S. De Pascale, "Interval Computations on the Spreadsheet", *This Volume.*

[9] M. E. Jerrell, "Applications of Interval Computations to Regional Economic Input-Output Models," *This Volume.*

[10] R. B. Kearfott, "A Review of Techniques in the Verified Solution of Constrained Global Optimization Problems," *This Volume.*

[11] L. J. Kohout and W. Bandler, "Fuzzy Interval Inference Utilizing the Checklist Paradigm and BK-Relational Products", *This Volume.*

[12] V. Kreinovich (ed.), *Reliable Computing*, 1995, Supplement (Extended Abstracts of APIC'95: International Workshop on Applications of Interval Computations, El Paso, TX, Febr. 23–25, 1995).

[13] R. E. Moore, *Automatic error analysis in digital computation*, Lockheed Missiles and Space Co. Technical Report LMSD-48421, Palo Alto, CA, 1959.

[14] R. E. Moore and C. T. Yang, *Interval analysis*, Lockheed Missiles and Space Co. Technical Report LMSD-285875, Palo Alto, CA, 1959.

[15] R. E. Moore, *Interval analysis*, Prentice Hall, Englewood Cliffs, NJ, 1966.

[16] H. T. Nguyen and V. Kreinovich, "Nested Intervals and Sets: Concepts, Relations to Fuzzy Sets, and Applications", *This Volume.*

[17] P. M. Pardalos and J. B. Rosen, *Constrained Global Optimization: Algorithms and Applications*, Springer-Verlag, New York, 1987.

[18] L. M. Rocha, V. Kreinovich, and R. B. Kearfott, "Computing Uncertainty in Interval Based Sets", *This Volume.*

[19] J. Rohn, "Linear Interval Equations: Computing Enclosures with Bounded Relative Or Absolute Overestimation is NP-Hard," *This Volume.*

[20] M. J. Schulte and E. E. Swartzlander, Jr., "Software and Hardware Techniques for Accurate, Self-Validating Arithmetic," *This Volume.*

[21] A. L. Semenov, "Solving Optimization Problems with Help of the UniCalc Solver," *This Volume.*

[22] G. W. Walster, "Stimulating Hardware and Software Support for Interval Arithmetic," *This Volume.*

# 2

# A REVIEW OF TECHNIQUES IN THE VERIFIED SOLUTION OF CONSTRAINED GLOBAL OPTIMIZATION PROBLEMS

**R. Baker Kearfott***

*Department of Mathematics*
*University of Southwestern Louisiana*
*U.S.L. Box 4-1010, Lafayette, LA 70504-1010, USA*
*email: rbk@usl.edu*

*This work was supported in part by National Science Foundation grant CCR-9203730.

## ABSTRACT

Elements and techniques of state-of-the-art automatically verified constrained global optimization algorithms are reviewed, including a description of ways of rigorously verifying feasibility for equality constraints and a careful consideration of the role of active inequality constraints. Previously developed algorithms and general work on the subject are also listed. Limitations of present knowledge are mentioned, and advice is given on which techniques to use in various contexts. Applications are discussed.

## 1 INTRODUCTION, BASIC IDEAS AND LITERATURE

We consider the constrained global optimization problem

$$\begin{array}{lrcll} \text{minimize} & \phi(X) & & & \\ \text{subject to} & c_i(X) & = & 0, & i=1,\ldots,m \\ & a_{i_j} \leq & x_{i_j} & \leq b_{i_j}, & j=1,\ldots,q, \end{array} \qquad (2.1)$$

where $X = (x_1, \ldots, x_n)^T$. A general constrained optimization problem, including inequality constraints $g(X) \leq 0$ can be put into this form by introducing slack variables $s$, replacing by $s + g(X) = 0$, and appending the bound constraint $0 \leq s < \infty$; see §2.2.



*R. B. Kearfott and V. Kreinovich (eds.), Applications of Interval Computations*, 23–59.

We wish to find *all* minimizers in problem 2.1, and to *verify* bounds on the local minimum. Because of this, the methods used must go beyond (and build upon) traditional methods utilizing descent, line searches, trust regions, etc. in floating-point arithmetic, such as those in the books [7], [13] [46], or in the software package of [4]: these "approximate minimization" methods find only one approximate minimizer per run, may terminate near points other than minimizers without indications that they have done so, may converge to local minimizers that are not global minimizers, etc. The efficiency of such approximate methods, however, makes them practical. They can be used as in [22] and [24] or [3] to make verification algorithms efficient.

The conditions $a_{i_j} \le x_{i_j} \le b_{i_j}$, $j = 1, \ldots, q$ represent actual bound constraints intrinsic to the problem formulation. In rigorous branch and bound algorithms, an *overall search region* $\mathbf{X}_0$ is generally defined through similar bounds:

$$\mathbf{X}_0 = \{(x_{0,1}, \ldots, x_{0,n}) \mid a_i \le x_{0,i} \le b_i, \quad 1 \le i \le n\}; \tag{2.2}$$

only those bounds corresponding to the index set $\{i_j\}_{j=1}^{q}$ should be treated as actual bound constraints.

Deterministic location of the global minima and all global minimizers of the non-convex constrained optimization problem 2.1 is in general very difficult (and, indeed, NP-complete; cf. §2.2 and §2.3). However, interval branch and bound methods have exhibited a degree of success in many instances. Various authors, including Moore [43] and Skelboe [65], Hansen ([17], [18] and [19]), Ichida [21], Ratschek and Rokne [52], Jansson and Knüppel ([23] and [24]), Caprani, Godthaab and Madsen [3] have contributed to the knowledge of such algorithms. See [52] for a coherent explanation of this type of algorithm, as well as for the requisite introduction to interval arithmetic[1].

The basic ideas behind versions of such algorithms for unconstrained optimization (where $m = 0$) can be stated with minimal notation and without reference to details of interval arithmetic. The principles include a *check on the range* of $\phi$ and a *computational existence / uniqueness* test. We let $\mathbf{X} = (\mathbf{x}_1, \ldots \mathbf{x}_n)^T = ([\underline{x}_1, \overline{x}_1], \ldots, [\underline{x}_n, \overline{x}_n])^T$ be the interval vector ("box") representing the search region $\underline{x}_i \le x_i \le \overline{x}_i$, $1 \le i \le n$. Branch and bound algorithms maintain one or more of these lists: a list $\mathcal{L}$ of boxes $\mathbf{X}$ to be processed, a list $\mathcal{U}$ of boxes the algorithm has reduced to small diameter, and a list $\mathcal{C}$ of boxes that have been verified to contain critical points. The general pattern is as follows.

[1] Other texts, such as [1], [15], [19], [43] and [48] also give good introductions to the subject, and contain techniques relevant to global optimization.

**Algorithm 1** (Abstract Branch-and-Bound Pattern)

1. *Initialize $\mathcal{L}$ by placing the initial search region $\mathbf{X}_0$ in it.*
2. DO WHILE $\mathcal{L} \neq \emptyset$.
   (a) *Remove the first*[2] *box* $\mathbf{X}$ *from* $\mathcal{L}$;
   (b) (Process $\mathbf{X}$) *Do one of the following:*
   - *reject* $\mathbf{X}$;
   - *reduce the size of* $\mathbf{X}$;
   - *determine that* $\mathbf{X}$ *contains a unique critical point, then find the critical point to high accuracy.*
   - *subdivide* $\mathbf{X}$ *to make it more likely to succeed at rejecting, reducing, or verifying uniqueness.*

   (c) *Insert one or more boxes derived from* $\mathbf{X}$ *onto* $\mathcal{L}$, $\mathcal{U}$ *or* $\mathcal{C}$, *depending on the size of the resulting box(es) from step 2b and whether the (possible) computational existence test in that step has determined a unique critical point.*

**End Algorithm 1**

Many details, such as stopping criteria and tolerances, are absent from Algorithm 1, which represents a general description. Such details differ in particular actual algorithms.

A combination of several techniques is used in state-of-the-art interval global optimization codes to do step 2b of Algorithm 1.

**Algorithm 2** (Range Check and Critical Point Verification)

1. *Input a box* $\mathbf{X}$ *and the current best rigorous upper bound* $\overline{\phi}$ *on the global minimum.*
2. (Feasibility check; for constrained problems only)
   (a) (Exit if infeasibility is proven) DO *for* $i = 1$ *to* $m$:

[2] The boxes in $\mathcal{L}$ are in general inserted in a particular order, depending on the actual algorithm

i. *Compute an enclosure* $c_i(\mathbf{X})$ *for the range of* $c_i$ *over* $\mathbf{X}$.

ii. IF $0 \notin c_i(\mathbf{X})$ THEN *discard* $\mathbf{X}$ *and* EXIT.

(b) *Prove computationally, if possible, that there exists at least one feasible point in* $\mathbf{X}$.

3. (Range check or "midpoint test")

   (a) *Compute a lower bound* $\underline{\phi}(\mathbf{X})$ *on the range of* $\phi$ *over* $\mathbf{X}$.

   (b) IF $\underline{\phi}(\mathbf{X}) > \overline{\phi}$ THEN *discard* $\mathbf{X}$ *and* EXIT.

4. (Update the upper bound on the minimum.) IF *the problem is unconstrained or feasibility was proven in step 2b,* THEN

   (a) *Use interval arithmetic to compute an upper bound* $\overline{\phi}(\mathbf{X})$ *of the objective function* $\phi$ *over* $\mathbf{X}$.

   (b) $\overline{\phi} \leftarrow \min\{\overline{\phi}, \overline{\phi}(\mathbf{X})\}$.

5. ("monotonicity test")

   (a) *Compute an enclosure* $\nabla\phi(\mathbf{X})$ *of the range of* $\nabla\phi$ *over* $\mathbf{X}$. *(Note: If* $\mathbf{X}$ *is "thin", i.e. if some bound constraints are active over* $\mathbf{X}$*, then a* reduced gradient *can be used; see §2.3 and [32].)*

   (b) IF $0 \notin \nabla\phi(\mathbf{X})$ THEN *discard* $\mathbf{X}$ *and* EXIT.

6. ("concavity test") *If the Hessian matrix*[3] $\nabla^2\phi$ *cannot be positive definite anywhere in* $\mathbf{X}$ THEN *discard* $\mathbf{X}$ *and* EXIT.

7. (Quadratic convergence and computational existence / uniqueness) *Use an interval Newton method*[4] *(with the Fritz–John conditions as in §3.5 in the constrained case) to possibly do one or more of the following:*

   - *reduce the size of* $\mathbf{X}$;
   - *discard* $\mathbf{X}$;
   - *verify that a unique critical point exists in* $\mathbf{X}$.

8. (Bisection or geometric tessellation) *If step 7 did not result in a sufficient change in* $\mathbf{X}$*, then bisect* $\mathbf{X}$ *along a coordinate direction (or otherwise tessellate* $\mathbf{X}$*), returning all resulting boxes for subsequent processing.*

**End Algorithm 2**

---

[3] or reduced Hessian matrix, as in step 5

[4] possibly in a subspace, as in steps 5 and 6

Since techniques for constrained problems are somewhat more involved, step 2, checking for infeasibility and verifying feasibility, will be explained separately in §2.1.

Step 3 is called the "midpoint test" because the upper bound $\overline{\phi}$ is often obtained by evaluating $\phi$ at the midpoint vector of $\mathbf{X}$, properly taking account of rounding errors for rigor. Of course, step 5 is called the "monotonicity test " since $\phi$ is monotone over $\mathbf{X}$ in the $i$-th variable if the $i$-th component of $\nabla\phi$ does not vanish over $\mathbf{X}$.

Improved techniques for carrying out step 6, checking non-convexity, are desirable. Presently, sufficient conditions, such as checking the sign of the diagonal entries of an interval evaluation $\nabla^2\phi(\mathbf{X})$, can be used. One method of verifying convexity appears as Theorem 14.1 in [15] and Lemma 2.7.2 in [54]. Also, Neumaier [49] has shown that every element matrix of an interval matrix $\mathbf{A}$ is positive-definite, provided some point matrix $A \in \mathbf{A}$ is positive definite and $\mathbf{A}$ is regular according to Definition 3 of §3 below. This result can be sharpened as follows.

**Theorem 1** (Shi, [64, Appendix B]) *If $A \in \mathbf{A}$ is symmetric, $A$ has $p$ negative eigenvalues and $\mathbf{A}$ is regular, then every symmetric point matrix in $\mathbf{A}$ has $p$ negative eigenvalues.*

Theorem 1 will allow a sharper concavity test.

Interval global optimization algorithms, viewed abstractly, are similar to branch and bound algorithms that do not explicitly use interval arithmetic to bound ranges, such as that in [50, Ch. 6]. Interval global optimization algorithms differ among themselves in the ordering of the lists $\mathcal{L}$, $\mathcal{U}$ and $\mathcal{C}$ in step 2c of Algorithm 1, and in how (and which) steps of Algorithm 2 are carried out.

## 1.1 Early and Simplified Algorithms

Early algorithms worked only with the list $\mathcal{L}$, without lists $\mathcal{U}$ and $\mathcal{C}$. Also, although the processes in steps 3 through 7 of Algorithm 2 make actual implementations practical and efficient, they are not an essential part of the branch and bound structure. In the early but well-known Moore / Skelboe algorithm, only the list $\mathcal{L}$ appears, and the boxes $\mathbf{X} \in \mathcal{L}$ are ordered in order of increasing $\underline{\phi}(\mathbf{X})$. In step 8 of Algorithm 2, $\mathbf{X}$ is bisected along the largest coordinate direction, and both progeny are placed in order in $\mathcal{L}$. Steps 3 through 7 of Algorithm 2 are absent. When the algorithm is terminated, the first box in the list is taken to approximate the global minimizer.

An algorithm attributed to Ichida [21] improves upon the Moore / Skelboe algorithm by including the midpoint test (step 3 of Algorithm 2) to avoid placing boxes generated during bisection onto $\mathcal{L}$ if they cannot contain optimizers. Additionally, the algorithm described in [21] contains a method for grouping together clusters of boxes corresponding to particular minimizers.

Hansen's algorithms, described in [17], [18] and [19] generally use second-order information (step 7 of Algorithm 2) and other sophisticated techniques. However an algorithm sometimes called "Hansen's algorithm" is a simplified version. In "Hansen's algorithm," $\mathcal{L}$ is ordered not in terms of the function, but in order of decreasing diameter (i.e. width of largest coordinate interval) of the $\mathbf{X}$. Furthermore, in the midpoint test, the entire list $\mathcal{L}$ is culled (and not just the boxes that have just been produced by bisection) whenever a new $\overline{\phi}$ is obtained. (We note that in Hansen's actual codes, as in the experiments in [66], the list is ordered such that the first box is the one with smallest lower bound on $\phi$. Hansen and Walster claim this is much better than ordering in terms of decreasing diameter.) Various modifications of the list ordering, such as that in [54, §2.2.5.1], have appeared more recently.

None of these simplified methods employs interval-Newton acceleration.

Convergence properties of the Moore / Skelboe, Ichida and Hansen algorithms, as well as some numerical experiments with Hansen's algorithm are analyzed in [44]. A wide-ranging survey that concentrates on these methods (but also mentions some newer techniques) is the book [52].

## 1.2 Recent Practical Algorithms

More recent algorithms and practical implementations usually involve interval Newton methods for computational existence and uniqueness of critical points and for quadratic convergence properties. However, some successful newer algorithms are derivative-free, and concentrate on use of approximate optimizers, order in which the list is searched, properties of the inclusion function, or parallelization.

A thorough exposition of background, starting with the elements of interval arithmetic, and of numerous techniques for interval unconstrained optimization along with a substantial number of careful numerical experiments appears in Ratz [54]. Some of these ideas are implemented in the Pascal-XSC code described in [15].

Ratz, continuing development of his algorithms, has concentrated on better choice of coordinate in the bisection process of step 8 of Algorithm 2, and on *splitting strategies*, cf. [55]. Regarding bisection strategies, Ratz claims better success when choosing the coordinate to bisect according to the scaling

$$\nabla\phi(\mathbf{X})(\mathbf{X} - X),$$

rather than merely bisecting along the longest coordinate direction of $\mathbf{X}$; cf. [54], pp. 41–42 and [6]. Convergence ofgeneralized bisection based global optimization algorithms with this coordinate selection strategy is also proven in [6]. This scheme is related to the *maximal smear* scheme introduced in [30]. Box splitting is a process, first discussed by Hansen in relation to the interval Gauss–Seidel method, by which extended interval arithmetic in the sense of Kahan is used to obtain two disjoint boxes. If applied wherever possible, too many boxes are produced, thus slowing the overall branch and bound algorithm. Coordinate choice in bisection, box-splitting strategies, and the ordering in the list $\mathcal{L}$ can be crucial in an overall global optimization algorithm.

Hansen's book [19] provides an informal description of many sophisticated techniques and heuristics for use in global optimization algorithms. Many examples are given, although thorough numerical experiments are absent. Good numerical experiments with an earlier optimization algorithm incorporating many of the same ideas appear in [66].

In [23] and [24], Jansson and Knüppel have presented a method without derivatives (no gradient test or interval Newton method), but with a sophisticated use of bisection and local optimization. In particular, a local optimization (to obtain an approximate optimizer) is performed at certain stages of the process, and the results are used to update $\overline{\phi}$. Though heuristic, the algorithm performs well on many reasonably complicated functions, including non-differentiable ones, such as maximizing the smallest singular value of a matrix. The report [24] also contains a collection of test examples, along with numerical results and three-dimensional graphs of those (numerous) test problems that are two-variable functions. Jansson [25, §2.5] proposes a variant in which derivatives are used only in an interval Newton method to verify and sharpen bounds for approximate optima. This variant is carefully tested on forty test problems in [26].

In [3] Caprani, Godthaab, and Madsen also propose use of an approximate minimizer obtained through a local method with floating point arithmetic. In their algorithm, an approximate local minimizer is found, then a box **X** is constructed about this minimizer. An interval Newton method is then applied to **X** to determine existence or uniqueness of a critical point. If existence can be proven, **X** is expanded as much as possible, subject to success of the interval Newton method in verifying uniqueness, then **X** is removed from the region by cutting the complement of **X** into remaining boxes to be processed. The minimizer-inflation technique is related to "$\epsilon$-inflation" as in [57, p. 58] or [41] used to provide error bounds on approximate solutions to linear and nonlinear systems. It is illustrated in [3] that the Caprani/Godthaab/Madsen method parallelizes well.

In [45], basic algorithms for non-differentiable and differentiable objective functions are reviewed, then a sophisticated coarse-grained algorithm for optimization on a distributed-memory multicomputer, implemented on a distributed system of workstations, is explained. In this algorithm, each processor shares a portion of the list $\mathcal{L}$. The load is dynamically balanced as the computations proceed. The algorithm was programmed in C++, based on a system for interval arithmetic developed by Leclerc; an encapsulated explanation appears in [40]. The numerical experiments feature a very difficult parameter-fitting problem.

In [10] and [11] Eriksson et al also study parallelization of an unconstrained global optimization algorithm, implemented on an Intel hypercube. Various load balancing strategies are compared on a set of six test problems, one of which was designed specifically to test different parallelization schemes.

In [32], experimental results are reported for a FORTRAN-77 code containing techniques for the monotonicity test and iteration / verification, as well as use of a local optimization process for computing $\overline{\phi}$ ("midpoint test "). The special preconditioners preconditioning techniques of [29] and [31] for the interval Gauss–Seidel method (a type of interval Newton method) in the optimization context, as well as a technique for handling bound constraints througl. the tessellation are studied there[5]. A more flexible Fortran-90 code utilizing the

[5]The latter two techniques are more fully explored in [54].

| Method / Authors | Midpoint Test | Monotonicity Test | Concavity Test | Interval Newton | Paralleli-sation | Use of Local Minimiser | Ref. |
|---|---|---|---|---|---|---|---|
| Moore / Skelboe | | | | | | | [43] and [65] |
| Ichida | yes | | | | | | [21] |
| "Hansen's algorithm" | yes, and to cull list | | | | | | [52] |
| Hansen's actual | yes | yes | yes | yes | | | [19] |
| Kearfott '92 | yes | yes | | yes | | yes | [32] |
| Ratz | yes | yes | yes | yes | | yes | [54] and [15] |
| Jansson / Knüppel | yes | | | | yes | yes | [24] |
| Caprani / Godthaab / Madsen | yes | | | yes | yes | yes | [3] |
| Hansen / Moore / Leclerc | yes | yes | yes | yes | subject of study | | [45] |
| Eriksson | yes | yes | | yes | subject of study | | [11] |
| Wolfe | yes | yes | | yes | | | [67] |

**Table 1** Summary attributes of various global optimization algorithms

system of [10] and including techniques for constrained problems is presently under development.

Theoretical and empirical consequences of the order of the interval extension used to obtain $\underline{\phi}(\mathbf{X})$ are studied in [8] and [37]. However, exhaustive studies on a practical algorithm do not appear there.

Some (but not all) of the attributes of the algorithms in this section and in §1.1 are summarized in Table 1. Here, the label "Kearfott '92" refers to the code of [32]. "Hansen's actual" is used to denote the most recent algorithms of Hansen, described in [19] and forming the basis of the work in [45] and [40]. Blank spaces in the table indicate that the feature is not present.

## 1.3 Notation

In the remainder of this paper, we will use the following notation. We will use boldface to denote intervals, lower case to denote scalar quantities, and upper case to denote vectors and matrices. We will use underscores to denote lower bounds of intervals and overscores to denote upper bounds of intervals. For components of vectors, we will use corresponding lower case letters. For example, we may have:

$$\mathbf{X} = [\mathbf{x}_1, \mathbf{x}_2, \ldots, \mathbf{x}_n]^T,$$

where $\mathbf{x}_i = [\underline{x}_i, \overline{x}_i]$. The notation $\check{X}$ will denote a representative point, usually in the interval vector $\mathbf{X}$. The ***magnitude*** of an interval is defined as $|\mathbf{x}| = \max\{|\underline{x}|, |\overline{x}|\}$.

The width of an interval $\mathbf{x}$ is denoted by $w(\mathbf{x}) = \overline{x} - \underline{x}$. The width of an interval vector $\mathbf{X}$, denoted $w(\mathbf{X})$, is defined component-wise. We use $w(\mathbf{X})$ in the context of $\|w(\mathbf{X})\| = \|w(\mathbf{X})\|_\infty$.

The symbol $\phi(\mathbf{X})$ denotes an interval extension of $\phi$ over $\mathbf{X}$.

Whenever $\|\cdot\|$ is used, it will mean $\|\cdot\|_\infty$.

We will use calligraphic letters such as $\mathcal{L}$, $\mathcal{U}$ and $\mathcal{C}$, as above, to denote stacks and lists of boxes.

Brackets $[\cdot]$ will be used to delimit both intervals and matrices and vectors. For example, we have the interval $[1, 2]$ and the interval vector $[[1, 2], [3, 4]]'$. Meanings should be clear from the context.

The notation $\mathrm{int}(\mathbf{X})$ will be used to denote the topological interior of an interval vector $\mathbf{X}$.

Interval arithmetic will not be reviewed here, as numerous introductions, such as those in [1], [43], [48], [52] or even [28], exist.

We will also assume familiarity with general concepts of numerical methods for constrained and unconstrained optimization, such as are found in [13].

# 2 ON CONSTRAINED OPTIMIZATION PROBLEMS

The book [52] contains an explanation of fundamental interval means of handling inequality constraints, while [19] contains an in-depth treatment of many interval techniques for both inequality and equality constraints. However, extensive numerical results using these techniques have not been published. (See Table 2; blank spaces mean the feature is absent.)

Handling of simple bound constraints through the tessellation process has been explored in [32] and [54]. In general, the computational effort for this technique

| Method / Authors | Bound Constraints | Inequality Constraints | Equality Constraints | Second Order | Use of approx. minimiser | Numerical Experiments |
|---|---|---|---|---|---|---|
| "Hansen's" | | | | | | unconstrained |
| Hansen's book | (as equality constraints) | yes | yes | yes | | unconstrained only |
| Opt. '92 | yes | | | | yes | yes |
| Rats | yes | | | | | yes |
| Jansson / Knüppel | yes | | | variant in [25, §2.5] | yes | unconstrained only |
| Caprani / Godthaab / Madsen | | | | | yes | unconstrained |
| Hansen / Moore / Leclere | | | | | | unconstrained only |
| Eriksson | | | | | | unconstrained |
| Wolfe | | yes | yes | yes | | yes |

**Table 2** **Summary of handling of constraints in various global optimization algorithms.**

increases exponentially with the number of bound constraints, but this may not be so for many specific problems. However, bound-constrained problems are intrinsically hard [51]. Although bound constraints can be handled as inequality constraints as in [19], it is unclear without published experimentation how such algorithms behave: it is possible that large numbers of small boxes, clustered on the boundaries of the constraints, are produced through the $n$-dimensional tessellation.

Alternately, inequality constraints can be handled as bound constraints by introduction of slack variables.

We elaborate on these concepts in the remainder of this section.

## 2.1 Checking Feasibility / Infeasibility

The following computations may be done with the constraints:

- using the constraints to delete portions of a region $\mathbf{X}$ that are infeasible;
- proving feasibility or infeasibility relative to inequality constraints;
- proving feasibility or infeasibility relative to equality constraints.

These possibilities are discussed in the remainder of this section.

In [19, §11.6], Hansen proposes heuristics for using inequality constraints to delete portions of a box $\mathbf{X}$ that cannot be feasible. Alternately, if the inequality constraints are converted into equality constraints first, the optimal preconditioner techniques of [29] and [31] in conjunction with the interval Gauss–Seidel method or interval Gaussian elimination may be used directly on the underdetermined $m$ by $n$ system of constraints. The latter provides a certain theoretical optimality not present in the heuristics of [19], with a smaller system than with the entire Fritz–John system; this appears in [36]. However, experiments in [36] indicate that the scheme reported in this section is usually better.

An algorithm for constructing large boxes within which inequality constraints of the form $g(x) < 0$ are rigorously verified appears in [39], along with numerical experimentation. We have not included this algorithm in our tables, however, since it is not a general global search algorithm, but a method of dealing with constraints.

As indicated in step 2a of Algorithm 2, an elementary check for infeasibility of an entire box $\mathbf{X}$ with respect to the equality constraints $c_i$ is to verify that $0 \notin c_i(\mathbf{X})$ for some $i$. There is a corresponding check if an inequality constraint $d(x) \leq 0$ is used: the region is infeasible if simply $d(\mathbf{X}) > 0$. Similarly, if only inequality constraints of the form $d_i \leq 0$ are present, feasibility is proven if $d_i(\mathbf{X}) \leq 0$ for each $i$. In this case, Hansen calls the region *certainly feasible.*

On the other hand, proving feasibility for problems with equality constraints $c_i(x) = 0$ (like problem 2.1), although necessary to get useful rigorous upper bounds $\overline{\phi}$ on the global minimum, is more difficult. In general, unless the number of constraints $m$ is less than the number of variables $n$, the problem is infeasible unless the constraints are linearly dependent. Walster and Hansen have mentioned work (in private communication) related to handling dependent constraints, but we are unaware of published material in this area. See [36] for some thoughts on dependent constraints; we will assume independent constraints and $m \leq n$ here.

In [67], Wolfe proposes an algorithm, based on a penalty function, for handling equality constraints. However, that algorithm considers feasibility to be rigorously proven only provided $c_i(x) \in [\epsilon, \epsilon]$ for each $i$ and some fixed $\epsilon$. In contrast, in a method proposed in Hansen [19, §12.3 ff] and in [36], $c_i(x) = 0$ is rigorously proven. Both these methods are based on applying an interval Newton method to the system $c_i(x) = 0$, $i = 1, \ldots, m$ with $n - m$ of the variables held fixed. However, the methods differ in application of the interval Newton method and in the choice variables to be held fixed. For details and numerical comparisons, see [36].

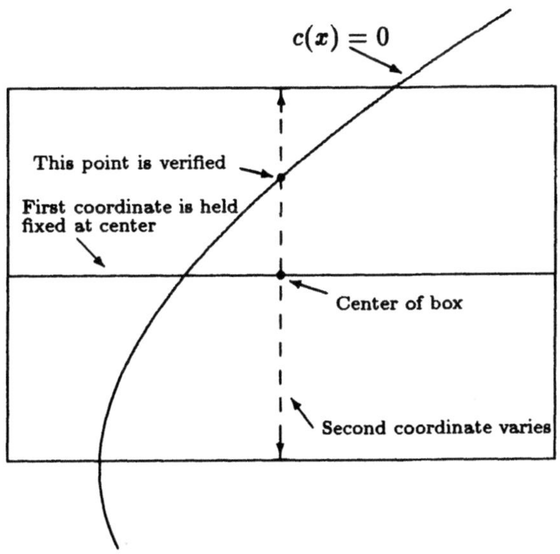


**Figure 1** Verifying a feasible point with Algorithm 3

The following general algorithm encompasses Hansen's and our proposed feasibility verification methods of [19] and [36]; the techniques differ in the details of step 3.

**Algorithm 3** (Prove Feasibility for Equality-Constrained Problems)

1. *Input an approximation* $\check{X}$ *to a feasible point, obtained through a conventional algorithm such as that of [4].*

2. *Let* $C(X) = (c_1(X), \ldots c_n(X))^T$ *and* $\nabla C(X)$ *represent the equality constraints and Jacobi matrix of the equality constraints, respectively (*$C : \mathbf{R}^n \rightarrow \mathbf{R}^m$*).*

3. *Choose coordinates* $\{p_k\}_{k=1}^m$ *to be varied in the interval Newton method in such a way that the resulting system is likely to be nonsingular.*

4. *Evaluate* $C(\check{X})$ *and* $\nabla \mathbf{C}(\hat{\mathbf{X}})$*, where* $\hat{\mathbf{X}}$ *has coordinates* $\hat{\mathbf{x}}_i = \check{x}_i$ *if* $i \neq p_k$ *for any* $k$ *and* $\hat{\mathbf{x}}_i = [\check{x}_i - \epsilon, \check{x}_i + \epsilon]$ *for* $i = p_k$ *for some* $\epsilon$*. That is, construct a box in an* $m$*-dimensional subspace that is, in a sense, most nearly perpendicular to the null space of* $\nabla C(\check{X})$*.*

5. *Let* $Y$ *be an approximate inverse of the* $m \times m$ *matrix consisting of columns* $p_1$ *through* $p_m$ *of the midpoint matrix of* $\nabla \mathbf{C}(\hat{\mathbf{X}})$*.*

6. *Apply an interval Newton method to* $Y\nabla \mathbf{C}(\hat{\mathbf{X}})(X-\check{X}) = -C(\check{X})$, *obtaining an image* $\hat{\mathbf{X}}_{new}$.

7. *If the result* $\hat{\mathbf{X}}_{new} \subseteq \text{int}(\hat{\mathbf{X}})$, *then the point* $\check{X}$ *is feasible. Find* $\phi(\check{X})$, *and take the upper bound as a rigorous upper bound for a minimum.*

**End Algorithm 3**

Here, the interval Newton method of step 6 may be taken as an operator $\mathbf{N}(\hat{\mathbf{X}}, F)$ operating on a function $F : \mathbb{R}^p \to \mathbb{R}^p$ defined on a box $\hat{\mathbf{X}}$, such that

1. the image $\mathbf{N}(\mathbf{X}, F)$ is a box in $\mathbb{R}^p$;
2. if $\mathbf{N}(\mathbf{X}, F)$ is contained in the interior of $\mathbf{X}$, then there is a unique solution to $F(X) = 0$ within $\mathbf{X}$;
3. $\mathbf{N}(\mathbf{X}, F)$ is straightforward to compute with interval arithmetic in an appropriate software environment.

See §3 for advice on interval Newton methods and for pointers to the abundant literature.

Full experiments on this algorithm, as well as comparisons, are reported in [36]. The general conclusions are that Algorithm 3 is effective, and that step 3 is effective, *provided a good approximation to a feasible point is centered in a box whose center is at* $\check{X}$. For a detailed explanation and additional illustrations, see [36].

## 2.2 On Equality, Inequality and Bound Constraints

The book [52] poses the optimization problem analogous to problem 2.1 with both equality constraints $c_i(x) = 0$ and inequality constraints $d_i(x) \leq 0$, while [19] contains separate chapters on inequality constrained problems and equality constrained problems. At first glance, inequality constrained problems seem easier than equality-constrained ones. This is because feasibility can sometimes be proven without use of an interval Newton method: We merely bound the range of each $d_j$ using interval arithmetic, then check that the upper bounds so obtained are all negative, i.e. by checking $\mathbf{d}_j(\mathbf{X}) \le 0$ for each $j$. Also, besides the sophisticated technique in [19, §11.6], similar verification that $\mathbf{d}_j(\mathbf{X}) > 0$ for each $j$ proves infeasibility over all of $\mathbf{X}$, allowing $\mathbf{X}$ to be eliminated from the global search region. Such techniques *should* probably be used in practical algorithms as additional tools to verify feasibility and to eliminate subregions. However, in such analyses, it is ignored that the inequality constraints can be *active*, that is, that they are in effect equality constraints.

Various theoretical results have been published in recent years showing that global optimization problems containing inequality constraints are NP-complete in the number of constraints. For example, it is shown in [51] that quadratic programming problems with one negative eigenvalue are NP-complete.

The possibility that computational effort can increase exponentially in the number of constraints becomes apparent if we examine the algorithm for bound-constrained problems in [32], reviewed in §2.3 below. However, as explained in §2.3, it is possible for heuristics to reduce the running time for specific problems below that predicted by the exponential worst-case bounds, without compromising rigor.

## 2.3 Handling Bound Constraints

In [32] and here, the bounds $a_{i_j} \le x_{i_j} \le b_{i_j}$ in problem 2.1 are viewed as bound constraints[6]. The bound constraints are then handled by separating the region into all possible subregions of lower dimensions, as is illustrated in Figure 2 for $n = 2$ These subregions are placed on the list $\mathcal{L}$ and processed as usual, except that reduced gradients and reduced Hessian matrices[7] are used in the interval Newton method on lower-dimensional regions. If all boxes are stored in $\mathcal{L}$, it is not difficult to see that the total number of boxes of all dimensions so obtained is $3^p$, where $p$ is the number of bound constraints. However, if a good upper bound $\overline{\phi}$ on the global minimum (as can sometimes be obtained with conventional algorithms such as that of [4]) is available, then many of the boxes can be rejected during the "peeling" process.

[6] in contrast to, say, [66], where these bounds are viewed as merely defining a search region in an unconstrained problem

[7] i.e. with rows and columns corresponding to variables held fixed deleted

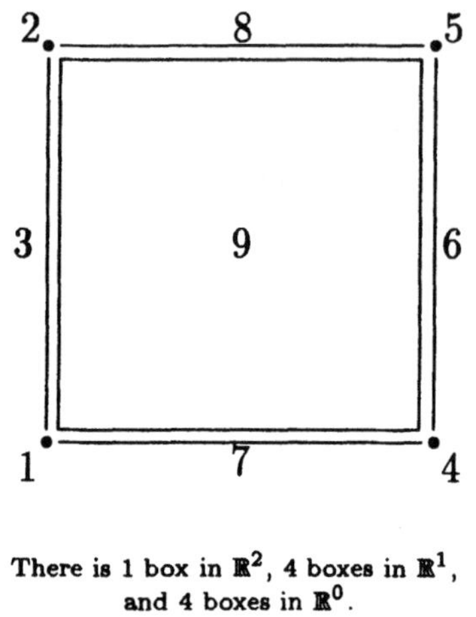


**Figure 2** "Peeling" a box to produce lower-dimensional boundary elements

The structure of the algorithm for producing the list of lower-dimensional boxes can be described simply in recursive form. We have:

**Algorithm 4** ("Peeling" the Boundary)

1. *Input the present coordinate index $i$, the box $\mathbf{X} = (\mathbf{x}_1, \ldots, \mathbf{x}_n)$, the list $\mathcal{I}$ of those coordinate indices for $\mathbf{X}$ that have already been considered, and the list $\mathcal{L}$ of boxes that have been already been generated for further processing.*
2. IF $i \notin \mathcal{I}$ THEN
   (a) (Process the lower boundary box.)
      i. *Set all coordinates of $\tilde{\mathbf{X}}$ but the $i$-th to corresponding coordinates of $\mathbf{X}$. Set the $i$-th coordinate of $\tilde{\mathbf{X}}$ to $\underline{x}_i$.*
      ii. *Set the index list $\mathcal{I}_{\text{new}}$ to $\mathcal{I}$ with $i$ appended.*
      iii. IF $i = n$
         THEN *store $\tilde{\mathbf{X}}$ in $\mathcal{L}$.*
         ELSE *execute this algorithm with $i+1$, $\tilde{\mathbf{X}}$, and $\mathcal{I}_{\text{new}}$ replacing $i$, $\mathbf{X}$, and $\mathcal{I}$, respectively.*
      END IF
   (b) (Process the upper boundary box.)
      i. *Set all coordinates of $\tilde{\mathbf{X}}$ but the $i$-th to corresponding coordinates of $\mathbf{X}$. Set the $i$-th coordinate of $\tilde{\mathbf{X}}$ to $\overline{x}_i$.*
      ii. *Set the index list $\mathcal{I}_{\text{new}}$ to $\mathcal{I}$ with $i$ appended.*

*iii.* IF $i = n$

THEN *store* $\tilde{\mathbf{X}}$ *in* $\mathcal{L}$.

ELSE *execute this algorithm with* $i+1$, $\tilde{\mathbf{X}}$, *and* $\mathcal{I}_{\text{new}}$ *replacing* $i$, $\mathbf{X}$, *and* $\mathcal{I}$, *respectively.*

END IF

*(c)* (Process the interior box.)

*i.* *Set* $\tilde{\mathbf{X}}$ *to* $\mathbf{X}$.

*ii.* *Set the index list* $\mathcal{I}_{\text{new}}$ *to* $\mathcal{I}$ *with* $i$ *appended.*

*iii.* IF $i = n$

THEN *store* $\tilde{\mathbf{X}}$ *in* $\mathcal{L}$.

ELSE *execute this algorithm with* $i+1$, $\tilde{\mathbf{X}}$, *and* $\mathcal{I}_{\text{new}}$ *replacing* $i$, $\mathbf{X}$, *and* $\mathcal{I}$, *respectively.*

END IF

**End Algorithm 4**

The numbering of the nine boxes of dimensions 2, 1, and 0 in Figure 2 represents the order they would appear in $\mathcal{L}$ if each box generated with $i - n$ in steps 2(a)iii, 2(b)iii, and 2(c)iii of Algorithm 4 were stored. The processing order in Algorithm 4 can be viewed as traversing a ternary tree, as in Figure 3. The levels of this tree correspond to the coordinates $i$, with the root at $i = 1$ at the top and the leaves at $i = n$ at the bottom. As drawn in Figure 3, the order the leaves eventually appear in $\mathcal{L}$ is from left to right.

Of course, actual implementations of Algorithm 4 have additional steps to

- eliminate the boxes $\mathbf{X}$ and $\tilde{\mathbf{X}}$ before further processing or storage in $\mathcal{L}$ by checking $\phi(\mathbf{X})$ or $\phi(\tilde{\mathbf{X}})$ and the reduced gradient of $\phi$ on $\mathbf{X}$ or $\tilde{\mathbf{X}}$;
- skip coordinates $i$ for which the bounds $a_i \le x_i \le b_i$ represent the extent of the search region, and not actual bound constraints for the problem.

These steps have been left out of the presentation of Algorithm 4 for clarity in the geometric process. However, they could be indispensable in reducing the number of boxes in $\mathcal{L}$ to a practical number. Observe that such steps can prune the tree in Figure 3 at a high level.

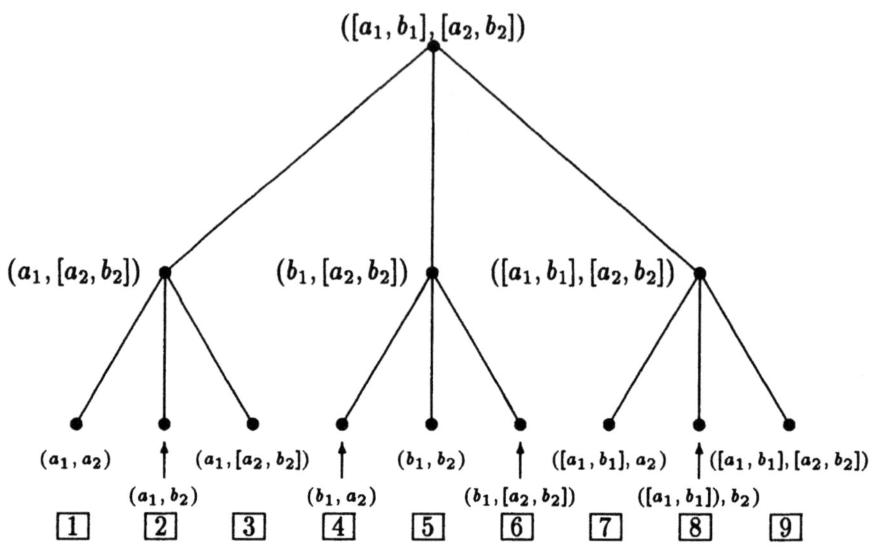


**Figure 3** "Peeling" the box into lower-dimensional boundary elements

# 3 ON USE OF INTERVAL NEWTON METHODS

An interval Newton method is used in step 6 of Algorithm 2, and also must be used to verify feasibility in step 6 of Algorithm 3. Thus, interval Newton methods are used in the analysis of nonlinear algebraic systems arising in global optimization

- to reduce the size of regions **X**, with quadratic convergence of the widths to zero;
- to computationally (but rigorously) prove existence of solutions within **X**;
- to prove that there is a unique solution within **X**;
- to prove that there can be no solutions in **X**.

The basics of interval Newton methods are described in [1, Ch. 19], [43, Ch. 5 and Ch. 6], [52, §2.9], [19, Ch. 4 and Ch. 8], and numerous papers, with extensive theoretical development reviewed in [48, Ch. 4 and Ch. 5]. Here we will concentrate on incorporating interval Newton methods into global optimization algorithms.

As mentioned in §2.1, an interval Newton method generally is an operator $\mathbf{N}(\mathbf{X}, F)$ associated with a function $F : \mathbf{R}^p \to \mathbf{R}^p$ defined on a box $\mathbf{X}$. Although actual derivations differ, we can intuitively think of interval Newton methods in terms of bounding the solution set to an interval linear system

$$\mathbf{A}(X - \check{X}) = -F(\check{X}). \tag{2.3}$$

The point $\check{X}$ is the midpoint of the box $\mathbf{X}$, an approximation to a solution to $F(X) = 0$, or some point that is otherwise appropriately chosen (usually, but not always, such that $\check{X} \in \mathbf{X}$). The interval matrix $\mathbf{A}$ can be chosen to be a *Lipschitz matrix* or a *slope matrix* for $F$; cf. [48] and definitions 1 and 2 in §3.3 below. In particular, an interval evaluation of the Jacobi matrix of $F$ will yield an $\mathbf{A}$ that is a Lipschitz matrix. This is appropriate for many purposes.

The interval Newton method is not applied directly to equation (2.3), but to the preconditioned equation

$$Y\mathbf{A}(X - \check{X}) = -YF(\check{X}), \tag{2.4}$$

where $Y$ is a $p$ by $p$ floating point matrix making the solution set to equation (2.4) easier to bound than that of equation 2.3. However, the solution set to equation (2.3) is in general a subset of the solution set to equation (2.4); cf. [48, §4.1].

Various ways of bounding the solution set to equation (2.3) or equation (2.4) appear in the literature. The interval hull, the smallest box or interval vector containing the solution set, can in principle be computed precisely. However, it may be impractical to compute the hull in practice, because the computation is NP complete in the order $p$ of the nonlinear system; cf. [56].

There are three classes of methods in common use to compute interval bounds on the solution set to equation (2.3) or equation (2.4). These are the Krawczyk method, the interval Gauss–Seidel method, and interval Gaussian elimination.

Let $\tilde{\mathbf{X}}$ represent the bounds obtained on the solution set to equation (2.4). Then, omitting some technical conditions[8] on the interval Newton methods, we have the following.

1. Let $\mathbf{A}$ be a slope matrix for $\mathbf{X}$ based at $\check{X}$ or a Lipschitz matrix over $\mathbf{X}$. If $\tilde{\mathbf{X}} \subset \text{int}(\mathbf{X})$ then there exists a solution of $F = 0$ in $\mathbf{X}$.
2. Let $\mathbf{A}$ be a slope matrix for $\mathbf{X}$ based at $\check{X}$ or a Lipschitz matrix over $\mathbf{X}$. If $\tilde{\mathbf{X}} \cap \mathbf{X} = \emptyset$ then there are no solutions to $F(X) = 0$ in $\mathbf{X}$.
3. Let $\mathbf{A}$ be a Lipschitz matrix over $\mathbf{X}$. If $\tilde{\mathbf{X}} \subset \text{int}(\mathbf{X})$ then there exists a unique solution of $F = 0$ in $\mathbf{X}$.
4. If $\mathbf{A}$ is a slope matrix for $\mathbf{X}$ based at $\check{X}$ or a Lipschitz matrix over $\mathbf{X}$, then any root of $F$ in $\mathbf{X}$ is also in $\tilde{\mathbf{X}}$. Furthermore, under certain conditions, replacing $\mathbf{X}$ by $\tilde{\mathbf{X}}$ leads to quadratic convergence of the widths of the components of $\mathbf{X}$ to zero. This allows subregions to be rigorously searched without excessive tessellation.

In addition, a compound algorithm can be used to show uniqueness when $\mathbf{A}$ is merely a slope matrix, and not a Lipschitz matrix. See [58] and the implementation description and experiments in [35].

Thus, performance of an interval Newton method can have differing goals (existence, uniqueness, nonexistence, or reduction in the size of $\mathbf{X}$). Also, interval Newton methods can be applied to different types of systems (general nonlinear system, Lagrange multiplier or Fritz–John system, etc.) Based on these and other considerations, different interval Newton methods are constructed. In summary, the items that can be varied are

- choice of method (direct hull computation, interval Gauss–Seidel, Krawczyk method, or interval Gaussian elimination);
- choice of preconditioner (if any);
- choice of the matrix $\mathbf{A}$ (slope matrix or Lipschitz matrix, and how it is computed);
- choice of the base point $\check{X}$.

We consider these choices in the context of global optimization algorithms in the remainder of this section.

[8] cf. [48] and more recent papers

## 3.1 Choice of Method

Algorithms for computation of the smallest interval vector containing the solution set to equation (2.3) are reviewed in [48, Ch. 6], while Rohn (who developed much of the theory of hull computation) reviews NP-hardness results in [56]. To our knowledge, computation of the exact interval hull is not used in any implementations of global optimization algorithms.

The Krawczyk method [38],

$$\mathbf{K}(\mathbf{X}, \check{X}) = \check{X} - YF(\check{X}) + [I - Y\mathbf{A}](\mathbf{X} - \check{X}). \tag{2.5}$$

appears most frequently in the literature, especially in early work. A good analysis of the verification and convergence properties of the Krawczyk method appears in [42], and it appears as *the* interval Newton method in [43, Ch. 5]. It is known that the Krawczyk method is not as sharp as the interval Gauss–Seidel method $\mathbf{GS}(\mathbf{X}, \check{X})$, i.e. that $\mathbf{GS}(\mathbf{X}, \check{X}) \subseteq \mathbf{K}(\mathbf{X}, \check{X})$, for a given preconditioner $Y$. However, convergence analysis is easier with the Krawczyk method. Furthermore, Rump, as reviewed in [58], has developed a method of both *inner* and *outer* estimations to the solution set of equation (2.3) for the Krawczyk method, and thus rigorous computational bounds on the overestimation of the hull by $\mathbf{K}(\mathbf{X}, \check{X})$. Such estimates, and hence the Krawczyk method, are useful when the widths of the components of $\mathbf{X}$ are small.

The interval Gauss–Seidel method shares properties with the classical Gauss–Seidel method. Its iteration scheme is best described with an algorithm.

**Algorithm 5** (Preconditioned interval Gauss–Seidel method)

DO *for* $i = 1$ *to* $p$.

1. (Update a coordinate) *Compute the* $i$*-th row* $Y_i$ *of the preconditioner.*
2. (Do the step) *Let* $\mathbf{A}_j$ *denote the* $j$*-th row of* $\mathbf{A}$. *Compute*

$$\tilde{\mathbf{x}}_i = \check{x}_i - \left[ Y_i F(\check{X}) + \sum_{j=1}^{i-1} Y_i \mathbf{A}_j (\tilde{\mathbf{x}}_j - \check{x}_j) + \sum_{j=i+1}^{p} Y_i \mathbf{A}_j (\mathbf{x}_j - \check{x}_j) \right] \Big/ (Y_i \mathbf{A}_i).$$

*3.* IF $\tilde{\mathbf{x}}_i \cap \mathbf{x}_i = \emptyset$,

THEN

EXIT, *signaling no solution.*

ELSE

*Replace* $\tilde{\mathbf{x}}_i$ *by* $\tilde{\mathbf{x}}_i \cap \mathbf{x}_i$.

END IF

END DO

**End Algorithm 5**

If Algorithm 5 does not terminate in step 3, then we label its output

$$\mathbf{GS}(\mathbf{X}, \check{X}) = \tilde{\mathbf{X}} = [\tilde{\mathbf{x}}_1, \ldots, \tilde{\mathbf{x}}_p]^T. \quad (2.6)$$

The analogous interval Jacobi method (each $\tilde{\mathbf{x}}_j$ is replaced by $\mathbf{x}_j$) and asynchronous parallel variants (each $\tilde{\mathbf{x}}_j$ is computed separately and made available for computations for the other $j$ in an unpredictable way) are also worthy of consideration.

For a fixed preconditioner matrix $Y$, the image under the Gauss–Seidel method is at least as narrow (and hence, at least as good) as the image under the Krawczyk method. Furthermore, a set of optimal preconditioners, in the sense that the widths of the coordinates of the image are as small as possible, has been developed; cf. §3.2.

The third method, interval Gaussian elimination,interval Gaussian elimination, sometimes gives better and sometimes worse results than the Krawczyk method or the interval Gauss–Seidel method; cf. [48, p. 158ff]. Although similar to the floating point Gaussian elimination algorithm, the *interval Gauss algorithm* is worth stating. In particular, although operations are done on intervals with non-zero widths during the elimination phase, it is rigorous to replace the elements being eliminated by exactly zero. Also, although the floating point Gauss algorithm has many variants, all of which are mathematically equivalent in exact arithmetic, these variants are not all the same when interval arithmetic is used. For this reason, we present the interval Gauss algorithm useful in our present context, as in [1, Ch. 15].

**Algorithm 6** (Interval Gaussian Elimination)

*1.* (Input the matrix and right-hand side)
*Input* $Y\mathbf{A} = \mathbf{G} = [\mathbf{g}_{i,j}]$ *and*[9] $-YF(\check{X}) = \mathbf{S} = [\mathbf{s}_1, \ldots, \mathbf{s}_n]^T$.

*2.* (Factorization phase) DO *for* $i = 1$ *to* $p$*:*

   *(a)* IF $0 \in \mathbf{g}_{i,i}$ THEN EXIT *with failure.*

   *(b)* DO *for* $j = i+1$ *to* $p$ *and for* $k = i+1$ *to* $p$*:*

      $\mathbf{g}_{j,k} \leftarrow \mathbf{g}_{j,k} - (\mathbf{g}_{j,i}/\mathbf{g}_{i,i})\, \mathbf{g}_{i,k}$.

   END DO

   *(c)* $\mathbf{s}_i \leftarrow \mathbf{s}_i - (\mathbf{g}_{j,i}/\mathbf{g}_{i,i})\, b_i$

END DO

*3.* (Solution phase)

   *(a)* $\mathbf{s}_{p,p} \leftarrow \mathbf{s}_p/\mathbf{g}_{p,p}$.

   *(b)* DO *for* $i = p-1$ *to* $1$ *by* $-1$*:*

      $\mathbf{s}_i \leftarrow \left(\mathbf{s}_i - \sum_{j=i+1}^{p} \mathbf{g}_{i,j}\mathbf{s}_j\right) / \mathbf{g}_{i,i}$

END DO

*4.* $\tilde{\mathbf{x}}_i \leftarrow \mathbf{s}_i + \check{x}_i$ *for* $i = 1$ *to* $p$.

**End Algorithm 6**

The main advantage of the interval Gauss algorithm is that ***starting bounds*** $\mathbf{X}_0$ ***are not required, except to compute the matrix*** $\mathbf{A}$. This is a crucial property used in the method proposed in [20] and [19, §10.6] for bounding Lagrange multipliers, and hence for applying interval Newton methods to constrained problems; see §3.5 below.

Pivoting is generally not done in the interval Gauss algorithm, but the preconditioner matrix $Y$ is designed so that $\mathbf{G}$ is like an identity matrix; see §3.2. However, Algorithm 6 can fail when the widths of the elements of $\mathbf{A}$ are large, when the preconditioner matrix $Y$ is ill-conditioned, or when some combination of these conditions occurs. Even when the Gauss algorithm fails, it may still be possible to use the partial factorization to reduce some coordinate widths, provided initial estimates $\mathbf{X}_0$ for the coordinate intervals are available.

[9] Although $-YF(\check{X})$ is a point, $\mathbf{S}$ is transformed into an interval vector in this algorithm.

## 3.2 Choice of Preconditioner

The most commonly seen preconditioner $Y$ in the the Krawczyk, interval Gauss–Seidel, and interval Gauss algorithms is the approximate inverse of the matrix formed from the midpoints of the elements of $\mathbf{A}$, termed the ***inverse midpoint preconditioner***. With this preconditioner, $\mathbf{G} = Y\mathbf{A}$ is[10] symmetric about the identity matrix. In certain contexts [58], this symmetry can be used to replace interval arithmetic by appropriately rounded floating point arithmetic. The preconditioner also has certain optimality properties that have been discussed in the literature in relation to all three methods. It is relatively simple to compute, its effect is relatively easy to analyze, and it is effective when the widths of $\mathbf{A}$ are small.

In [29], preconditioner rows for the interval Gauss–Seidel method based upon certain ***width optimality*** conditions were proposed. Since their computation requires solution of linear programming problems, these preconditioners are termed ***optimal LP preconditioners***. Alternate optimality conditions and reformulation to a smaller LP problem appeared in [31]. The goal in [29] and [31] was to find $Y$ so that the widths of the components $\tilde{\mathbf{X}}$ were minimized, and thus enable or speed convergence of interval iteration to a point.

As seen in [29], LP preconditioners can be effective provided the structure of the corresponding LP problems is utilized in their solution. Otherwise, the cost of solving these problems could overwhelm the advantages of an optimal optimal preconditioner. It is still uncertain how useful these preconditioners can be in optimization. They are most advantageous when the relative widths of elements in different rows of $\mathbf{A}$ differ widely.

OptimalLP preconditioners as originally formulated may not be appropriate for verification. This is because minimization of $w(\tilde{\mathbf{x}}_i)$ may not be consistent with $\tilde{\mathbf{x}}_i \subset \mathrm{int}(\mathbf{x}_i)$. However, it is possible to formulate the LP problem in terms of maximization of the possibility that $\tilde{\mathbf{x}}_i \subset \mathrm{int}(\mathbf{x}_i)$. It is also possible to formulate optimal preconditioners for interval Gaussian elimination, and even for general systems[11] $\mathbf{A}X = B$. The inverse midpoint preconditioner is applicable to such systems while some such formulations lead to larger and hence more expensive-to-solve linear programming problems. Such formulations have not yet been fully investigated.

---

[10] approximately, if $Y$ is an approximate inverse

[11] The interval Newton equation (2.3) is special because both the right-hand-side and the unknown $X - \check{X}$ approximate the zero vector.

## 3.3 Choice of Derivative Matrix

The matrices $\mathbf{A}$ for equation (2.3) generally satisfy one of the following two properties.

**Definition 1** *(*[48, p. 174], etc.) *The matrix* $\mathbf{A}$ *is said to be a* Lipschitz matrix *for* $F$ *over* $\mathbf{X}$ *provided, for every* $X \in \mathbf{X}$ *and* $Y \in \mathbf{X}$, $F(X) - F(Y) = A(X - Y)$ *for some* $A \in \mathbf{A}$.

Generally, uniqueness can be proven with an interval Newton method provided $Y$ is nonsingular and $\mathbf{A}$ is a Lipschitz matrix; cf. [48].

An alternate, weaker property commonly discussed in the literature is the following.

**Definition 2** *The matrix* $\mathbf{A}$ *is said to be an* interval slope matrix *for* $F$ *over* $\mathbf{X}$ *and centered on the interval vector* $\check{\mathbf{X}}$, *provided, for every* $X \in \mathbf{X}$ *and* $\check{X} \in \check{\mathbf{X}}$, $F(X) - F(\check{X}) = A(X - \check{X})$ *for some* $A \in \mathbf{A}$.

Note that every Lipschitz matrix is a slope matrix, but not visa versa. As seen below, slope matrices can be computed whose entries are narrower than computed Lipschitz matrices.

The following concept is also relevant.

**Definition 3** *An interval matrix* $\mathbf{A}$ *is said to be* regular *provided every* $A \in \mathbf{A}$ *is nonsingular.*

For example, regularity of $\mathbf{A}$ is implied by $\tilde{\mathbf{X}} \subset \mathbf{X}$ in equation (2.3) or equation (2.4). See [58] and [63] for methods in which values or ranges of $F$, but only the elements of $\mathbf{A}$, appear in the computational formula.

We will illustrate various choices for $\mathbf{A}$ by considering the second-order information for the objective function $\phi(x_1, x_2) = x_1^4/12 + x_1(x_2^3/3) + x_1^2/2 + x_2^2/2$, whose Hessian matrix is

$$\nabla^2\phi(x_1, x_2) = \begin{bmatrix} x_1^2 + 1 & x_2^2 \\ x_2^2 & 2x_1x_2 + 1 \end{bmatrix}. \tag{2.7}$$

In these examples, $F$ denotes $\nabla\phi$, so the derivative matrix $\mathbf{A}$ of $F$ corresponds to the Hessian matrix $\nabla^2\phi$. We will examine matrices $\mathbf{A}$ for $\phi$ valid over the interval vector $\mathbf{X} = ([-.5, .5], [-.5, .5])^T$.

Using interval arithmetic to evaluate equation (2.7), we obtain:

$$[\text{Interval Hessian}] = \begin{bmatrix} [1, 1.25] & [0, .25] \\ [0, .25] & [.5, 1.5] \end{bmatrix}. \tag{2.8}$$

The interval Hessian matrix is a Lipschitz matrix for $F$ over $\mathbf{X}$. It can thus be used flexibly in interval Newton algorithms. For example, when it is used to bound the solution set to equation (2.4) using the Krawczyk, interval Gauss–Seidel, or interval Gauss algorithms, uniqueness is implied by $\tilde{\mathbf{X}} \subset \text{int}(\mathbf{X})$, regardless of the choice of base point $\check{X}$, as long as $\check{X} \in \mathbf{X}$. The drawback is that the widths of the entries of an interval Hessian matrix may be so large that the resulting image box $\tilde{\mathbf{X}}$ is not useful.

In [16] and later in [19, §6.3-6.4], Hansen observes that, when evaluating the Hessian matrix for interval Newton methods, not all entries need be intervals. In particular, the $j$-th component of $\mathbf{X}$ in evaluation of $\partial^2\phi(X)/\partial x_i \partial x_k$ in the $i$-th row[12] of the matrix may be replaced by the $j$-th component of $\check{X}$ for $j > k$. Theory of uniqueness verification using this scheme has not been published. The matrix $\mathbf{A}$ so obtained is no longer a Lipschitz matrix, so the general uniqueness theory in [48] is not valid. Nonetheless, such $\mathbf{A}$ are slope matrices, andexistence can be verified. Such matrices are appropriate for (at least the initial steps of) interval Newton iteration. The drawback is that the matrix must be recomputed with each new $\check{X}$.

Hansen's technique applied to our example problem with $\check{X} = [0, 0]^T$ and a specified ordering[13] to $\partial\phi/\partial x_2$ gives the matrix:

$$[\text{Hansen Hessian}] = \begin{bmatrix} [1, 1.25] & [0, 0.25] \\ [0, 0.25] & 1 \end{bmatrix}. \tag{2.9}$$

An alternate point-based way of computing matrices $\mathbf{A}$, more well-known than Hansen's scheme, is utilization of ***interval slopes***. Simple one-dimensional examples show that, for $\check{\mathbf{X}}$ a point, uniqueness is ***not*** implied if $\tilde{\mathbf{X}} \subset \text{int}(\mathbf{X})$ when

[12] Other orderings are also possible; cf. [19, §6.4].

[13] cf. [19, §6.3-6.4]. In this example, we are holding $x_1$ fixed in computing $\mathbf{a}_{2,2}$.

**A** is formed from interval slopes. However, such **A** are slope matrices in the sense of Definition 2, so existence is implied. Furthermore, as introduced in [58], uniqueness can be verified over a large box **X** by first implying existence over a small box $\check{\mathbf{X}}$ centered at a point $\check{X}$, then showing that the slope matrix over a large box **X** and centered on $\check{\mathbf{X}}$ is regular.

As described by various authors, slope matrices may be computed by operator overloading with techniques similar to automatic differentiation. A sharp scheme, applicable when **X** is large but subject to roundout error for small **X**, appears in [58]. An alternate scheme that can be combined with that of [58] appears in [48]. With the scheme of [58] applied to the exact gradient vector $\nabla\phi = [x_1^3/3 + x_1 + x_2^3/3, x_1x_2^2 + x_2]^T$ and $\check{\mathbf{X}} = \check{X} = [0,0]^T$, the slope matrix for our example is:

$$[\text{Computed slope}] = \begin{bmatrix} [1, 1.0834] & [0, 0.08334] \\ [0, 0.25] & [.75, 1.25] \end{bmatrix}. \tag{2.10}$$

Actually, Hansen's technique and automatic computation of slopes can be combined. In particular, if components of $\check{\mathbf{X}}$ are substituted for components of **X** as in [19, §6.4] and an automatic slope computation scheme is used, the resulting matrix is a slope matrix in the sense of Definition 2. Thus, existence verification, as well as uniqueness verification over large boxes with the two-stage scheme of [58] can proceed with such **A**.

The matrix **A** obtained for our example by combining slope computation with Hansen's idea is:

$$[\text{Hansen–slope}] = \begin{bmatrix} [1, 1.0834] & [0, 0.08334] \\ [0, 0.25] & 1 \end{bmatrix}. \tag{2.11}$$

Based on this present state-of-the-art, we have the following recommendations:

- Hansen's idea combined with slopes should be used:
  - if **A** is to be used only for interval iteration over relatively large boxes.
  - if existence verification only is desired;
  - if uniqueness verification only in as large a box **X** as possible is desired (for example, for deleting subregions in a global search, as in [3]);
  - if non-existence verification only is desired.

- any combination of the above.

- An interval Jacobi matrix should be used:
  - for interval iteration over small boxes where an interval Jacobi matrix is adequate for convergence, especially if $\check{X}$ must change frequently.
  - if uniqueness verification is required over small boxes $\mathbf{X}$.

If existence only is required over a small box, such as for checking feasibility in Algorithm 3 of §2.1, Hansen's technique combined with slopes is theoretically sharper than interval Hessian matrices. That is, the Hansen-slope scheme is in general preferable in $\epsilon$-inflation algorithms that determine existence. However, depending on the implementation of slopes, the computed slopes may actually be wider than corresponding interval derivatives[14]. Also, other considerations, such as ease of implementation, may enter into consideration.

## 3.4 Choice of Base Point $\check{X}$

Little appears in the literature concerning selection of the point $\check{X}$. Centered forms for bounding the range, satisfying optimality conditions, were given in [2]. These forms, based on particular choice of point $\check{X}$, can be used when iterating interval Newton methods.

Common choices of $\check{X}$ include the center of the box $\mathbf{X}$, a root of $F(X) = 0$, or both. Some analysis of the case when $\check{X}$ is both a root and the center of the box appears in [63]. This choice, natural in $\epsilon$-inflation when one wishes to verify bounds on a solution obtained through a conventional floating-point method, appears to make verification more likely. An early, abstract analysis of verification tests in [27] indicates that verification is easier if the root of $F(X) = 0$ is near the center of the box. Pictures in [36] also illustrate that it is wise for $\check{X}$ in Algorithm 3 to be a good approximation to a feasible point (corresponding to a root of $F$), and for the box $\hat{\mathbf{X}}$ in that algorithm to be centered on such a feasible point. Nonetheless, additional analysis of both choosing $\check{X}$ to be a root of $F$ and choosing $\check{X}$ to be centered in the box $\mathbf{X}$ can be done.

[14] The divided differences in [58] should *not* be used for boxes $\mathbf{X}$ that are small in relation to the computational precision.

A graphical examination of the case $F : \mathbf{R}^1 \rightarrow \mathbf{R}^1$ suggests that verification of existence or non-existence using slopes can sometimes be more effective when $\check{X}$ is chosen so $\check{X} \notin \mathbf{X}$. Further investigation of this may be useful.

In summary, the present state of knowledge indicates that both roots of $F(X) = 0$ and $\check{X}$ should be centered in $\mathbf{X}$, where possible, if existence or uniqueness verification is desired.

## 3.5 The Fritz–John Conditions

In certain circumstances, we wish to simultaneously verify feasibility and local optimality. For instance, a conventional floating point constrained optimizer may have been used to obtain an approximate global optimum $\check{X}$. We may then wish to verify uniqueness of a critical point in as large a box $\mathbf{X}$ as possible about $\check{X}$, so that $\mathbf{X}$ may be excluded from further consideration in the exhaustive search[15]. The system of equations to be used in the interval Newton method must therefore embody the necessary conditions for constrained optima. General conditions, not requiring "constraint qualification" assumptions, are the Fritz–John conditions.

The Fritz–John conditions have been advocated by Hansen *et al* in [20] and [19]. Their use is thoroughly explained in those works, although empirical results are lacking. Here, for convenience, we present the function $F$ and derivative matrix $\nabla F$ corresponding to the Fritz–John conditions for our formulation in equation (2.1).

The variables in our system are $X = (x_1, \ldots, x_n)$, $V = (v_1, \ldots, v_m)$, and $u_0$, for a total of $n + m + 1$ variables. We will write $W = (X, u_0, V)$. The function $F$ is:

$$F(W) = \begin{bmatrix} u_0 \nabla\phi(X) + \sum_{i=1}^{m} v_i \nabla c_i(X) \\ c_1(X) \\ \vdots \\ c_m(X) \\ \left[u_0 + \sum_{i=1}^{m} v_i [1, 1+\epsilon]\right] - 1 \end{bmatrix} = 0, \qquad (2.12)$$

The $v_i$ are unconstrained and represent Lagrange multipliers for the equality constraints $c_i = 0$. The last equation is a normalization condition suggested and justified in [20]; $\epsilon$ is on the order of the computational precision.

[15] This is done in the unconstrained case in [3] and for general nonlinear systems in [35].

By not including the bound constraints $a_{i_j} \le x_{i_j} \le b_{i_j}$ in this function, we reduce the size of the system by $2q$. Furthermore, it is more flexible to include the bound constraints through the process reviewed in §2.3. Thus, equation (2.12) is applicable for points $X$ when none of the bound constraints are active. When one or more bound constraints are active, an analogue of equation (2.12) in the appropriate lower-dimensional subspace is used. In fact, it can be shown that, if the entire space is used and the bound constraints are included in the Fritz–John function $F$, then the Jacobi matrix of $F$ must be singular; cf. [36] for a precise statement of this fact.

The Jacobi matrix of $F$, used to form the matrix $\mathbf{A}$ for interval Newton methods, is:

$$H(W) = \left[\begin{array}{c|c|c|c|c} u_0\nabla^2\phi(X) + \sum_{i=1}^m v_i\nabla^2 c_i(X) & \nabla\phi(X) & \nabla c_1(X) & \ldots & \nabla c_m(X) \\ \hline \nabla c_1(X) & 0 & 0 & \ldots & 0 \\ \vdots & \vdots & \vdots & \vdots & \vdots \\ \nabla c_m(X) & 0 & 0 & \ldots & 0 \\ \hline 0 & 1 & [1, 1+\epsilon] & \ldots & [1, 1+\epsilon] \end{array}\right]. \tag{2.13}$$

If Hansen's technique as explained in §3.3 is used to evaluate $H$ for the matrix $\mathbf{A}$ in the interval Newton method, then only (guess) point values of $u_0$ and the $v_i$ enter. Thus, if Gaussian elimination is used to solve equation (2.4), initial bounds on the Lagrange multipliers $u_0$ and $v_i$ are not required. This procedure, first suggested in [20], is recommended by us, although further experimentation is needed to ascertain what are reasonable guesses for the base points for $u_0$ and the $v_i$. In verifying points obtained by floating point constrained optimization software, approximate Lagrange multiplier values may be available along with the approximate optimizers.

# 4 APPLICATIONS

Actual successful applications of constrained global optimization codes have to date been limited by general lack of availability of such codes to engineers and scientists. A notable exception is the work in [39], in which an algorithm to determine feasibility of constraints is presented and used in structural design analysis with composite laminates. There has also been some success at applying related general nonlinear equation codes or associated techniques. For example, Schnepper and Stadtherr [59], [60] have modified the TOMS algorithm INTBIS [30] for efficient analysis of chemical process-based flowsheeting. K. Okumura, (private communication), tailoring the techniques to the problem, has applied interval methods for nonlinear systems to analysis of linear and nonlinear electrical networks. In fact, not only did Okumura achieve rigor, but the method proved to be orders of magnitude faster than a Monte Carlo simulation.

In [14], Hager has used techniques in common with interval global optimization algorithms to determine when certain constraints are violated, for use in robot sensing and manipulation, etc.

Csendes and Ratz [6, problem Ex-2, §4] have recently applied Ratz' unconstrained global optimization code successfully to a parameter estimation problem related to respiratory mechanical model identification.

More generally, there has been a steady, and recently increasing interest in interval nonlinear equations technology and interval computations in general for robust geometric computations in computer-aided geometric design. An early review of such techniques is [47]. Computation of the solution of polynomial systems of equations in this context appears in [62]. Experiments with an interval algorithm for intersection of curves appear in [61]. A clever application of interval arithmetic to solution of the nonlinear systems involved in interval ray tracing, allowing substantial speedup of the process, appears in [9].

Our own examination of the realistic test problems in [12] leads us to believe that there is much potential for practical application of interval global optimization algorithms in chemical engineering and process design.

# 5 SUMMARY AND PRESENT WORK

Various researchers have produced codes for verified global optimization. The books [52] and [19] are devoted to the subject, while Arnold Neumaier is preparing a third. Additionally, the book [48] treats verified solution of nonlinear algebraic systems, while several books are devoted to the underlying interval arithmetic techniques.

While techniques for constrained problems have been published, high-quality experimental reports appear to exist only for unconstrained optimization. Also, with the possible exception of the codes of [15] or [54], most verified optimization programs to date have been research codes, or have been written in non-portable programming languages.

Besides reviewing this work, we have carefully proposed a method for handling both equality and inequality constraints, and have proposed a variant of a method of Hansen for rigorously verifying feasibility. We have advocated a particular structure when handling inequality, equality, and bound constraints. We have also suggested combining interval slopes with an idea of Hansen for derivative matrices.

We have been developing a portable Fortran-90 environment to support global optimization and numerical nonlinear algebra. The environment, described in [10], is based on the portable FORTRAN-77 package INTLIB of [9]. We intend to complete and test a comprehensive optimization code in this system. This code will be based on the ideas in this work and on the principle that the global algorithm should build on solutions of approximate optimizers that have been verified.

## Acknowledgements

I wish to thank George Corliss, Eldon Hansen, Dietmar Ratz, Vladik Kreinovich and other researchers of this subject for their encouragement, careful reading, and suggestions.

## REFERENCES

[1] G. Alefeld and J. Herzberger, *Introduction to Interval Computations*, Academic Press, New York, 1983.

[2] E. Baumann, "Optimal Centered Forms", *BIT*, 1988, Vol. 28, No. 1, pp. 80–87.

[3] O. Caprani, B. Godthaab, and K. Madsen, "Use of a Real-Valued Local Minimum in Parallel Interval Global Optimization", *Interval Computations*, 1993, No. 2, pp. 71–82.

[4] A. R. Conn, N. Gould, and Ph. L. Toint, *LANCELOT: A Fortran Package for Large-Scale Nonlinear Optimization*, Springer-Verlag, New York, 1992.

[5] T. Csendes, "Nonlinear Parameter Estimation by Global Optimization – Efficiency and Reliability", *Acta Cybernetica*, 1988, Vol. 8, No. 4, pp. 361–370.

[6] T. Csendes and D. Ratz, *Subdivision Direction Selection in Interval Methods for Global Optimization*, preprint, 1994.

[7] J. E. Dennis and R. B. Schnabel, *Numerical Methods for Unconstrained Optimization and Nonlinear Least Squares*, Prentice-Hall, Englewood Cliffs, NJ, 1983.

[8] Kaisheng Du and R. B. Kearfott, "The Cluster Problem in Global Optimization The Univariate Case", *Computing* (Suppl.), 1992, Vol. 9, pp. 117–127.

[9] W. Enger, "Interval Ray Tracing – A Divide and Conquer Strategy for Realistic Computer Graphics", *The Visual Computer*, 1992, Vol. 9, pp. 91–104.

[10] J. Eriksson, *Parallel Global Optimization using Interval Analysis*, Ph.D. dissertation, University of Umeå, Institute of Information Processing, 1991.

[11] J. Eriksson and P. Lindstrœm, "A Parallel Interval Method Implementation for Global Optimization Using Dynamic Load Balancing", *Reliable Computing*, Vol. 1, No. 1, pp. 77–92.

[12] C. A. Floudas and P. M. Pardalos, *A Collection of Test Problems for Constrained Global Optimization Algorithms*, Springer-Verlag, New York, 1990.

[13] P. E. Gill, W. Murray, and M. Wright, *Practical Optimization*, Academic Press, New York, 1981.

[14] G. D. Hager, *Solving Large Systems of Nonlinear Constraints with Application to Data Modeling*, preprint, 1993.

[15] R. Hammer, M. Hocks, U. Kulisch, and D. Ratz, *Numerical Toolbox for Verified Computing I*, Springer-Verlag, New York, 1993.

[16] E. R. Hansen, "Interval Forms of Newton's Method", *Computing*, 1978, Vol. 20, pp. 153–163.

[17] E. R. Hansen, "Global Optimization Using Interval Analysis: The One-Dimensional Case", *J. Optim. Theory Appl.*, 1979, Vol. 29, No. 3, pp. 331-344.

[18] E. R. Hansen, "Global Optimization Using Interval Analysis: the Multidimensional Case", *Numer. Math.*, 1980, Vol. 34, No. 3, pp. 247–270.

[19] E. R. Hansen, *Global Optimization Using Interval Analysis*, Marcel Dekker, Inc., New York, 1992.

[20] E. R. Hansen and G. W. Walster, "Bounds for Lagrange Multipliers and Optimal Points", *Comput. Math. Appl.*, 1993, Vol. 25, No. 10, pp. 59–ff.

[21] K. Ichida and Y. Fujii, "An Interval Arithmetic Method for Global Optimization", *Computing*, 1979, Vol. 23, No. 1, pp. 85–97.

[22] C. Jansson, *A Global Minimization Method: The One-Dimensional Case*, Technical Report No. 91.2, 1991.

[23] C. Jansson, "A Global Optimization Method Using Interval Arithmetic", In: L. Atanassova and J. Herzberger (eds.), *Computer Arithmetic and Enclosure Methods. Proc. Third International IMACS-GAMM Symposium on Computer Arithmetic and Scientific Computing*, North–Holland, Amsterdam, Netherlands, 1992, pp. 259–268.

[24] C. Jansson and O. Knüppel, *A Global Minimization Method: The Multi-Dimensional Case*, preprint, 1992.

[25] C. Jansson, "On Self-Validating Methods for Optimization Problems", In: J. Herzberger (ed.), *Topics in Validated Computations*, North-Holland, Amsterdam, Netherlands, 1994, pp. 381–439.

[26] C. Jansson and O. Knüppel, *Numerical Results for a Self-Validating Global Optimization Method*, Technical Report, 1994.

[27] R. B. Kearfott, "Abstract Generalized Bisection and a Cost Bound", *Math. Comp.*, 1987, Vol. 49, No. 179, pp. 187–202.

[28] R. B. Kearfott, "Interval Arithmetic Techniques in the Computational Solution of Nonlinear Systems of Equations: Introduction, Examples, and Comparisons", In: E. L. Allgower and K. Georg (eds.), *Computational Solution of Nonlinear Systems of Equations (Lectures in Applied Mathematics, Vol. 26)*, American Mathematical Society, Providence, RI, 1990, pp. 337–358.

[29] R. B. Kearfott, "Preconditioners for the Interval Gauss–Seidel Method", *SIAM J. Numer. Anal.*, 1990, Vol. 27, No. 3, pp. 804–822.

[30] R. B. Kearfott and M. Novoa, "INTBIS, A Portable Interval Newton/Bisection Package (Algorithm 681)", *ACM Trans. Math. Software*, 1990, Vol. 16, No. 2, pp. 152–157.

[31] R. B. Kearfott, C. Y. Hu, and M. Novoa III, "A Review of Preconditioners for the Interval Gauss–Seidel Method", *Interval Computations*, 1991, No. 1, pp. 59–85.

[32] R. B. Kearfott, "An Interval Branch and Bound Algorithm for Bound Constrained Optimization Problems", *Journal of Global Optimization*, 1992, Vol. 2, pp. 259–280.

[33] R. B. Kearfott, M. Dawande, K.-S. Du, and C.-Y. Hu, "Algorithm 737: INTLIB: A Portable FORTRAN 77 Interval Standard Function Library" *ACM Trans. Math. Software*, 1994, Vol. 20, No. 4, pp. 447–459.

[34] R. B. Kearfott, "A Fortran 90 Environment for Research and Prototyping of Enclosure Algorithms for Constrained and Unconstrained Nonlinear Equations", *ACM Trans. Math. Software*, 1995, Vol. 21, No. 1, pp. 63–78.

[35] R. B. Kearfott, *Empirical Evaluation of Innovations in Interval Branch and Bound Algorithms for Nonlinear Algebraic Systems*, preprint, 1994.

[36] R. B. Kearfott, *On Verifying Feasibility in Equality Constrained Optimization Problems*, preprint, 1994.

[37] R. B. Kearfott and K. Du, "The Cluster Problem in Multivariate Global Optimization", *Journal of Global Optimization*, 1994, Vol. 5, pp. 253–265.

[38] R. Krawczyk, "Newton-Algorithmen zur Bestimmung von Nullstellen mit Fehlershranken", *Computing*, 1969, Vol. 4, pp. 187–201.

[39] B. P. Kristinsdottir, Z. B. Zabinsky, T. Csendes, M. E. Tuttle, "Methodologies for Tolerance Intervals", *Interval Computations*, 1993, No. 3, pp. 133–147.

[40] A. Leclerc, "Parallel Interval Global Optimization in C++", *Interval Computations*, 1993, No. 3, pp. 148–163.

[41] G. Mayer, *Epsilon-Inflation in Verification Algorithms*, preprint, 1993.

[42] R. E. Moore, "A Test for Existence of Solutions to Nonlinear Systems", *SIAM J. Numer. Anal.*, 1977, Vol. 14, No. 4, pp. 611–615.

[43] R. E. Moore, *Methods and Applications of Interval Analysis*, SIAM, Philadelphia, 1979.

[44] R. E. Moore and H. Ratschek, "Inclusion Functions and Global Optimization II", *Math. Prog.*, 1988, Vol. 41, No. 3, pp. 341–356.

[45] R. E. Moore, E. Hansen, and A. Leclerc, "Rigorous Methods for Parallel Global Optimization", In: A. Floudas and P. Pardalos (eds.), *Recent Advances in Global Optimization* Princeton Univ. Press, Princeton, N.J., 1992, pp. 321–342.

[46] J. J. Moré and S. J. Wright, *Optimization Software Guide*, SIAM, Philadelphia, 1993.

[47] S. P. Mudur and P. A. Koparkar, "Interval Methods for Processing Geometric Objects", *IEEE Comput. Graphics and Appl.*, 1984, Vol. 4, No. 2, pp. 7–17.

[48] A. Neumaier, *Interval Methods for Systems of Equations*, Cambridge University Press, Cambridge, England, 1990.

[49] A. Neumaier, *Second-Order Sufficient Optimality Conditions for Local and Global Nonlinear Programming*, preprint, 1994.

[50] P. M. Pardalos and J. B. Rosen, *Constrained Global Optimization: Algorithms and Applications*, Springer-Verlag, New York, 1987.

[51] P. M. Pardalos and S. A. Vavasis, "Quadratic Programming with One Negative Eigenvalue is NP-Hard", *Journal of Global Optimization*, 1992, Vol. 1, No. 1.

[52] H. Ratschek and J. Rokne, *New Computer Methods for Global Optimization*, Wiley, New York, 1988.

[53] H. Ratschek and R. L. Voller, "Global Optimization over Unbounded Domains", *SIAM J. Control Optim.*, 1990, Vol. 28, No. 3, pp. 528–539.

[54] D. Ratz, *Automatische Ergebnisverifikation bei globalen Optimierungsproblemen*, Ph.D. dissertation, Universität Karlsruhe, 1992.

[55] D. Ratz, "Box-Splitting Strategies for the Interval Gauss–Seidel Step in a Global Optimization Method", *Computing*, 1994, Vol. 53, pp. 337–354.

[56] J. Rohn, *NP-Hardness Results for Linear Algebraic Problems with Interval Data*, preprint, 1994.

[57] S. M. Rump, *Kleine Fehlerschranken bei Matrixproblemen*, Ph.D. dissertation, Universität Karlsruhe, 1980.

[58] S. M. Rump, "Verification Methods for Dense and Sparse Systems of Equations", In: J. Herzberger (ed.), *Topics in Validated Computations*, North–Holland, Amsterdam, 1994, pp. 63–135.

[59] C. A. Schnepper, *Large Grained Parallelism in Equation-Based Flowsheeting Using Interval Newton / Generalized Bisection Techniques*, Ph.D. dissertation, University of Illinois, Urbana, Department of Chemical Engineering, 1992.

[60] C. A. Schnepper and M. A. Stadtherr, "Application of a Parallel Interval Newton/Generalized Bisection Algorithm to Equation-Based Chemical Process Flowsheeting", *Interval Computations*, 1993, No. 4, pp. 40–64.

[61] T. W. Sederberg and S. R. Parry, "Comparison of Three Curve Intersection Algorithms", *Comput. Aided Des.*, 1986, Vol. 18, No. 1, pp. 58–63.

[62] E. C. Sherbrooke and N. M. Patrikalakis, *Computation of the Solutions of Nonlinear Polynomial Systems*, preprint, 1993.

[63] X. Shi and R. B. Kearfott, *Some Results on the Regularity of an Interval Matrix*, preprint, 1994.

[64] X. Shi, *Intermediate Expression Preconditioning and Verification for Rigorous Solution of Nonlinear Problems*, Ph.D. dissertation, University of Southwestern Louisiana, Department of Mathematics, August 1995.

[65] S. Skelboe, "Computation of Rational Interval Functions", *BIT*, 1974, Vol. 14, pp. 87–95.

[66] G. W. Walster, E. R. Hansen, and S. Sengupta, "Test Results for a Global Optimization Algorithm", In: P. T. Boggs, R. H. Byrd, and R. B. Schnabel (eds.), *Numerical Optimization 1984*, SIAM, Philadelphia, 1985, pp. 272–287.

[67] M. A. Wolfe, "An Interval Algorithm for Constrained Global Optimization", *J. Comput. Appl. Math.*, 1994, Vol. 50, pp. 605–612.

# 3

# THE SHAPE OF THE SYMMETRIC SOLUTION SET


**Götz Alefeld*, Vladik Kreinovich**, and Günter Mayer*****

**Institut für Angewandte Mathematik, Universität Karlsruhe, D-76128Karlsruhe, Germany, email goetz.alefeld@mathematik.uni-karlsruhe.de*

***Department of Computer Science, University of Texas at El Paso, El Paso, TX 79968, email vladik@cs.utep.edu*

****Fachbereich Mathematik, Universität Rostock, D–18051 Rostock, Germany, email guenter.mayer@mathematik.uni-rostock.de*



## ABSTRACT

We give a new deduction of the set of inequalities which characterize the solution set $S$ of real linear systems $Ax = b$ with the $n \times n$ coefficient matrix $A$ varying between a lower bound $\underline{A}$ and an upper bound $\overline{A}$, and with $b$ similarly varying between $\underline{b}$ and $\overline{b}$. The idea of this deduction can also be used to construct a set of inequalities which describe the so–called symmetric solution set $S_{\text{sym}}$, i.e., the solution set of $Ax = b$ with $A = A^T$ varying between the bounds $\underline{A} = \underline{A}^T$ and $\overline{A} = \overline{A}^T$. This is the main result of our paper. We show that in each orthant $S_{\text{sym}}$ is the intersection of $S$ with sets of which the boundaries are quadrics.


## 1 INTRODUCTION

In [2], [3], [4], [5], [6], [7], and others, the intersection $S \cap O$ of the *solution set*

$$S := \{x \in \mathbf{R}^n \mid Ax = b,\ A \in [A],\ b \in [b]\} \tag{3.1}$$

with an orthant $O$ is characterized by a set of inequalities which describe half–spaces. Here, $[A]$ denotes an $n \times n$ matrix with real compact intervals as entries, $[b]$ is a given vector with $n$ real compact intervals as components, and $[A]$ is assumed to be *regular*, i.e., it contains only regular matrices as elements. In [2] it is also shown that for regular *symmetric* interval matrices $[A] = [A]^T$ the




*R. B. Kearfott and V. Kreinovich (eds.), Applications of Interval Computations*, 61–79.

*symmetric solution set*

$$S_{\text{sym}} := \{x \in \mathbf{R}^n \mid Ax = b,\ A = A^T \in [A],\ b \in [b]\} \subseteq S \tag{3.2}$$

is compact and connected. For $2 \times 2$ interval matrices, the intersection $S_{\text{sym}} \cap O$ was characterized as an intersection of $S$ with sets of which the boundaries are quadrics. The proof could, however, not be generalized to the $n \times n$ case, $n > 2$. In the present paper, we handle this general problem by another technique based on results of Tarski [9] and Seidenberg [8]. In addition, we dispense with the regularity of $[A]$ in the definition of $S$ and $S_{\text{sym}}$. It is obvious that these sets then no longer need to be bounded. First, we apply the new technique to obtain the description of $S \cap O$ by the above–mentioned set of inequalities. Then we show that the same technique can be used to characterize $S_{\text{sym}} \cap O$ as intersection of $S$ with sets of which the boundaries are quadrics. This is the generalization of the results from [2] to the case $n > 2$. The proof is constructive, i.e., all the inequalities can be explicitly derived. However, their number $n_{\#}$ increases tremendously with the number $n$ of rows and columns of $[A]$. Therefore, we restrict ourselves to describing the way in which to derive them. An optimal bound for $n_{\#}$ is given in the particular case of symmetric tridiagonal $n \times n$ matrices $[A]$, which can also be viewed as generalizations of $2 \times 2$ matrices.

We conclude our paper with several examples which illustrate the theory.

# 2 NOTATION

By $\mathbf{R}^n$, $\mathbf{R}^{n\times n}$, $\mathbf{IR}$, $\mathbf{IR}^n$, $\mathbf{IR}^{n\times n}$ we denote the set of real vectors with $n$ components, the set of real $n \times n$ matrices, the set of intervals, the set of interval vectors with $n$ components and the set of $n \times n$ interval matrices, respectively. By 'interval' we always mean a real compact interval. Interval vectors and interval matrices are vectors and matrices, respectively, with interval entries. We write intervals in brackets with the exception of degenerate intervals (so–called ***point intervals***) which we identify with the element being contained, and we proceed similarly with interval vectors and interval matrices. As usual, we denote the lower and upper bound of an interval $[a]$ by $\underline{a}$ and $\overline{a}$, respectively. Similarly, we use $\underline{A}$ and $\overline{A}$ for the bounds of interval matrices $[A] \in \mathbf{IR}^{n\times n}$. We write $[A] = [\underline{A}, \overline{A}] = ([a_{ij}]) = ([\underline{a}_{ij}, \overline{a}_{ij}]) \in \mathbf{IR}^{n\times n}$ simultaneously, without further reference. By $[A]^T$ we mean the transposed matrix of $[A]$. We mention that $[A] = [A]^T$ is equivalent to $\underline{A} = \underline{A}^T$ and $\overline{A} = \overline{A}^T$.

For computations with interval quantities we refer to [1].

We denote any orthant of $\mathbf{R}^n$ by $O$, and the first orthant by $O_1$.

# 3 RESULTS

## 3.1 A Fundamental Lemma

We start this section with the following lemma, which goes back to results of Tarski [9] and Seidenberg [8]. This lemma is basic for all our subsequent considerations. In particular, it contains the constructive procedure which leads to the inequalities mentioned in Section 1.

**Lemma 1.** *Let $L_i = L_i(z_1, \ldots, z_n)$, $i = 1, \ldots, m$, be $m$ given linear functions of $n$ real variables $z_1, \ldots, z_n$:*

$$L_i(z_1, \ldots, z_n) = c_i + \sum_{j=1}^{n} l_{ij} z_j \ , \quad i \in \{1, \ldots, m\}.$$

*Without loss of generality, let*

$$\begin{aligned} l_{i1} &= 0 \quad \text{if} \quad i \in M^0 := \{1, \ldots, m_1\}, \\ l_{i1} &< 0 \quad \text{if} \quad i \in M^- := \{m_1 + 1, \ldots, m_2\}, \\ l_{i1} &> 0 \quad \text{if} \quad i \in M^+ := \{m_2 + 1, \ldots, m\}, \end{aligned}$$

*where at least one of the sets $M^0$, $M^-$, $M^+$ is not empty. If $M^0 = \emptyset$ we set $m_1 := 0$, if $M^- = \emptyset$ we set $m_2 := m_1$, if $M^+ = \emptyset$ we set $m_2 := m$.*

*Let $\hat{m} := m_1 + (m_2 - m_1)(m - m_2)$. If $\hat{m} > 0$ then construct linear functions $\hat{L}_i := \hat{L}_i(z_2, \ldots, z_n)$, $i = 1, \ldots, \hat{m}$, in the following way:*

*i) If $M^0 \neq \emptyset$ then let*

$$\hat{L}_i := L_i, \ i = 1, \ldots, m_1.$$

*ii) If $M^- \neq \emptyset$ and $M^+ \neq \emptyset$ then for $p \in M^-$, $q \in M^+$ and $i = p + (q - m_2 - 1)(m_2 - m_1)$ (i.e., $p = m_1 + 1, \ldots, m_2$, $q = m_2 + 1, \ldots, m$, and therefore $i = m_1 + 1, \ldots, m_2 + (m - m_2 - 1)(m_2 - m_1) = \hat{m}$) let*

$$\hat{L}_i := \frac{c_q}{l_{q1}} - \frac{c_p}{l_{p1}} + \sum_{j=2}^{n} \left( \frac{l_{qj}}{l_{q1}} - \frac{l_{pj}}{l_{p1}} \right) z_j .$$

*No $\hat{L}_i$ is constructed if $\hat{m} = 0$.*

*Let the real numbers $\tilde{z}_2, \ldots \tilde{z}_n$ be given. Then the following two assertions are equivalent.*

*a) There exists a real number $\tilde{z}_1$ such that*

$$L_i(\tilde{z}_1, \ldots, \tilde{z}_n) \leq 0, \quad i = 1, \ldots, m.$$

*b) The inequalities*

$$\hat{L}_i(\tilde{z}_2, \ldots, \tilde{z}_n) \leq 0, \quad i = 1, \ldots, \hat{m}, \tag{3.3}$$

*hold, where we define (3.3) to be true in the case $\hat{m} = 0$, i.e., if no condition $\hat{L}_i(\tilde{z}_2, \ldots, \tilde{z}_n) \leq 0$ exists.*

*Remark.* The procedure of constructing the functions $\hat{L}_i$ starting from the functions $L_i$ is called an *elimination of* $z_1$.

*Proof of Lemma 1.*
a) $\Rightarrow$ b) : Let $L_i(\tilde{z}_1, \ldots, \tilde{z}_n) \leq 0$ hold for $i = 1, \ldots, m$. Then $\hat{L}_i(\tilde{z}_2, \ldots, \tilde{z}_n) = L_i(\tilde{z}_1, \ldots, \tilde{z}_n) \leq 0$ for $i = 1, \ldots, m_1$. If $M^- = \emptyset$ and $M^+ = \emptyset$, respectively, then $m_2 = m_1$ and $m = m_2$, respectively; hence $\hat{m} = m_1$, and b) follows.

Assume now $M^- \neq \emptyset$ and $M^+ \neq \emptyset$. Let $p \in M^-$. Then $l_{p1} < 0$; hence $L_p(\tilde{z}_1, \ldots, \tilde{z}_n) \leq 0$ is equivalent to

$$-\frac{c_p}{l_{p1}} - \sum_{j=2}^{n} \frac{l_{pj}}{l_{p1}} \tilde{z}_j \leq \tilde{z}_1 . \tag{3.4}$$

Analogously,

$$\tilde{z}_1 \leq -\frac{c_q}{l_{q1}} - \sum_{j=2}^{n} \frac{l_{qj}}{l_{q1}} \tilde{z}_j \quad \text{for } q \in M^+. \tag{3.5}$$

Combining (3.4) with (3.5) results in

$$-\frac{c_p}{l_{p1}} - \sum_{j=2}^{n} \frac{l_{pj}}{l_{p1}} \tilde{z}_j \leq -\frac{c_q}{l_{q1}} - \sum_{j=2}^{n} \frac{l_{qj}}{l_{q1}} \tilde{z}_j \quad \text{for } p \in M^-,\ q \in M^+, \tag{3.6}$$

which is equivalent to

$$\hat{L}_i(\tilde{z}_2, \ldots, \tilde{z}_n) \leq 0 \quad \text{for } i = p + (q - m_2 - 1)(m_2 - m_1).$$

This shows b).

b) $\Rightarrow$ a) : Let $\hat{L}_i(\tilde{z}_2, \ldots, \tilde{z}_n) \le 0$ hold for $i = 1, \ldots, \hat{m}$. Then, by definition, $L_i(\tilde{z}_1, \ldots, \tilde{z}_n) \le 0$ for $i = 1, \ldots m_1$ and for any $\tilde{z}_1 \in \mathbf{R}$.
If $M^- = \emptyset$ choose any $\tilde{z}_1$ such that

$$\tilde{z}_1 \le \min_{q \in M+} \left\{ -\frac{c_q}{l_{q1}} - \sum_{j=2}^{n} \frac{l_{qj}}{l_{q1}} \tilde{z}_j \right\} .$$

Then $L_q(\tilde{z}_1, \ldots, \tilde{z}_n) \le 0$ for each $q \in M^+$.
If $M^+ = \emptyset$ choose any $\tilde{z}_1$ such that

$$\max_{p \in M^-} \left\{ -\frac{c_p}{l_{p1}} - \sum_{j=2}^{n} \frac{l_{pj}}{l_{p1}} \tilde{z}_j \right\} \le \tilde{z}_1 .$$

which is equivalent to $L_p(\tilde{z}_1, \ldots, \tilde{z}_n) \le 0$ for each $p \in M^-$.
If $M^- \neq \emptyset$ and $M^+ \neq \emptyset$ then $\hat{L}_i(\tilde{z}_2, \ldots, \tilde{z}_n) \le 0$ for $i = m_1 + 1, \ldots, \hat{m}$. By the construction of $\hat{L}_i$ this means

$$-\frac{c_p}{l_{p1}} - \sum_{j=2}^{n} \frac{l_{pj}}{l_{p1}} \tilde{z}_j \le -\frac{c_q}{l_{q1}} - \sum_{j=2}^{n} \frac{l_{qj}}{l_{q1}} \tilde{z}_j \quad \text{for each } \ p \in M^-,\ q \in M^+ .$$

This implies

$$c_{\max} := \max_{p \in M^-} \left\{ -\frac{c_p}{l_{p1}} - \sum_{j=2}^{n} \frac{l_{pj}}{l_{p1}} \tilde{z}_j \right\} \le c_{\min} := \min_{q \in M+} \left\{ -\frac{c_q}{l_{q1}} - \sum_{j=2}^{n} \frac{l_{qj}}{l_{q1}} \tilde{z}_j \right\} .$$

Choose $\tilde{z}_1 \in [c_{\max}, c_{\min}]$; then $L_i(\tilde{z}_1, \ldots \tilde{z}_n) \le 0$ for $i = m_1 + 1, \ldots, m$. This proves b). □

## 3.2 Characterization of the Solution Set $S$ by Inequalities

Using Lemma 1, we will characterize the solution set S by inequalities. To this end let $[A] \in \mathbf{IR}^{n \times n}$, $[b] \in \mathbf{IR}^n$ and $Ax = b$ with $A \in [A]$ and $b \in [b]$. Without loss of generality, let $x \in O_1$. Then

$$\begin{aligned} \underline{a}_{ij} x_j \le a_{ij} x_j \le \overline{a}_{ij} x_j \ , &\qquad i, j = 1, \ldots n, \\ \underline{b}_i \le \sum_{j=1}^{n} a_{ij} x_j \le \overline{b}_i \ , &\qquad i = 1, \ldots n. \end{aligned} \tag{3.7}$$

With

$$\tilde{z}_{ij} := a_{ij}x_j \ , \quad i,j = 1,\ldots n,$$

the $n^2+n$ double inequalities (3.7) are equivalent to the $2(n^2+n)$ inequalities

$$\left.\begin{array}{lll} \left.\begin{array}{ll} \underline{a}_{ij}x_j - \tilde{z}_{ij} & \leq 0 \\ -\overline{a}_{ij}x_j + \tilde{z}_{ij} & \leq 0 \end{array}\right\} & j=1,\ldots,n & \\ \underline{b}_i - \sum_{j=1}^{n} \tilde{z}_{ij} \quad \leq 0 & & \\ -\overline{b}_i + \sum_{j=1}^{n} \tilde{z}_{ij} \quad \leq 0 & & \end{array}\right\} \quad i=1,\ldots,n\,. \tag{3.8}$$

It is obvious that (3.8) is in the form prescribed by Lemma 1a) (not yet ordered in the way described there) with $\tilde{z}_{ij}$ instead of $\tilde{z}_k$. Therefore, if $\tilde{z}_{11}, \tilde{z}_{12}, \tilde{z}_{13}, \ldots, \tilde{z}_{nn}$ satisfy (3.8) then $\tilde{z}_{12}, \tilde{z}_{13}, \ldots, \tilde{z}_{nn}$ satisfy a system $\hat{L}_i(\tilde{z}_{12}, \tilde{z}_{13}, \ldots, \tilde{z}_{nn}) \leq 0$ with $\hat{L}_i$ from Lemma 1. Note that for fixed $i$ and $j$ the term $\tilde{z}_{ij}$ occurs exactly once in each line of (3.8), i.e., four times for each $i$. In the notation of Lemma 1, the values of the coefficients $l_{ij}$ are only $-1$, 0, or 1, and $M^- \neq \emptyset$, $M^+ \neq \emptyset$. Elimination of $\tilde{z}_{11}$ means replacing the inequalities in (3.8)

$$\begin{array}{rl} \underline{a}_{11}x_1 - \tilde{z}_{11} & \leq 0 \\ -\overline{a}_{11}x_1 + \tilde{z}_{11} & \leq 0 \\ \underline{b}_1 - \sum_{j=1}^{n} \tilde{z}_{1j} & \leq 0 \\ -\overline{b}_1 + \sum_{j=1}^{n} \tilde{z}_{1j} & \leq 0 \end{array}$$

by

$$\left.\begin{array}{rcl} \underline{a}_{11}x_1 & \leq & \overline{a}_{11}x_1 \\ \underline{a}_{11}x_1 & \leq & \overline{b}_1 - \sum_{j=2}^{n} \tilde{z}_{1j} \\ \underline{b}_1 - \sum_{j=2}^{n} \tilde{z}_{1j} & \leq & \overline{a}_{11}x_1 \\ \underline{b}_1 - \sum_{j=2}^{n} \tilde{z}_{1j} & \leq & \overline{b}_1 - \sum_{j=2}^{n} \tilde{z}_{1j} \Longleftrightarrow \underline{b}_1 \leq \overline{b}_1 \end{array}\right\} \tag{3.9}$$

while keeping the remaining ones. (Cf. the process of getting (3.6) from (3.4) and (3.5) in the proof of (3.3) in Lemma 1.) The first and the fourth inequality of (3.9) always hold; therefore we delete them. Then (3.9) is equivalent to

$$\begin{aligned} \underline{b}_1 - \overline{a}_{11}x_1 - \sum_{j=2}^{n} \tilde{z}_{1j} &\leq 0\,, \\ -\overline{b}_1 + \underline{a}_{11}x_1 + \sum_{j=2}^{n} \tilde{z}_{1j} &\leq 0\,. \end{aligned}$$

Together with the inequalities of (3.8) which do not contain $\tilde{z}_{11}$ yields a system of inequalities for which the arguments above can be repeated for $\tilde{z}_{12}, \tilde{z}_{13}, \ldots \tilde{z}_{nn}$. After a total of $n^2$ elimination steps we end at the following $n_{\#} = 2n$ relations.

$$\left.\begin{aligned} \underline{b}_i \leq \sum_{j=1}^{n} \overline{a}_{ij}x_j \\ \sum_{j=1}^{n} \underline{a}_{ij}x_j \leq \overline{b}_i \end{aligned}\right\} \quad i = 1, \ldots, n. \tag{3.10}$$

Repeated application of Lemma 1 shows that the statements '$x \in S$' and '$x$ satisfies (3.10)' are equivalent for $x \in O_1$. Note that (3.10) is just the well-known characterization of $S$ which can be found, e.g., in [2].

We assumed $x \in O_1$. If $x \notin O_1$ then the bounds in the first group of inequalities in (3.7) are changed for those $j$ for which $x_j < 0$. For $x_j < 0$, they are $\overline{a}_{ij}x_j \leq a_{ij}x_j \leq \underline{a}_{ij}x_j$ . The elimination process for the $\tilde{z}_{ij}$ proceeds now in an analogous manner. The final result is summarized in the following theorem (cf. [2]). Equivalent formulations can be found in [3], [4], [5], [6] and [7].

**Theorem 1.** *Let $[A] \in \mathbf{IR}^{n\times n}$ and let $[b] \in \mathbf{IR}^n$. Then $x \in S$ if and only if $x$ satisfies*

$$\begin{aligned} \sum_{j=1}^{n} \check{a}_{ij}x_j \leq \overline{b}_i \quad & \text{with } \check{a}_{ij} := \begin{cases} \overline{a}_{ij} & \text{if } x_j < 0 \\ \underline{a}_{ij} & \text{if } x_j \geq 0 \end{cases} \\ \underline{b}_i \leq \sum_{j=1}^{n} \hat{a}_{ij}x_j \quad & \text{with } \hat{a}_{ij} := \begin{cases} \underline{a}_{ij} & \text{if } x_j < 0 \\ \overline{a}_{ij} & \text{if } x_j \geq 0 \end{cases} \end{aligned} \tag{3.11}$$

*for $i = 1, \ldots, n$.*

## 3.3 Characterization of the Symmetric Solution Set $S_{\mathrm{sym}}$ by Inequalities

Let the symmetric interval matrix $[A] = [A]^T$ be given. Of course, the solution set $S$ of $[A]$ is characterized by Theorem 1 also in this particular case. We are now going to describe the symmetric solution set $S_{\mathrm{sym}}$.

We want to apply Lemma 1 to show that the symmetric solution set $S_{\mathrm{sym}}$ can be characterized as an intersection of closed half–spaces and closed sets of which the boundary is formed by quadrics. This result generalizes that in [2] which was only valid for $2 \times 2$ matrices.

Due to the equality of $a_{ij}$ and $a_{ji}$ the numbers $\tilde{z}_{ij} := a_{ij}x_j$ and $\tilde{z}_{ji} := a_{ji}x_i$ are no longer independent if $x_i$ and $x_j$ are given. Therefore, the procedure for characterizing $S$ must be modified in order to get a set of inequalities which describe $S_{\mathrm{sym}}$. Instead of considering $a_{ij}x_j$ we will now use the $\dfrac{n(n+1)}{2}$ numbers

$$\tilde{z}_{ij} := a_{ij}x_ix_j, \quad i,j = 1,\dots n, \quad i \le j.$$

They can be introduced by multiplying (3.7) by $x_i$ taking into account the sign of $x_i$. For simplicity we restrict ourselves (without loss of generality) to the first orthant. We then get the inequalities

$$\left.\begin{array}{ll} \underline{a}_{ij}x_ix_j \le a_{ij}x_ix_j = \tilde{z}_{ij} \le \overline{a}_{ij}x_ix_j, & i,j = 1,\dots n,\ i \le j \\ \underline{b}_i x_i \le \sum\limits_{j=1}^{n} a_{ij}x_ix_j = \sum\limits_{j=1}^{n} \tilde{z}_{ij} \le \overline{b}_i x_i, & i = 1,\dots n, \\ & \text{with } \tilde{z}_{ij} := \tilde{z}_{ji} \text{ in the case } j < i \end{array}\right\} \tag{3.12}$$

which are equivalent to

$$\left.\begin{array}{ll} \left.\begin{array}{lcl} \underline{a}_{ij}x_ix_j - \tilde{z}_{ij} & \le & 0 \\ -\overline{a}_{ij}x_ix_j + \tilde{z}_{ij} & \le & 0 \end{array}\right\} & j = i,\dots,n \\ \underline{b}_i x_i - \sum\limits_{j=1}^{n} \tilde{z}_{ij} \quad \le \quad 0 & \\ -\overline{b}_i x_i + \sum\limits_{j=1}^{n} \tilde{z}_{ij} \quad \le \quad 0 & \end{array}\right\} \quad i = 1,\dots,n. \tag{3.13}$$

with $\tilde{z}_{ij} = \tilde{z}_{ji}$. Lemma 1 can now be applied as it was in the proof of Theorem 1. Thus, if $\tilde{z}_{11}, \tilde{z}_{22}, \dots, \tilde{z}_{nn}$ are eliminated first, the number of non–trivial inequalities in (3.13) reduces from $2\left(\frac{n(n+1)}{2}+n\right) = n(n+1)+2n$ to $n(n+1)$. The inequalities now read

$$\left.\begin{array}{l} \left.\begin{array}{lll} \underline{a}_{ij}x_ix_j - \tilde{z}_{ij} & \leq & 0 \\ -\overline{a}_{ij}x_ix_j + \tilde{z}_{ij} & \leq & 0 \end{array}\right\} \quad j = i+1,\ldots,n \\ \underline{b}_ix_i - \overline{a}_{ii}x_i^2 - \sum_{\substack{j=1\\ j\neq i}}^{n} \tilde{z}_{ij} \quad \leq \quad 0 \\ -\overline{b}_ix_i + \underline{a}_{ii}x_i^2 + \sum_{\substack{j=1\\ j\neq i}}^{n} \tilde{z}_{ij} \quad \leq \quad 0 \end{array}\right\} \quad i = 1,\ldots,n. \tag{3.14}$$

In order to show what happens when eliminating $\tilde{z}_{ij} = \tilde{z}_{ji}$ for $i \neq j$ we will choose $\tilde{z}_{12}$ as the next candidate. Here, we have to replace the six inequalities

$$\left.\begin{array}{rcl} \underline{a}_{12}x_1x_2 - \tilde{z}_{12} & \leq & 0 \\ -\overline{a}_{12}x_1x_2 + \tilde{z}_{12} & \leq & 0 \\ \underline{b}_1x_1 - \overline{a}_{11}x_1^2 - \sum_{j=2}^{n} \tilde{z}_{1j} & \leq & 0 \\ -\overline{b}_1x_1 + \underline{a}_{11}x_1^2 + \sum_{j=2}^{n} \tilde{z}_{1j} & \leq & 0 \\ \underline{b}_2x_2 - \overline{a}_{22}x_2^2 - \sum_{\substack{j=1\\ j\neq 2}}^{n} \tilde{z}_{2j} & \leq & 0 \\ -\overline{b}_2x_2 + \underline{a}_{22}x_2^2 + \sum_{\substack{j=1\\ j\neq 2}}^{n} \tilde{z}_{2j} & \leq & 0 \end{array}\right\} \tag{3.15}$$

by the six non–trivial inequalities

$$\begin{array}{rcl} \underline{a}_{12}x_1x_2 & \leq & \overline{b}_1x_1 - \underline{a}_{11}x_1^2 - \sum_{j=3}^{n} \tilde{z}_{1j} \\ \underline{a}_{12}x_1x_2 & \leq & \overline{b}_2x_2 - \underline{a}_{22}x_2^2 - \sum_{j=3}^{n} \tilde{z}_{2j} \\ \underline{b}_1x_1 - \overline{a}_{11}x_1^2 - \sum_{j=3}^{n} \tilde{z}_{1j} & \leq & \overline{a}_{12}x_1x_2 \\ \underline{b}_2x_2 - \overline{a}_{22}x_2^2 - \sum_{j=3}^{n} \tilde{z}_{2j} & \leq & \overline{a}_{12}x_1x_2 \end{array}$$

$$\underline{b}_1 x_1 - \overline{a}_{11} x_1^2 - \sum_{j=3}^{n} \tilde{z}_{1j} \leq \overline{b}_2 x_2 - \underline{a}_{22} x_2^2 - \sum_{j=3}^{n} \tilde{z}_{2j}$$

$$\underline{b}_2 x_2 - \overline{a}_{22} x_2^2 - \sum_{j=3}^{n} \tilde{z}_{2j} \leq \overline{b}_1 x_1 - \underline{a}_{11} x_1^2 - \sum_{j=3}^{n} \tilde{z}_{1j}$$

which are equivalent to

$$\left.\begin{aligned}
\underline{b}_1 x_1 - \overline{a}_{11} x_1^2 - \overline{a}_{12} x_1 x_2 - \sum_{j=3}^{n} \tilde{z}_{1j} &\leq 0 \\
-\overline{b}_1 x_1 + \underline{a}_{11} x_1^2 + \underline{a}_{12} x_1 x_2 + \sum_{j=3}^{n} \tilde{z}_{1j} &\leq 0 \\
\underline{b}_2 x_2 - \overline{a}_{22} x_2^2 - \overline{a}_{12} x_1 x_2 - \sum_{j=3}^{n} \tilde{z}_{2j} &\leq 0 \\
-\overline{b}_2 x_2 + \underline{a}_{22} x_2^2 + \underline{a}_{12} x_1 x_2 + \sum_{j=3}^{n} \tilde{z}_{2j} &\leq 0 \\
\underline{b}_1 x_1 - \overline{b}_2 x_2 - \overline{a}_{11} x_1^2 + \underline{a}_{22} x_2^2 - \sum_{j=3}^{n} (\tilde{z}_{1j} - \tilde{z}_{2j}) &\leq 0 \\
-\overline{b}_1 x_1 + \underline{b}_2 x_2 + \underline{a}_{11} x_1^2 - \overline{a}_{22} x_2^2 + \sum_{j=3}^{n} (\tilde{z}_{1j} - \tilde{z}_{2j}) &\leq 0 \, .
\end{aligned}\right\} \qquad (3.16)$$

The remaining inequalities of (3.14) are unchanged. Continuing the reduction process of $\tilde{z}_{ij}$ along the lines of Lemma 1 ends up with a set of inequalities each of which describes a quadric, if '$\leq 0$' is replaced by '$= 0$'. Among this set one finds the inequalities

$$\left.\begin{aligned}
\underline{b}_i x_i - \sum_{j=1}^{n} \overline{a}_{ij} x_i x_j &\leq 0 \\
-\overline{b}_i x_i + \sum_{j=1}^{n} \underline{a}_{ij} x_i x_j &\leq 0
\end{aligned}\right\} \quad i = 1, \ldots, n \, . \qquad (3.17)$$

(For a proof use the first four inequalities of (3.16) and the first two sets of inequalities from (3.13).) For $x_i > 0$ we can divide (3.17) by $x_i$ to end up with (3.10). This means that the set of inequalities characterizing $S$ must necessarily be satisfied for $x \in S_{\text{sym}}$ – a trivial statement. Nevertheless curvilinear boundaries normally will occur for $S_{\text{sym}}$. They arise from the last two inequalities of (3.16). If $x_i = 0$, the division by $x_i$ is forbidden in (3.17). Since $x = (x_i) \in S_{\text{sym}}$ implies $x \in S$, the inequalities (3.10) are satisfied also in this particular case $x_i = 0$.

As with the non-symmetric solution set $S$, repeated application of Lemma 1 shows that the statements '$x \in S_{\text{sym}}$' and '$x$ satisfies the set of inequalities constructed by the above-mentioned procedure' are equivalent for $x \in O_1$. Here again, dispensing with the restriction $x \in O_1$ means changing the bounds in (3.12) which involve $x_i$, $x_j$. Thus we obtain

$$\overline{a}_{ij} x_i x_j \le \tilde{z}_{ij} \le \underline{a}_{ij} x_i x_j \quad \text{if } x_i x_j < 0,$$

and

$$\overline{b}_i x_i \le \sum_{j=1}^{n} \tilde{z}_{ij} \le \underline{b}_i x_i \quad \text{if } x_i < 0.$$

Due to our previous considerations we get the following theorem.

**Theorem 2.** *Let* $[A] = [A]^T \in \mathbf{IR}^{n \times n}$ *and let* $[b] \in \mathbf{IR}^n$. *Then in each orthant the symmetric solution set* $S_{\text{sym}}$ *can be represented as the intersection of the unsymmetric solution set* $S$ *and sets with quadrics as boundaries.*

## 3.4 On the Number of Inequalities to Describe $S_{\text{sym}}$ for Tridiagonal Matrices

We want to derive an upper bound $\overline{n}$ for the number $n_{\#}$ of the inequalities which are necessary to describe $S_{\text{sym}}$ for tridiagonal matrices $[A] = [A]^T \in \mathbf{IR}^{n \times n}$. In this case, we use the elements $\tilde{z}_{ij}$ only for $0 \le j - i \le 1$, and we start with the inequalities

$$\left.\begin{array}{l}
\left.\begin{array}{rcl}
\underline{a}_{i,i+1} x_i x_{i+1} - \tilde{z}_{i,i+1} & \le & 0 \\
-\overline{a}_{i,i+1} x_i x_{i+1} + \tilde{z}_{i,i+1} & \le & 0
\end{array}\right\} \quad i = 1, \ldots, n-1, \\
\begin{array}{rcl}
\underline{b}_1 x_1 - \overline{a}_{11} x_1^2 - \tilde{z}_{12} & \le & 0 \\
-\overline{b}_1 x_1 + \underline{a}_{11} x_1^2 + \tilde{z}_{12} & \le & 0
\end{array} \\
\left.\begin{array}{rcl}
\underline{b}_i x_i - \overline{a}_{ii} x_i^2 - \tilde{z}_{i-1,i} - \tilde{z}_{i,i+1} & \le & 0 \\
-\overline{b}_i x_i + \underline{a}_{ii} x_i^2 + \tilde{z}_{i-1,i} + \tilde{z}_{i,i+1} & \le & 0
\end{array}\right\} \quad i = 2, \ldots, n-1, \\
\begin{array}{rcl}
\underline{b}_n x_n - \overline{a}_{nn} x_n^2 - \tilde{z}_{n-1,n} & \le & 0 \\
-\overline{b}_n x_n + \underline{a}_{nn} x_n^2 + \tilde{z}_{n-1,n} & \le & 0
\end{array}
\end{array}\right\} \qquad (3.18)$$

for $S_{\text{sym}} \cap O_1$ which we get after having eliminated the diagonal elements $\tilde{z}_{ii}$, $i = 1, \ldots, n$ (cf. (3.14)).

Consider a fixed element $\tilde{z}_{k,k+1}$. The elimination process shows that in each step of this process half of the inequalities containing $\tilde{z}_{k,k+1}$ has the form

$$r + \tilde{z}_{k,k+1} \le 0 \qquad (3.19)$$

while the other half has the form

$$\hat{r} - \tilde{z}_{k,k+1} \le 0 \tag{3.20}$$

with $r$, $\hat{r}$ independent of $\tilde{z}_{k,k+1}$. Each inequality of the form (3.19) has a counterpart (3.20) which one gets by changing plus to minus and the upper bars to lower ones, and vice versa.

We will eliminate $\tilde{z}_{k,k+1}$ in the order $k = 1, 2, \ldots n-1$. In the $k$–th of these elimination steps we will find the following situation.

**Lemma 2.** *Let $[A] = [A]^T \in \mathbf{IR}^{n \times n}$ be a tridiagonal interval matrix and let $[b] \in \mathbf{IR}^n$. Assume that the off–diagonal element $\tilde{z}_{k,k+1}$ is eliminated in the $k$–th elimination step where the $n$ preceding elimination steps for the diagonal elements $\tilde{z}_{ii}$, $i = 1, \ldots n$, have already been performed (and therefore are not counted here).*

*a) Before the $k$–th elimination step $\tilde{z}_{k,k+1}$ occurs in $4 + 2k$ inequalities of which $2 + 2k = 2(k+1)$ contain $\tilde{z}_{k,k+1}$ as the only element $\tilde{z}_{i,i+1}$ while the remaining two inequalities contain $\tilde{z}_{k,k+1}$ together with $\tilde{z}_{k+1,k+2}$. (In the case $k = n-1$ we define $\tilde{z}_{n,n+1} := 0$.)*

*b) Before the $k$–th elimination step $\tilde{z}_{i,i+1}$, $i = k+1, \ldots, n-1$, occurs in six inequalities. With the exception of the two inequalities which contain $\tilde{z}_{k,k+1}$ together with $\tilde{z}_{k+1,k+2}$, these inequalities are the original ones (which have not been changed after having eliminated the elements $\tilde{z}_{ii}$, $i = 1, \ldots, n$).*

*c) Right after the $k$–th elimination step, $k < n-1$, there are $k^2 + k$ final inequalities, i.e., inequalities which are no more changed during the succeeding elimination process. If $k = n-1$ this number increases by $2n$.*

*Proof.* The assertion is trivial for $k = 1$. Let it hold for some fixed $k < n-1$. Those inequalities which contain $\tilde{z}_{k,k+1}$ as the only element $\tilde{z}_{i,i+1}$ combine to final inequalities. Combining an inequality with its counterpart results in an inequality which always holds. (Keeping track of the history of an inequality $L_s \le 0$, one can see that $L_s$ can be expressed as sum of some of the terms $+\underline{b}_i|x_i|$, $-\overline{b}_i|x_i|$, $+\underline{a}_{ij}|x_i| \cdot |x_j|$, $-\overline{a}_{ij}|x_i| \cdot |x_j|$ and those $\tilde{z}_{ij}$ which have not yet been eliminated. Therefore $L_s$ and its counterpart combine to a final inequality, say $\hat{L}_t \le 0$, in which $\hat{L}_t$ is the sum of some of the terms $-d([b]_i)|x_i|$, $-d([a]_{ij})|x_i| \cdot |x_j|$, where $d([a]) := \overline{a} - \underline{a}$ denotes the diameter of an interval $[a]$. Hence $\hat{L}_t \le 0$ is trivially true.) We will omit such trivial inequalities in our further considerations. In the $k$–th elimination step we thus get

$(k+1)^2 - (k+1) = k^2 + k$ final inequalities and $2(k+1)$ new ones. These new inequalities contain $\tilde{z}_{k+1,k+2}$ as the only element $\tilde{z}_{i,i+1}$. All other inequalities remain unchanged. Among them there are two inequalities which contain only $\tilde{z}_{k+1,k+2}$ and two which contain $\tilde{z}_{k+1,k+2}$ together with $\tilde{z}_{k+2,k+3}$. The increase of $2n$ in the case $k = n-1$ results from the $2n$ new inequalities for $\tilde{z}_{n,n+1}$ which are final ones since we defined $\tilde{z}_{n,n+1}$ to be zero. This proves the lemma. □

Adding all the final inequalities results in

$$\begin{aligned} \overline{n} &= \sum_{k=1}^{n-1} k^2 + \sum_{k=1}^{n-1} k + 2n \\ &= \frac{(n-1)n(2n-1)}{6} + \frac{n(n-1)}{2} + 2n = \frac{n}{3}(n^2+5). \end{aligned}$$

The elimination process for tridiagonal matrices shows, that $n_{\#} = \overline{n}$ may be possible, i.e., the above bound $\overline{n}$ is *sharp*. There are also cases in which $n_{\#} < \overline{n}$ holds, as can be seen from any diagonal matrix.

We state our result in the following theorem.

**Theorem 3.** *Let* $[A] = [A]^T \in \mathbf{IR}^{n\times n}$ *be a tridiagonal interval matrix, let* $[b] \in \mathbf{IR}^n$ *and let* $n_{\#}$ *be the number of inequalities obtained by the elimination process described above for the intersection of* $S_{\text{sym}}$ *with an orthant. Then* $n_{\#}$ *is bounded by* $\overline{n} = \frac{n}{3}(n^2+5)$. *This bound is sharp.*

## 3.5 Particular Cases

In the particular case of 2×2 matrices we end up with the following 6 inequalities for $S_{\text{sym}} \cap O_1$ which can be deduced from (3.18).

$$\begin{aligned} \underline{b}_1 x_1 - \overline{a}_{11} x_1^2 - \overline{a}_{12} x_1 x_2 &\leq 0 \\ -\overline{b}_1 x_1 + \underline{a}_{11} x_1^2 + \underline{a}_{12} x_1 x_2 &\leq 0 \\ \underline{b}_2 x_2 - \overline{a}_{22} x_2^2 - \overline{a}_{12} x_1 x_2 &\leq 0 \\ -\overline{b}_2 x_2 + \underline{a}_{22} x_2^2 + \underline{a}_{12} x_1 x_2 &\leq 0 \\ \underline{b}_1 x_1 - \overline{b}_2 x_2 - \overline{a}_{11} x_1^2 + \underline{a}_{22} x_2^2 &\leq 0 \\ -\overline{b}_1 x_1 + \underline{b}_2 x_2 + \underline{a}_{11} x_1^2 - \overline{a}_{22} x_2^2 &\leq 0\,. \end{aligned}$$

Dividing the first two inequalities by $x_1$, the next two by $x_2$ yields the inequalities for the unsymmetric solution set $S \cap O_1$. The last two inequalities just characterize the sets $C^+$ and $C^-$ in [2].

For tridiagonal $3 \times 3$ matrices $S_{\mathrm{sym}} \cap O_1$ is characterized by the 7 inequalities

$$\left.\begin{array}{rcl} \underline{b}_1 x_1 - \overline{a}_{11} x_1^2 - \overline{a}_{12} x_1 x_2 & \leq & 0 \\ \underline{b}_2 x_2 - \overline{a}_{12} x_1 x_2 - \overline{a}_{22} x_2^2 - \overline{a}_{23} x_2 x_3 & \leq & 0 \\ \underline{b}_3 x_3 - \overline{a}_{23} x_2 x_3 - \overline{a}_{33} x_3^2 & \leq & 0 \\ \underline{b}_1 x_1 - \overline{b}_2 x_2 - \overline{a}_{11} x_1^2 + \underline{a}_{22} x_2^2 + \underline{a}_{23} x_2 x_3 & \leq & 0 \\ \underline{b}_1 x_1 - \overline{b}_2 x_2 + \underline{b}_3 x_3 - \overline{a}_{11} x_1^2 + \underline{a}_{22} x_2^2 - \overline{a}_{33} x_3^2 & \leq & 0 \\ \underline{b}_1 x_1 - (\overline{b}_2 - \underline{b}_2) x_2 - \overline{a}_{11} x_1^2 - \overline{a}_{12} x_1 x_2 - (\overline{a}_{22} - \underline{a}_{22}) x_2^2 & \leq & 0 \\ \underline{b}_2 x_2 - \overline{b}_3 x_3 - \overline{a}_{12} x_1 x_2 - \overline{a}_{22} x_2^2 + \underline{a}_{33} x_3^2 & \leq & 0 \end{array}\right\} \quad (3.21)$$

and their 7 counterparts, which can easily be deduced from (3.21). The first three inequalities and their counterparts describe the unsymmetric solution set $S \cap O_1$. The sixth inequality equals the first one if $[b_2]$ and $[a_{22}]$ are point intervals.

Without proof we mention that the number of inequalities for $S_{\mathrm{sym}} \cap O$ increases to 44 for a dense $3 \times 3$ system.

# 4 EXAMPLES

In this section we want to illustrate some aspects of the results in Section 3.

**Example 1.** Let

$$[A] = \begin{pmatrix} 1 & [-1,1] \\ [-1,1] & -1 \end{pmatrix}, \qquad [b] = \begin{pmatrix} 2 \\ 2 \end{pmatrix}.$$

The matrix $[A]$ contains the singular matrix $A_1 = \begin{pmatrix} 1 & 1 \\ -1 & -1 \end{pmatrix}$, for which no solution of the system $A_1 x = b$ exists, and the singular matrix $A_2 = \begin{pmatrix} 1 & -1 \\ 1 & -1 \end{pmatrix}$, for which the solutions of $A_2 x = b$ are given by $x_1 - x_2 = 2$. However, $[A]$ contains no singular *symmetric* matrix, since $\det \begin{pmatrix} 1 & s \\ s & -1 \end{pmatrix} = -1 - s^2 \leq -1$. The solution sets $S$ and $S_{\mathrm{sym}}$ can be represented by

$$S_{\mathrm{sym}} = \left\{ \frac{2}{1+s^2} \begin{pmatrix} 1+s \\ -1+s \end{pmatrix} \,\middle|\, -1 \leq s \leq 1 \right\} \subseteq O_4$$

and

$$S = \left\{ \frac{2}{1+st} \begin{pmatrix} 1+s \\ -1+t \end{pmatrix} \middle| -1 \le s,t \le 1,\ st \ne -1 \right\} \cup \left\{ \begin{pmatrix} x_1 \\ x_2 \end{pmatrix} \middle| x_1 - x_2 = 2 \right\},$$

where $O_4$ denotes the fourth quadrant of $\mathbf{R}^2$. The inequalities (3.11) characterizing $S \cap O_4$ read

$$\begin{aligned} 2 - x_1 + x_2 &\le 0 \\ -2 + x_1 + x_2 &\le 0 \\ -2 - x_1 - x_2 &\le 0 \end{aligned}$$

where we have dropped the fourth inequality since it coincides with the first one.

For $S_{\text{sym}} = S_{\text{sym}} \cap O_4$ we need, in addition, the inequalities

$$\begin{aligned} -(x_1^2 + x_2^2 - 2x_1 + 2x_2) &\le 0 \\ x_1^2 + x_2^2 - 2x_1 + 2x_2 &\le 0 \end{aligned}$$

which we combine to the equality

$$x_1^2 + x_2^2 - 2x_1 + 2x_2 = 0$$

or, equivalently,

$$(x_1 - 1)^2 + (x_2 + 1)^2 = 2. \tag{3.22}$$

Thus, $S_{\text{sym}}$ is the half–circle which results as the intersection of the circle (3.22) with $O_4$, while $S$ is the union of the half–strip in Figure 1 and the straight line $x_1 - x_2 = 2$.

**Example 2.** Let

$$[A] = \begin{pmatrix} -2 & 2 & 0 \\ 2 & 2 & [3,8] \\ 0 & [3,8] & -2 \end{pmatrix},\ A_t = \begin{pmatrix} -2 & 2 & 0 \\ 2 & 2 & t \\ 0 & t & -2 \end{pmatrix}, 3 \le t \le 8,$$

$$\text{and } b = [b] = \begin{pmatrix} 0 \\ 16 \\ 12 \end{pmatrix}.$$

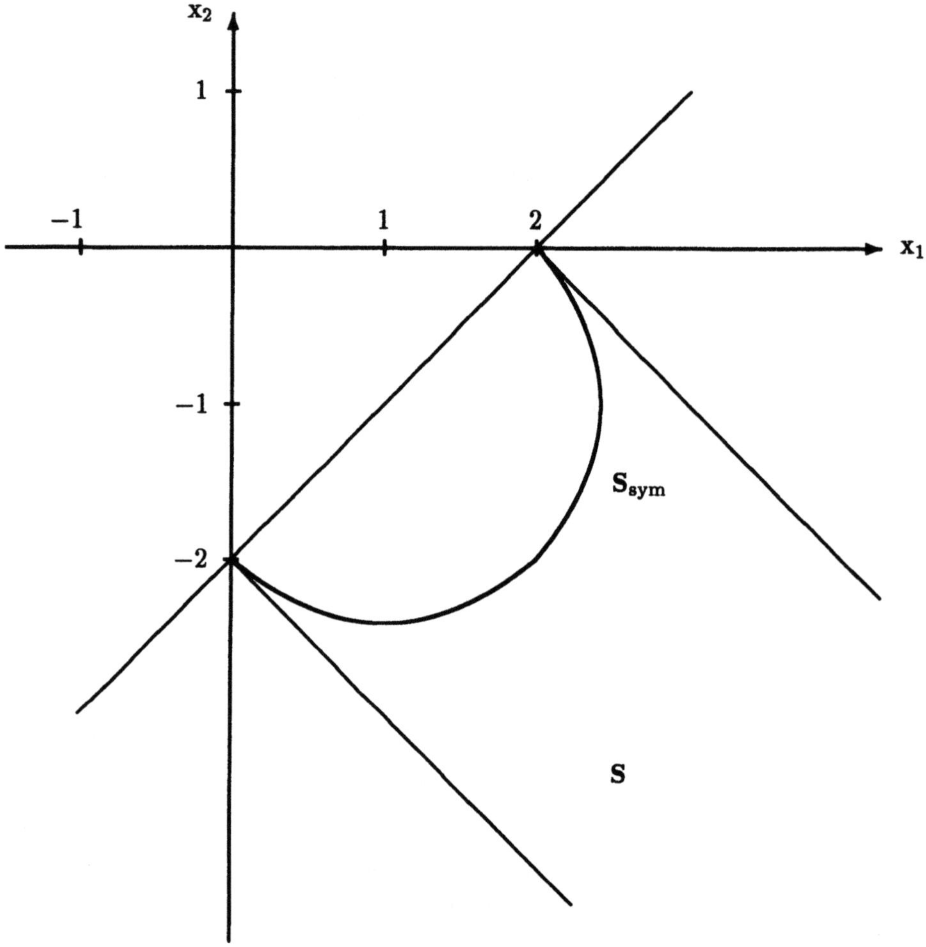


**Fig. 1** The shape of the solution sets $S$, $S_{\text{sym}}$ in Example 1.

From $A_t x = b$ we get

$$x_1 = x_2 = \frac{12t + 32}{t^2 + 8} \geq 0, \quad x_3 = \frac{16t - 48}{t^2 + 8} \geq 0\ . \qquad (3.23)$$

In particular, $A_t$ is regular and $S_{\text{sym}} = S_{\text{sym}} \cap O_1$. The inequalities in (3.21) yield

$$\begin{aligned} x_1 - x_2 &= 0, \\ 3x_3 \le 16 - 2x_1 - 2x_2 &\le 8x_3, \\ 3x_2 \le 12 + 2x_3 &\le 8x_2, \end{aligned} \tag{3.24}$$

which characterize $S \cap O_1$, and

$$\begin{aligned} 3x_2x_3 \le 16x_2 - 2x_1^2 - 2x_2^2 &\le 8x_2x_3, \\ x_1^2 + (x_2-4)^2 + (x_3+3)^2 &= 25, & (3.25) \\ -\left(\frac{x_1-8}{2}\right)^2 + \left(\frac{x_1-8}{2} + x_2\right)^2 + (x_3+3)^2 &= 9, & (3.26) \end{aligned}$$

which select $S_{\text{sym}}$ from $S \cap O_1$. Note that the sixth inequality of (3.21) coincides here with the first one (after having divided by $x_1$); therefore it has been omitted. The intersection of the sphere (3.25) with center $(0,4,-3)$ and radius 5 with the plane (3.24) is the circle $C$ on this plane with center $(2,2,-3)$ and radius $\sqrt{17}$. Since $S_{\text{sym}}$ is connected (see [2] or prove it directly from (3.23)) it must be an arc of $C$. A closer look at (3.23) reveals that $S_{\text{sym}}$ is the arc of $C$ which lies in $O_1$ between $(4,4,0)$ and $(\frac{16}{9}, \frac{16}{9}, \frac{10}{9})$. One obtains these endpoints as the solutions of $A_t x = b$ for the values $t = 3$ and $t = 8$, respectively. The equation (3.26) describes a distorted one–sheeted hyperboloid. $S_{\text{sym}}$ is illustrated in Figure 2.

The unsymmetric solution set $S$ is the triangle with the corners $(4,4,0)$, $(\frac{17}{8}, \frac{17}{8}, \frac{5}{2})$, $(\frac{16}{9}, \frac{16}{9}, \frac{10}{9})$ which lies in the plane (3.24) and in $O_1$, and which contains $S_{\text{sym}}$.

## Acknowledgements

The work of the second author was partially supported by NSF Grants No. CDA-9015006 and EEC-9322370, and by NASA Grant No. NAG 9-757.

**Fig. 2** The shape of the solution set $S_{\rm sym}$ in Example 2.

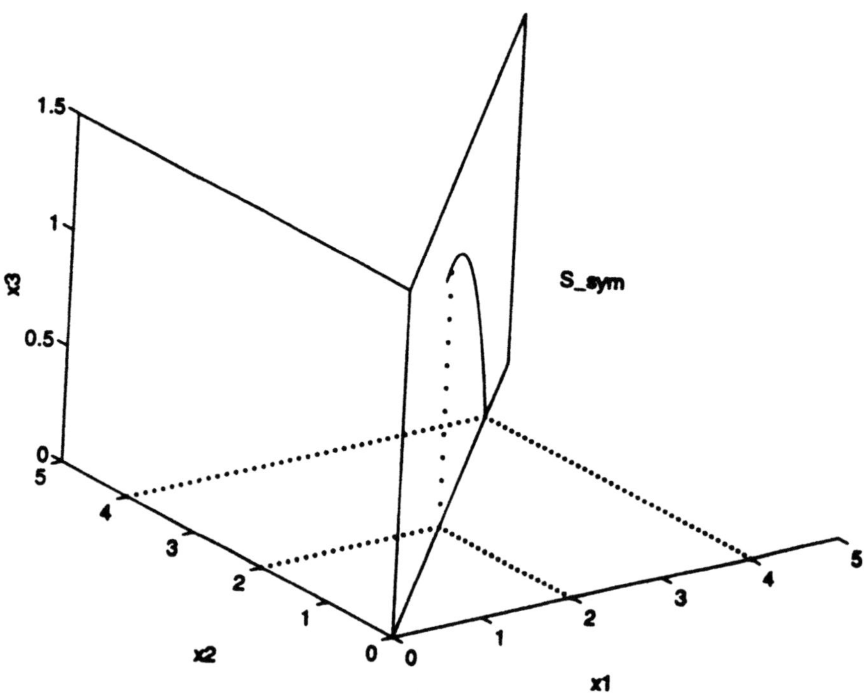

# REFERENCES

[1] G. Alefeld and J. Herzberger, *Introduction to Interval Computations*, Academic Press, N.Y., 1983.

[2] G. Alefeld and G. Mayer, "On the Symmetric and Unsymmetric Solution Set of Interval Systems", *SIAM J. Matrix Anal. Appl.*, 1995 (to appear).

[3] H. Beeck, "Über Struktur und Abschätzungen der Lösungsmenge von linearen Gleichungssystemen mit Intervallkoeffizienten", *Computing*, 1972, Vol. 10, pp. 231 - 244.

[4] D. J. Hartfiel, "Concerning the Solution Set of $Ax = b$ where $P \leq A \leq Q$ and $p \leq b \leq q$", *Numer. Math.*, 1980, Vol. 35, p. 355–359.

[5] A. Neumaier, *Interval Methods for Systems of Equations*, Cambridge University Press, Cambridge, 1990.

[6] W. Oettli and W. Prager, "Compatibility of Approximate Solution of Linear Equations with Given Error Bounds for Coefficients and Right-hand Sides", *Numer. Math.*, 1964, Vol. 6, pp. 405–409.

[7] J. Rohn, "Interval Linear Systems", *Freiburger Intervall-Berichte*, 1984, Vol. 84/7, pp. 33–58.

[8] A. Seidenberg, "A New Decision Method for Elementary Algebra", *Annals of Math.*, 1954, Vol. 60, pp. 365–374.

[9] A. Tarski, *A Decision Method for Elementary Algebra and Geometry*, 2nd ed., Berkeley and Los Angeles, 1951.

# 4

# LINEAR INTERVAL EQUATIONS: COMPUTING ENCLOSURES WITH BOUNDED RELATIVE OVERESTIMATION IS NP-HARD

**Jiří Rohn**

*Faculty of Mathematics and Physics, Charles University, Malostranské nám. 25, 118 00 Prague, Czech Republic*

*and*

*Institute of Computer Science, Academy of Sciences, Pod vodárenskou věží 2, 182 07 Prague, Czech Republic*

## ABSTRACT

It is proved that if there exists a polynomial-time algorithm for enclosing solutions of linear interval equations with relative overestimation better than $\frac{4}{n^2}$ (where $n$ is the number of equations), then P=NP. The result holds for the symmetric case as well.

## 1 INTRODUCTION

For a system of linear interval equations

$$A^I x = b^I \tag{4.1}$$

($A^I$ square), *enclosure* is defined as an interval vector $[\underline{y}, \overline{y}]$ satisfying

$$X \subseteq [\underline{y}, \overline{y}]$$

where $X$ is the solution set:

$$X = \{x;\ Ax = b \text{ for some } A \in A^I,\ b \in b^I\}.$$

Various enclosure methods can be found in Alefeld and Herzberger [2] or Neumaier [7]. If $A^I$ is regular, then there exists the narrowest (or: optimal) enclosure $[\underline{x}, \overline{x}]$ given by

$$\underline{x}_i = \min_X x_i,$$



*R. B. Kearfott and V. Kreinovich (eds.), Applications of Interval Computations, 81–89.*

$$\overline{x}_i = \max_X x_i$$

for each $i$. Computing $[\underline{x}, \overline{x}]$ was proved to be NP-hard (Rohn and Kreinovich [12]; also, Kreinovich, Lakeyev and Noskov [6] for the rectangular case). In this paper we show that the same is true for computing "sufficiently accurate" enclosures (Theorem 1), even in the symmetric case (Theorem 2).

# 2 THE RESULT

**Theorem 1.** *Suppose there exists a polynomial-time algorithm which for each strongly regular $n \times n$ interval matrix $A^I$ and each $b^I$ (both with rational bounds) computes a rational enclosure $[\underline{y}, \overline{y}]$ of $X$ satisfying*

$$\left| \frac{\overline{y}_i - \overline{x}_i}{\overline{x}_i} \right| \leq \frac{4}{n^2} \tag{4.2}$$

*for each $i$ with $\overline{x}_i \neq 0$. Then P=NP.*

**Comments.**

1) $A^I = [A_c - \Delta, A_c + \Delta]$ is called strongly regular if $\varrho(|A_c^{-1}|\Delta) < 1$ (a well-known sufficient regularity condition).

2) P and NP are the well-known complexity classes. The conjecture that P$\neq$NP, although unproved, is widely believed to be true (Garey and Johnson [3]).

3) Hence, the problem of computing sufficiently accurate enclosures is very difficult: an existence of a polynomial-time algorithm yielding the accuracy (4.2) would imply polynomial-time solvability of all problems in the class NP. At the current stage of the complexity theory (conjecture P$\neq$NP) this possibility cannot be excluded, but must be considered highly unlikely.

*Proof.* 1) Denote $e = (1, 1, \ldots, 1)^T \in R^n$ and $Z = \{z \in R^n;\ |z_i| = 1 \text{ for each } i\}$, so that $Z$ is the set of all $\pm 1$-vectors. We shall use matrix norms

$$\|M\|_s = e^T|M|e = \sum_i \sum_j |m_{ij}|$$

and

$$\|M\|_{\infty,1} = \max\{\|Mz\|_1;\ z \in Z\} \tag{4.3}$$

(where $\|x\|_1 = \sum_i |x_i|$; cf. [4]). $\lfloor \alpha \rfloor$ denotes the integer part of a real number $\alpha$.

2) A real symmetric $n \times n$ matrix $M = (m_{ij})$ is called an $MC$-matrix if it is of the form

$$m_{ij} \begin{cases} = n & \text{if } i = j \\ \in \{0, -1\} & \text{if } i \neq j \end{cases}$$

$(i, j = 1, \ldots, n)$. For an $MC$-matrix $M$ we obviously have

$$n \leq e^T M e \leq \|M\|_{\infty,1} \leq \|M\|_s \leq n(2n-1). \tag{4.4}$$

Also,

$$z_i(Mz)_i > 0 \tag{4.5}$$

holds for each $z \in Z$ and each $i \in \{1, \ldots, n\}$. We shall essentially use the fact that computing $\|M\|_{\infty,1}$ is NP-hard for $MC$-matrices [10, Corollary 7]. In the sequel we shall construct, for a given $n \times n$ $MC$-matrix $M$, a linear interval system with interval matrix of size $3n \times 3n$ such that if $\overline{y}_i$ satisfies (4.2), then

$$\|M\|_{\infty,1} = \lfloor \|M\|_s + 1 - \frac{1}{\overline{y}_i} \rfloor.$$

Hence, if such a $\overline{y}_i$ can be computed in polynomial time, then $\|M\|_{\infty,1}$ can also be computed in polynomial time and since this is an NP-hard problem, P=NP will follow.

3) For a given $n \times n$ $MC$-matrix $M$ (which is diagonally dominant and therefore nonsingular), consider a linear interval system

$$A^I x = b^I \tag{4.6}$$

with $A^I = [A_c - \Delta, A_c + \Delta]$, $b^I = [b_c - \delta, b_c + \delta]$ given by

$$A_c = \begin{pmatrix} 0 & -I & 0 \\ -I & 0 & M^{-1} \\ 0 & M^{-1} & M^{-1} \end{pmatrix},$$

$$\Delta = \begin{pmatrix} 0 & 0 & 0 \\ 0 & 0 & 0 \\ 0 & 0 & \beta e e^T \end{pmatrix}$$

(all the blocks are $n \times n$, $I$ is the unit matrix),

$$b_c = \begin{pmatrix} 0 \\ 0 \\ 0 \end{pmatrix},$$

$$\delta = \begin{pmatrix} 0 \\ 0 \\ \beta e \end{pmatrix}$$

(all the blocks are $n \times 1$) and

$$\beta = \frac{1}{\|M\|_s + 1}. \tag{4.7}$$

We shall first prove that $A^I$ is strongly regular. Since

$$A_c^{-1} = \begin{pmatrix} M^{-1} & -I & I \\ -I & 0 & 0 \\ I & 0 & M \end{pmatrix}$$

(as it can be easily verified), we have

$$|A_c^{-1}|\Delta = \begin{pmatrix} 0 & 0 & \beta ee^T \\ 0 & 0 & 0 \\ 0 & 0 & \beta |M| ee^T \end{pmatrix}.$$

This matrix has eigenvalues $\lambda = 0$ (multiple) and $\lambda = \beta\|M\|_s$. Hence $\varrho(|A_c^{-1}|\Delta) = \beta\|M\|_s < 1$ due to (4.7), and $A^I$ is strongly regular.

4) For the linear interval system (4.6), consider a solution $x$ of the linear system $\tilde{A}x = \tilde{b}$ for some $\tilde{A} \in A^I$, $\tilde{b} \in b^I$. If we decompose $x$ as

$$x = \begin{pmatrix} x^{(1)} \\ x^{(2)} \\ x^{(3)} \end{pmatrix},$$

then we have

$$\begin{aligned} x^{(1)} &= M^{-1}x^{(3)} \\ x^{(2)} &= 0 \\ M'x^{(3)} &= b' \end{aligned}$$

for some $M', b'$ satisfying $|M^{-1} - M'| \leq \beta ee^T$ and $|b'| \leq \beta e$, hence $x^{(3)}$ is a solution of the linear interval system

$$[M^{-1} - \beta ee^T, M^{-1} + \beta ee^T]x' = [-\beta e, \beta e] \tag{4.8}$$

whose matrix is obviously again strongly regular. From [8, Thm. 2.2] we have that for each $z \in Z$ the equation

$$M^{-1}x = \beta(\|x\|_1 + 1)z \tag{4.9}$$

has a unique solution $x_z$. A direct substitution shows that the solution has the form

$$x_z = \frac{\beta}{1-\beta\|Mz\|_1}Mz.$$

Now, from the same Theorem 2.2 in [8] we have that each solution of (4.8) belongs to the convex hull of the $x_z$'s, hence also

$$x^{(3)} \in \mathrm{Conv}\{\frac{\beta}{1-\beta\|Mz\|_1}Mz;\ z \in Z\}$$

which implies

$$x^{(1)} = M^{-1}x^{(3)} \in \mathrm{Conv}\{\frac{\beta}{1-\beta\|Mz\|_1}z;\ z \in Z\}.$$

Thus for each $i \in \{1,\ldots,n\}$ we have

$$x_i^{(1)} \le \frac{\beta}{1-\beta\max\{\|Mz\|_1;\ z\in Z\}} = \frac{\beta}{1-\beta\|M\|_{\infty,1}}$$

and the upper bound is obviously achieved at some $x_z$ which, due to (4.9) and (4.5), solves the equation

$$(M^{-1} - \beta zz^T)x_z = \beta z. \tag{4.10}$$

Hence for the $3n$-dimensional solution $x$ of (4.6) we have

$$\overline{x}_i = \overline{x}_i^{(1)} = \frac{\beta}{1-\beta\|M\|_{\infty,1}} \tag{4.11}$$

for each $i \in \{1,\ldots,n\}$ (cf. [12]).

5) Let $i \in \{1,\ldots,n\}$. Due to (4.4), (4.7) and (4.11) we have

$$\beta \ge \frac{1}{n(2n-1)+1} = \frac{1}{2n^2-n+1}$$

and

$$\overline{x}_i \ge \frac{\beta}{1-\beta n} \ge \frac{\frac{1}{2n^2-n+1}}{1-\frac{n}{2n^2-n+1}} = \frac{1}{2n^2-2n+1} \ge \frac{1}{2n^2}.$$

Hence, if an enclosure $[\underline{y},\overline{y}]$ of the solution set of (4.6) satisfies (4.2), then

$$0 \le \frac{\overline{y}_i - \overline{x}_i}{\overline{x}_i} \le \frac{4}{(3n)^2} < \frac{1}{2n^2} \le \overline{x}_i \le \overline{y}_i,$$

which implies

$$0 \leq \frac{1}{\overline{x}_i} - \frac{1}{\overline{y}_i} < 1. \tag{4.12}$$

Now, from (4.11) we have

$$\|M\|_{\infty,1} = \frac{1}{\beta} - \frac{1}{\overline{x}_i}$$

and adding this to (4.12), we obtain

$$\|M\|_{\infty,1} \leq \frac{1}{\beta} - \frac{1}{\overline{y}_i} < \|M\|_{\infty,1} + 1.$$

Since $\|M\|_{\infty,1}$ is integer for an $MC$-matrix $M$ (due to (4.3)), the last result implies

$$\|M\|_{\infty,1} = \lfloor \frac{1}{\beta} - \frac{1}{\overline{y}_i} \rfloor = \lfloor \|M\|_s + 1 - \frac{1}{\overline{y}_i} \rfloor.$$

Thus, if $\overline{y}_i$ satisfying (4.2) can be computed by a polynomial-time algorithm, then the same is true for $\|M\|_{\infty,1}$ and since computing $\|M\|_{\infty,1}$ is NP-hard for $MC$-matrices [10], P=NP follows. □

## 3 THE SYMMETRIC CASE

Let $A^I = [A_c - \Delta, A_c + \Delta]$ be a symmetric interval matrix (i.e., the bounds $A_c - \Delta$ and $A_c + \Delta$ are symmetric) and let $X^s$ be the set of solutions of (4.1) corresponding to systems with symmetric matrices only:

$$X^s = \{x;\ Ax = b \text{ for some } A \in A^I,\ b \in b^I,\ A \text{ symmetric}\}.$$

Again, $[\underline{y}, \overline{y}]$ is called an enclosure of $X^s$ if $X^s \subseteq [\underline{y}, \overline{y}]$ holds. Enclosure methods for the symmetric case were given by Jansson [5] and Alefeld and Mayer [1]. The narrowest enclosure is $[\underline{x}^s, \overline{x}^s]$, where

$$\underline{x}_i^s = \min_{X^s} x_i,$$

$$\overline{x}_i^s = \max_{X^s} x_i$$

for each $i$. We have an analogous result:

**Theorem 2.** *Suppose there exists a polynomial-time algorithm which for each strongly regular symmetric $n \times n$ interval matrix $A^I$ and each $b^I$ (both with* *rational bounds) computes a rational enclosure $[\underline{y}, \overline{y}]$ of $X^s$ satisfying*

$$\left| \frac{\overline{y}_i - \overline{x}_i^s}{\overline{x}_i^s} \right| \leq \frac{4}{n^2}$$

*for each $i$ with $\overline{x}_i^s \neq 0$. Then P=NP.*

*Proof.* The system (4.6) constructed in the proof of Theorem 1 has a symmetric interval matrix $A^I$ and each $\overline{x}_i$, $i = 1, \ldots, n$, is achieved at the solution of a system whose matrix is of the form

$$\begin{pmatrix} 0 & -I & 0 \\ -I & 0 & M^{-1} \\ 0 & M^{-1} & M^{-1} - \beta z z^T \end{pmatrix}$$

(Eq. (4.10)), hence it is symmetric (since an $MC$-matrix $M$ is symmetric). Thus we have

$$\overline{x}_i = \overline{x}_i^s \tag{4.13}$$

for $i = 1, \ldots, n$, and the proof of Theorem 1 applies to this case as well. □

In particular, (4.13) in view of (4.11) and [10, Corollary 7] implies that computing the narrowest enclosure $[\underline{x}^s, \overline{x}^s]$ is NP-hard. Hence, taking symmetry into account does not help to overcome the NP-hardness of computing the narrowest enclosure established in [12]. A related result [9] says that checking nonsingularity of all *symmetric* matrices contained in a symmetric interval matrix is NP-hard.

# 4 CONCLUDING REMARK

Small modifications in Eq. (4.7) and in the part 5) of the proof show that Theorem 1 also holds true if the relative overestimation bound (4.2) is replaced by the absolute overestimation bound

$$\overline{y}_i \leq \overline{x}_i + \frac{1}{4n^4} \tag{4.14}$$

for each $i$ (see [11]). This form seems to be less appropriate than (4.2) since the term $\frac{1}{4n^4}$ in (4.14) is not related to the magnitude of $\overline{x}_i$.

## Acknowledgements

This work was supported by the Czech Republic Grant Agency under grant GAČR 201/93/0429. I am indebted to Arnold Neumaier, Vladik Kreinovich, Anatoly Lakeyev and Sergey Shary for their justified critical remarks on an earlier draft of this paper that resulted in an improvement of the main result.

## REFERENCES

[1] G. Alefeld and G. Mayer, "On the symmetric and unsymmetric solution set of interval systems," to appear in *SIAM J. Matr. Anal. Appl.*

[2] G. Alefeld and J. Herzberger, *Introduction to Interval Computations*, Academic Press, N. Y. 1983.

[3] M. E. Garey and D. S. Johnson, *Computers and Intractability: A Guide to the Theory of NP-Completeness*, Freeman, San Francisco 1979.

[4] G. H. Golub and C. F. van Loan, *Matrix Computations*, The Johns Hopkins University Press, Baltimore, 1983.

[5] C. Jansson, "Interval linear systems with symmetric matrices, skew-symmetric matrices and dependencies in the right hand side," *Computing*, 1991, Vol. 46, pp. 265–274.

[6] V. Kreinovich, A. V. Lakeyev and S. I. Noskov, "Approximate linear algebra is intractable," *Lin. Alg. Appls.*, 1995 (to appear).

[7] A. Neumaier, *Interval Methods for Systems of Equations*, Cambridge University Press, Cambridge 1990.

[8] J. Rohn, "Systems of linear interval equations," *Lin. Alg. Appls.*, 1989, Vol. 126, pp. 39–78.

[9] J. Rohn, "Checking positive definiteness or stability of symmetric interval matrices is NP-hard," *Commentat. Math. Univ. Carolinae*, 1994, Vol. 35, pp. 795–797.

[10] J. Rohn, "NP-hardness results for some linear and quadratic problems," Technical Report No. 619, Institute of Computer Science, Academy of Sciences of the Czech Republic, Prague 1995, 11 pp.

[11] J. Rohn, ***Linear interval equations: computing sufficiently accurate enclosures is NP-hard***, Technical Report No. 621, Institute of Computer Science, Academy of Sciences of the Czech Republic, Prague 1995, 7 pp.

[12] J. Rohn and V. Kreinovich, "Computing exact componentwise bounds on solutions of linear systems with interval data is NP-hard," *SIAM J. Matr. Anal. Appl.*, 1995, Vol. 16, pp. 415–420.

# 5

# QUALITY IMPROVEMENT VIA OPTIMIZATION OF TOLERANCE INTERVALS DURING THE DESIGN STAGE


**Sevgui Hadjihassan*, Eric Walter*, and Luc Pronzato****

** Laboratoire des Signaux et Systèmes, CNRS/ESE, Plateau de Moulon, 91192 Gif sur Yvette, France, email {sevgui,walter}@lss.supelec.fr*

*** Laboratoire I3S, CNRS URA-1376, 250 av. A. Einstein, Sophia Antipolis, 66560 Valbonne, France, email pronzato@mimosa.unice.fr*



## ABSTRACT

A deterministic model-based approach to quality improvement is proposed, along Taguchi's ideas for off-line quality control. This new approach takes into account fluctuations in the factors; these fluctuations are characterized in terms of tolerance intervals.

The case of an *a priori* known model (i.e., known dependency of the performance characteristics on the factors) is considered first. The optimal factor design is chosen by worst-case optimization.

If the model's parameters are unknown, a bounded-error approach is used to characterize their uncertainty. A min-max optimization is then performed, taking into account the fluctuations of the quality of the product components as well as the uncertainty on the model parameters.


## 1 INTRODUCTION

### 1.1 Quality Control is Important

In a world where competition for markets is ever more intense, ensuring a satisfactory quality to products at the lowest possible cost is a vital issue. This is the main reason why industry finds statistical quality control increasingly



*R. B. Kearfott and V. Kreinovich (eds.), Applications of Interval Computations*, 91–131.

appealing. The development of a product consists of three successive steps, namely the design of the product, the design of the production process and the manufacturing, see Figure 1.

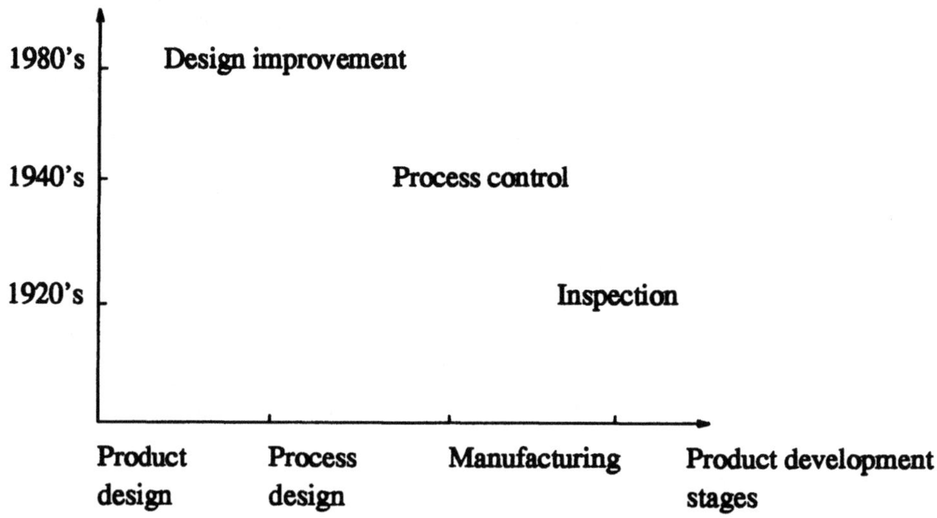


FIG 1: Evolution of quality control concepts [14].

The first efforts in quality control in the 20's were performed by inspection of the final product at the end of the manufacturing stage. Later, in the 40's, it was realized that process control and monitoring of the production line could reduce manufacturing imperfections and the cost of rejecting unsatisfactory products. During the 80's, efforts to control quality were extended to the design stage. Design improvement methods can further reduce manufacturing imperfections and make the product performance less sensitive to environmental variables and aging.

## 1.2 Quality Improvement in the Design Stage: an Example

Suppose the performance characteristic of interest of some electrical circuit is its output voltage, with a target value $y_0$. Assume that this output voltage depends on some factor $x$, for example the resistance of a resistor. Suppose also that the effect of $x$ on the output voltage is nonlinear, as shown in Figure 2. In order to obtain an output voltage $y_0$, the product designer must select the nominal value of the resistance. The factor $x$ should take either the value $x_0$ or the value $x_1$. Assume that at the production stage there is some fluctuation of $x$. For instance, the resistance may have a tolerance of 10% around its nominal value. If the deviation of the resistance is as shown on Figure 2, for a nominal value $x_1$ the output voltage will have a large interval of variation. On the other hand, if the circuit designer selects the nominal value of the resistance to be $x_0$, then the output voltage will have a much smaller interval of variation.

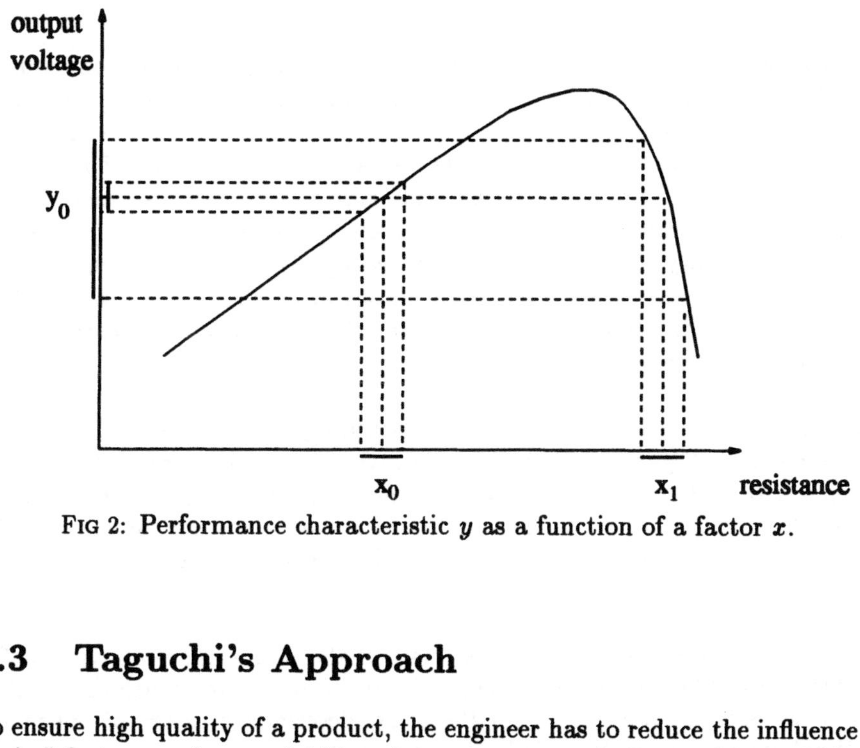


FIG 2: Performance characteristic $y$ as a function of a factor $x$.

## 1.3 Taguchi's Approach

To ensure high quality of a product, the engineer has to reduce the influence of "noise" factors, such as variability of components, environmental perturbations and aging.

The first elements of a scientific approach to identifying product and process design specifications that reduce the degree of performance variation have been proposed by G. Taguchi [33]. Taguchi divides all the factors that influence the performance into ***design*** and ***noise*** factors.

For example, for a resistor:

- its nominal resistance is a *design* factor, while
- such factors as error in the value of the resistance (i.e., the difference between the actual and the nominal values), temperature, etc, are *noise* factors.

If there were no noise factors, then one would simply choose the values of the design factors $\vec{x} = (x_1, ..., x_n)$ that optimize the value of the performance characteristic. However, in real life, due to the presence of noise factors, for the same values $\vec{x} = (x_1, x_2, ...)$ of design factors, we can get different values of the performance characteristic depending on the values of the noise factors. Therefore, to be able to choose a design $\vec{x}$, one must somehow select a measure $j(\vec{x})$ of the quality of this design. Taguchi has proposed several such measures; for example, we can take the mean square deviation of the actual performance characteristic from its target value $y^*$. He calls these measures *signal-to-noise ratios* (for performance variation).

To estimate the value $j(\vec{x})$ of a chosen measure $j$ for a given choice of design parameters $\vec{x}$, Taguchi proposes to run several (numerical or real-life) simulations with different possible values of noise factors, suitably chosen under controlled conditions. From the results of these experiments, an *estimate* $\tilde{j}(\vec{x})$ for the chosen measure can be computed; this estimate is called a *performance statistic.*

To choose the best design $\vec{x}$, the estimate of the performance statistic $\tilde{j}(\vec{x})$ is evaluated for different possible designs $\vec{x}$, and the design for which this statistic takes the optimal value (e.g., for the above-described measure of optimality, optimal means the smallest) is finally selected.

## 1.4 Successes and Problems with Taguchi's Approach

Taguchi's approach has been successfully applied to a number of problems in Japan, USA, and Europe, and has led to the development of high-quality products.

In spite of its good engineering ideas, some criticisms have been expressed about several aspects of the approach. They include the following:

- the criterion used is not always clearly related to the problem to be solved;
- too many experiments are required;
- there is no explicit optimization, due to the absence of a mathematical model of the performance statistics as a function of the design parameters;
- interactions between factors are difficult to handle.

## 1.5 An Approach Combining Taguchi's Ideas with the Response Surface Methodology

To overcome some of the above-mentioned drawbacks of Taguchi's approach, I. Vuchkov and L. Boyadjieva [36, 37, 38] have proposed to combine Taguchi's ideas with the so-called *response surface methodology*. The resulting approach consists of three steps:

- First, a *model* is constructed that describes the dependency of the desired performance characteristic on the design and noise factors. The graph of this model is called the *response surface*. One of the methods to determine the response surface is to measure the values of the performance characteristic in specially controlled environments, where the actual values of the factors can be measured accurately (with negligible error), and then to use the results of these measurements to compute the parameters of the model.

  For resistors, it will mean that we measure the actual resistance and the actual temperature, and that we control the temperature and other noise factors so that their values will not change during one experiment.

- Second, this model and moments of the noise distribution (these moments are supposed to be known *a priori*) are used to determine two new models that describe the dependency of the *mean* and *variance* of the performance characteristics at the production stage on the design factors.
- Finally, these last two models are used to compute *optimal values* for the design factors, e.g., values that minimize the predicted variance while keeping the predicted mean value on target.

This method requires fewer experiments than Taguchi's, and optimizes a criterion clearly related to the problem to be solved. The optimization is explicit, and possible interactions between factors are easily taken into account.

## 1.6 The Need for Deterministic (Interval) Quality Improvement Techniques

The model-based approach proposed by Vuchkov and Boyadjieva is *statistical*, and based on the assumption that information is available on the statistical distribution of noise factors.

Most often, however, the only available information on the fluctuations of the characteristics of components is expressed in terms of the intervals of their possible values. These intervals are called *tolerance intervals*.

> For example, for resistors, usually, tolerance intervals are given; they are specified in terms of percentages of the nominal values (e.g., $\pm 10\%$).

In such situations, to improve quality on the design stage, we need a *deterministic* (*interval*) counterpart of the above-described statistical approach.

In this chapter, we describe this new approach on the example of simplest realistic quality models.

## 1.7 The Structure of the Chapter

In Section 2, we describe basic models, and their origins.

In Section 3, we consider the case when a model of the dependency of the performance characteristic on the factors is assumed to be already available.

Section 4 considers the case when the structure of the model is known, but the parameters $\theta_1, ..., \theta_N$ of this model structure are determined from measurements; in these measurements, the factors are *controlled* (i.e., they are known with negligible errors). Because the model of the performance characteristic is always an approximation of reality, errors between the actual performance characteristic and that predicted by the model cannot be neglected. A special *parameter bounding* methodology is used to build the set $S_N$ of all values of $\vec{\theta}$ that are consistent with these experimental data. The optimization criterion that is used to choose the best design takes into account, in a guaranteed way, the fluctuations of the quality of the product components at the production stage as well as the uncertainty on the model parameters.

Section 5 addresses the situation where for the measurements that determine the values $\theta_k$, the errors with which we measure the values of the factors are also not negligible. In this section, we show that when the model of the performance characteristic is quadratic in the value of the factors, the set $S_N$ of possible values of the parameters can be included in a finite union of convex polyhedra. This allows the replacement of the set $S_N$ by an approximated one guaranteed to contain it.

A simple pedagogical example illustrates Sections 3–5. Finally, a realistic example of a temperature controller is presented in Section 6.

# 2 SOME BASIC MODELS, AND THEIR ORIGINS

## 2.1 General Description

A typical quality control problem is to attain certain target values $y_c^*$ of given performance characteristics $y_c$, $1 \leq c \leq C$.

The dependency of the performance characteristic $y_c$ on some factors $f_1, ..., f_n$ is described by a mathematical model: $y_c = \eta_c(f_1, ..., f_n)$. We do not completely control these factors. In what follows, we assume that the actual value of $f_i$ in mass production is the sum of the desired value $x_i$ (design factor) and some error (noise factor), which corresponds to the "inner noise" of Taguchi's method. We know what combinations of factors are possible for a given design $\vec{x} = (x_1, ..., x_n)$. In mathematical terms, we know a *function* $F(\vec{x})$ that assigns to every design $\vec{x}$, a set $F(\vec{x}) \subseteq R^n$ of possible values of the factor vector $\vec{f} = (f_1, ..., f_n)$.

Our goal is to minimize the losses caused by the difference $\Delta_c = y_c^* - \eta_c(\vec{f})$ between the desired $y_c^*$ and the predicted $\eta_c(\vec{f})$ values of the performance characteristics. For this purpose, we must be able, given the differences $\Delta_c$, to calculate the loss $L = L(\Delta_1, ..., \Delta_C)$ caused by these differences. To emphasize the fact that this loss depends on $\vec{f}$, the resulting loss $L(y_1^* - \eta_1(\vec{f}), ..., y_C^* - \eta_C(\vec{f}))$ will be denoted by $h(\vec{f})$.

For a given design $\vec{x}$, the only information that we have about the possible values of the factor vector $\vec{f}$ is that it belongs to the set $F(\vec{x})$. Our goal is to *guarantee* the quality of the design, i.e., guarantee that the losses would never exceed a certain number $L_{\max}$. Since we have no information about the frequency of difference values $\vec{f} = F(\vec{x})$, the only way to guarantee the quality is to guarantee that even in the worst case, the resulting losses are limited by a given number, because it could be that this worst case is actually the most frequent one. Therefore, as a natural criterion $j(\vec{x})$ for choosing a design $\vec{x}$, we can take the worst-case loss associated with this design $\vec{x}$:

$$j(\vec{x}) = \max_{\vec{f} \in F(\vec{x})} h(\vec{f}).$$

So, to describe a quality control problem in mathematical terms, we must know three things:

- how the performance characteristics $\eta_c$ depend on the factors $f_1, ..., f_n$,
- how the set $F(\vec{x})$ of possible values of factors depends on the design,
- how the loss $L$ depends on the differences $\Delta_c = y_c^* - \eta_c(\vec{f})$ between the desired and the predicted values of performance characteristics.

In the following subsection, we will describe the corresponding models.

## 2.2 The Main Model: the Dependency of the Performance Characteristic on the Factors

In this chapter, we will consider the simplest case when there is only one performance characteristic $y$.

Let us denote all the factors on which this characteristic depends by $f_1, ..., f_n$. The possible range of these factors is usually reasonably small, so, we can expand the dependency of $y$ on $f_i$ into power series, and neglect terms of sufficiently high order. The simplest case would be when we neglect quadratic and higher order terms, and assume that the dependency of $y$ on $f_k$ is linear. In many practical cases, this is, however, an oversimplification. Moreover, nonlinearity is essential to make it possible to reduce the fluctuation of the performance characteristics due to the noise factor (see Figure 2). To get a more realistic model $\eta$ of the dependency of $y$ on $f_k$, we must therefore retain quadratic terms as well. Such a quadratic model provides a pretty good description of many real-life situations. In general, the quadratic model can be written as:

$$\eta(\vec{\theta}, \vec{x}) = \theta_0 + \sum_{i=1}^{n} \theta_i f_i + \sum_{i=1}^{n} \theta_{ii} f_i^2 + \sum_{i=1}^{n-1} \sum_{j=i+1}^{n} \theta_{ij} f_i f_j. \qquad (5.1)$$

In the following text, the coefficients $\theta_0$, $\theta_1$, ..., $\theta_n$, $\theta_{11}$, ..., $\theta_{nn}$ will be called *parameters* of this model.

We can simplify this formula if we use the following vector notation:

- By $\vec{f} = (f_1, ..., f_n)$, we denote the $n$–dimensional vector consisting of all the factors.
- By $q_0$, we denote the value of $\theta_0$.
- By $\vec{q} = (\theta_1, ..., \theta_n)$, we denote the $n$–dimensional vector consisting of the coefficients of the part of the model that is linear in $\vec{f}$.
- By $Q$, we denote a symmetric $n \times n$ matrix whose diagonal coefficients are equal to $\theta_{ii}$, and off-diagonal to $(1/2)\theta_{ij}$ or $(1/2)\theta_{ji}$, so that

$$\sum_{i=1}^{n} \theta_{ii} f_i^2 + \sum_{i=1}^{n-1} \sum_{j=i+1}^{n} \theta_{ij} f_i f_j = \frac{1}{2} \vec{f}^T Q \vec{f}.$$

- By $\vec{\theta}$, we denote a $p$–dimensional vector ($p = 1 + n + n(n+1)/2$) that consists of all the parameters: $\vec{\theta} = (\theta_0, \theta_1, ..., \theta_n, \theta_{11}, ..., \theta_{nn})$.

In these terms, the quadratic model takes the form

$$\eta(\vec{\theta}, \vec{f}) = q_0 + \vec{q}^T \vec{f} + \frac{1}{2} \vec{f}^T Q \vec{f}.$$

Square matrices are generically non-degenerate, so, we may safely assume that the real-life matrix $Q$ is non-degenerate (invertible). This additional assumption is often made in quality control problems.

## 2.3 How Tolerance Intervals Depend on the Design

In this chapter, we consider the simplest case when for each parameter $f_i$, there is a *manufacturing guaranteed accuracy* $e_{fi}(x_i)$ so that if the nominal value of this parameter is $x_i$, then the set of its possible actual values is an interval $[x_i - e_{fi}(x_i), x_i + e_{fi}(x_i)]$.

For example, if a tolerance is 10% around the nominal value, then $e_{fi}(x_i) = 0.1 \cdot x_i$.

In this case, the set $F(\vec{x})$ of possible values of $\vec{f}$ is simply a box

$$F(\vec{x}) = [x_1 - e_{f1}(x_1), x_1 + e_{f1}(x_1)] \times ... \times [x_n - e_{fn}(x_n), x_n + e_{fn}(x_n)].$$

In the following text, for simplicity, we will write $e_{fi}$ instead of $e_{fi}(x_i)$.

## 2.4 Choice of a Loss Function

For a good quality design, the difference $\Delta$ between the desired and the predicted values of the performance characteristic must be small. Therefore, to get a good approximation to the dependency of the loss $L$ in $\Delta$, we can assume that the loss is analytical, expand it into the power series: $L = L(0) + L'(0)\Delta + ...$, and keep only the main term in this series.

Losses are defined in such a way that in the ideal case $\Delta = 0$, the loss is also 0. So, this function has no zero order terms.

Since losses are the smallest when $\Delta = 0$, the function $L$ has a minimum at 0. Hence, $L'(0) = 0$, and the function $L$ has no linear terms either. So, the smallest non-zero term in this expansion is const $\cdot \Delta^2$.

We are not interested in the *absolute* value of a loss, we are only interested in choosing the design for which the loss is the smallest. For this purpose, we can as well change the unit for measuring losses, and thus switch to a new loss function $\tilde{L}(\Delta) = L(\Delta)/L(1)$. This new loss function has an even simpler form: $\tilde{L}(\Delta) = \Delta^2$. This is the loss function that we will be using in this chapter.

For this loss function,

$$h(\vec{f}) = (y^* - \eta(\vec{f}))^2,$$

and

$$j(\vec{x}) = \max_{\vec{f} \in F(\vec{x})} (y^* - \eta(\vec{f}))^2 = \max_{\vec{f} \in F(\vec{x})} h(\vec{f}).$$

# 3 MODEL OF PERFORMANCE CHARACTERISTIC IS KNOWN BEFOREHAND

In this section, we will assume that the model of the performance characteristic $\eta$ is described by the formula (5.1) with known values $\theta_i$ (this model may, for instance, have been constructed according to the response surface methodology). We also assume that the criterion to be minimized is:

$$j(\vec{x}) = \max_{\vec{f} \in F(\vec{x})} (y^* - \eta(\vec{\theta}, \vec{f}))^2 = \max_{\vec{f} \in F(\vec{x})} h(\vec{f}), \tag{5.2}$$

where $y^*$ is the target value for the performance characteristic, and $F(\vec{x}) \subset \mathcal{R}^n$ is the vector interval (i.e. the axis-aligned box) defining the tolerances in mass production around the nominal value $\vec{x}$ of the factors (Figure 3).

A straightforward (brute force) way to solve this problem would be to find the minimum of $j(\vec{x})$ by trying all possible values of $\vec{x}$ on a dense grid. Since computing $j(\vec{x})$ for each $\vec{x}$ also involves solving a complicated (constrained) optimization problem (with respect to $\vec{f}$), this approach would lead to extensive computations even for problems of modest size.

Other possibilities are:

- to apply a min-max optimization algorithm based on a relaxation procedure [32];
- to exploit the specific properties of the criterion $j$ and/or of the set $F(\vec{x})$.

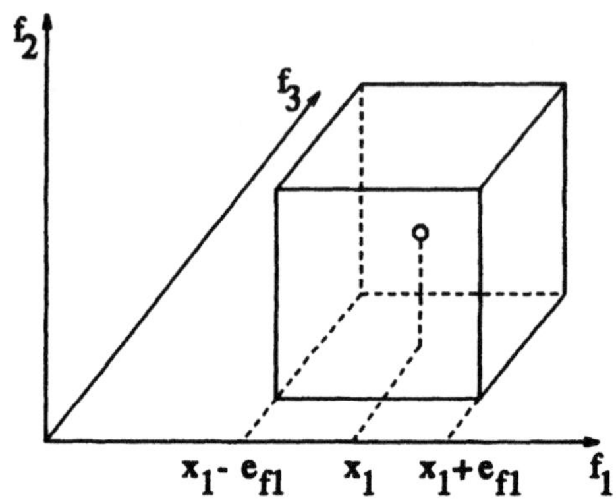


FIG 3: Tolerance in mass production around the nominal value for the factors.

- to apply techniques of interval analysis [13, 22] (including the use of Gauss decomposition [4]); or

In this section, we will analyze the last two possibilities.

## 3.1 Enumeration of all Stationary Points

Let us first try to use the specific features of the criterion $j(\vec{x})$ and of the set $F(\vec{x})$. These features will be used to simplify the computation of $j(\vec{x})$ for a given $\vec{x}$. To find the *optimal* design, i.e., the design $\vec{x}$ for which $j(\vec{x}) \to \min$, we can use, e.g., adaptive random search [29].

Let us describe how, for each fixed value $\vec{x}$, we search for the maximum with respect to $\vec{f}$. For our quadratic model, the performance characteristic can be written as

$$\eta(\vec{\theta}, \vec{f}) = q_0 + \vec{q}^T \vec{f} + \frac{1}{2} \vec{f}^T Q \vec{f}.$$

To simplify the exposition, we have assumed in what follows that the matrix $Q$ is invertible. The maximum of $h(\vec{f})$ is obtained at the points where the modified criterion

$$h'(\vec{f}) = y^* - q_0 - \vec{q}^T \vec{f} - \frac{1}{2} \vec{f}^T Q \vec{f} \tag{5.3}$$

attains the smallest or the largest value.

If this smallest or largest value is attained in the interior of the box $F(\vec{x})$, then it is attained at the stationary point of $h'$, i.e., at the point $\vec{f}^* = -Q^{-1}\vec{q}$.

If this smallest or largest value of $h'(\vec{f})$ is attained at the border or in one of the vertices of the box $F(\vec{x})$, then it is attained at a point $\vec{f}$ in which the $h'(\vec{f}) \to \min$ under the condition that equalities that describe the corresponding face of the box are satisfied. These points can be found by using the Lagrange multipliers method. According to this method, the minima and maxima are the stationary points of the modified criterion

$$h''(\vec{f}) = y^* - q_0 - \vec{q}^T \vec{f} - \frac{1}{2}\vec{f}^T Q \vec{f} + \vec{\lambda}^T (A\vec{f} - \vec{b}), \tag{5.4}$$

where $A\vec{f} = \vec{b}$ corresponds to the saturated constraints and $\vec{\lambda}$ is a vector of Lagrange multipliers. For $k$ saturated constraints, $\dim \vec{b} = k \times 1$, $\dim A = k \times n$ and $\dim \vec{\lambda} = k \times 1$. Since $\vec{f}$ is a stationary point of $h''$, the first derivatives of the modified criterion with respect to the components of $\vec{f}$ must be equal to 0:

$$\frac{\partial h''(\vec{f})}{\partial \vec{f}} = -(\vec{q} + Q\vec{f}) + A^T \vec{\lambda} = \vec{0}.$$

This equality is satisfied when

$$\vec{f} = \vec{f}^* = Q^{-1}(A^T \vec{\lambda}^* - \vec{q}). \tag{5.5}$$

Since $A\vec{f}^* = \vec{b}$, (5.5) implies

$$AQ^{-1}A^T \vec{\lambda}^* - AQ^{-1}\vec{q} = \vec{b}.$$

The optimal value of $\vec{\lambda}$ is thus

$$\vec{\lambda}^* = (AQ^{-1}A^T)^{-1}(AQ^{-1}\vec{q} + \vec{b}).$$

Carrying $\vec{\lambda}^*$ back into (5.5), one gets the final expression for the "conditional" stationary point:

$$\vec{f}^* = Q^{-1}A^T(AQ^{-1}A^T)^{-1}(AQ^{-1}\vec{q} + \vec{b}) - Q^{-1}\vec{q}. \tag{5.6}$$

So, the list of possible candidates for $\vec{f}$ includes:

- The vertices of the box $F(\vec{x})$.
- The stationary point of the function $h'(\vec{f})$ (if it is inside the box $F(\vec{x})$).
- Those "conditional" stationary points that belong to the boundary of the box $F(\vec{x})$.

## 3.2 Example 1

Let us illustrate this method on a simple pedagogical example, in which $n = 1$, the model of the performance characteristic is:

$$\eta(f) = 2 - 2f + f^2,$$

the target value is $y^* = 1.25$, and $F(x) = [x - e_f, x + e_f]$. The results are presented in Table 1.

| $e_f$ | $y^*$ | $x^*$ | $j(x^*)$ |
|---|---|---|---|
| 0.01 | 1.25 | 0.5001 | $9.996.10^{-5}$ |
| | | 1.4999 | $9.996.10^{-5}$ |
| 0.1 | 1.25 | 0.5101 | 0.0096 |
| | | 1.4898 | 0.0096 |

Table 1. Optimization results for Example 1 (no uncertainty in the parameters).

In the ideal case of no noise ($e_f = 0$ and $f = x$), the target 1.25 corresponds to the factor levels $x = f = 0.5$ and $x = f = 1.5$, see Figure 4. When the noise

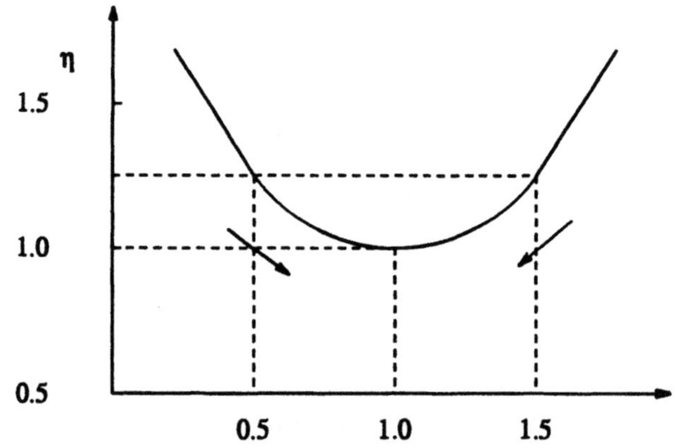


FIG 4: When the error on the factor increases, the optimal nominal value for the factor moves towards the stationary point.

(characterized by the value $e_f$) increases, the optimal value for the factor moves towards the stationary point. This is already apparent for small tolerances $e_f$, and becomes clearer when the tolerances get larger, see Table 1.

## 3.3 Gauss Decomposition

### *A Need for a Fast Enclosure Method*

The method described in a previous subsection always produces the precise value of $j(\vec{x})$, but requires that we analyze all the cases when some of $2n$ inequalities that define a box $F(\vec{x})$ turn into equalities. To describe each case, we must specify which of $n$ factors $f_1, ..., f_n$ attains the boundary values; therefore, the total number of such cases is equal to the total number $2^n$ of subsets of an $n$−element set $\{1, ..., n\}$. For big $n$, this number of cases can exceed computational abilities. So, in addition to the above-described method of computing the exact value of $j(\vec{x})$, we would like to have a method that may give approximate estimates for $j(\vec{x})$, but that completes its computations in reasonable time.

Our goal is to guarantee a certain quality, i.e., to guarantee that the worst-case loss $j(\vec{x})$ does not exceed a certain threshold $L_{\max}$. This goal can be attained if we have an estimating algorithm that provides an *enclosure* $\tilde{j}(\vec{x})$, i.e., a number for which $j(\vec{x}) \leq \tilde{j}(\vec{x})$. In this case, if $\tilde{j}(\vec{x}) \leq L_{\max}$, we can guarantee that $j(\vec{x}) \leq L_{\max}$.

*Interval computations* is a technique that provides us with such enclosures.

### *The Idea of Interval Computations*

One possibility is to apply interval computations to the original formula (5.1) for $\eta(\vec{\theta}, \vec{x})$. Namely, since we know that the actual values of $f_i$ belong to the intervals $[f_i^-, f_i^+] = [x_i - e_{fi}, x_i + e_{fi}]$, we can conclude that

$$\eta(\vec{\theta}, \vec{x}) = \theta_0 + \sum_{i=1}^{n} \theta_i f_i + \sum_{i=1}^{n} \theta_{ii} f_i^2 + \sum_{i=1}^{n-1} \sum_{j=i+1}^{n} \theta_{ij} f_i f_j \in [\eta^-, \eta^+],$$

where

$$[\eta^-, \eta^+] =$$

$$\theta_0 + \sum_{i=1}^{n} \theta_i [f_i^-, f_i^+] + \sum_{i=1}^{n} \theta_{ii} [f_i^-, f_i^+]^2 + \sum_{i=1}^{n-1} \sum_{j=i+1}^{n} \theta_{ij} [f_i^-, f_i^+] \cdot [f_j^-, f_j^+].$$

Here, the operations with intervals are defined as usual:

- the product of a real number $c$ and an interval $[a^-, a^+]$ is defined as follows:
  - If $c \geq 0$, then $c \cdot [a^-, a^+] = [c \cdot a^-, c \cdot a^+]$.
  - If $c < 0$, then $c \cdot [a^-, a^+] = [c \cdot a^+, c \cdot a^-]$.
- The sum of two intervals is defined as
$$[a^-, a^+] + [b^-, b^+] = [a^- + b^-, a^+ + b^+].$$
- The square of an interval $[a^-, a^+]$ is defined as follows:
  - If $0 \leq a^- \leq a^+$, then $[a^-, a^+]^2 = [(a^-)^2, (a^+)^2]$.
  - If $a^- \leq a^+ \leq 0$, then $[a^-, a^+]^2 = [(a^+)^2, (a^-)^2]$.
  - If $a^- \leq 0 \leq a^+$, then $[a^-, a^+]^2 = [0, \max(|a^-|, a^+)^2]$.
- The product of two intervals is defined as
$$[a^-, a^+] \cdot [b^-, b^+] =$$
$$[\min(a^-b^-, a^-b^+, a^+b^-, a^+b^+), \max(a^-b^-, a^-b^+, a^+b^-, a^+b^+)].$$

(the only difference with the usual interval operations is the formula for the square).

## *Interval Computations Sometimes Overestimate*

If we apply interval computations to the original quadratic expression for $\eta(\vec{x})$, then, for big $n$, we often get a huge overestimate $\tilde{j}(\vec{x}) \gg j(\vec{x})$.

The reason for such an overestimation is so-called *interval dependency*: Crudely speaking, the above formulas are exact if the intervals are independent. For example, the formula for the sum of intervals is exact if all pairs $(a, b)$ with $a \in [a^-, a^+]$ and $b \in [b^-, b^+]$ are possible. If the intervals to which we apply these formulas are, in turn, obtained from other intervals, then the values $a$ and $b$ could be dependent. Therefore, not all pairs $(a, b)$ with $a \in [a^-, a^+]$ and $b \in [b^-, b^+]$ are possible, and interval methods lead to an overestimate.

For example, if we apply interval computations to estimate the range of the function $f(x) = x - x^2$ for $x \in [x^-, x^+] = [0, 1]$, then we get an exact interval of possible values for $y = x^2$: $[y^-, y^+] = [0, 1]$, but when we apply the formula of interval computation to compute the interval of possible values of the difference

$f(x) = x + (-1)y$, we get $[0,1] + (-1) \cdot [0,1] = [0,1] + [-1,0] = [-1,1]$. In using this formula, we did not take into consideration the fact that the values of $x$ and $y$ are actually related ($y = x^2$), and so, not all pairs $(x, y)$ with $x \in [0,1]$ and $y \in [0,1]$ are possible. As a result, we get a correct enclosure $[-1,1]$ that is a gross overestimation of the actual range $[0, 0.25]$.

## *To Decrease this Overestimation, We may Want to Rewrite the Expression*

Overestimation is caused by applying interval operations to intervals $[a^-, a^+]$ and $[b^-, b^+]$ that are *dependent* in the sense that some data used in computations leading to $[a^-, a^+]$ were also used to compute $[b^-, b^+]$. So, a natural way to decrease overestimation is to decrease the number of such operations.

The expression (1) can be represented as a sum of $2 + n(n-1)/2$ ($\sim n^2$) parts for each of which interval computation leads to the exact interval. Namely:

- Each variable $f_i$ appears only once in the expression $\theta_0 + \sum \theta_i f_i$; therefore, according to known properties of interval computations, the resulting interval $\theta_0 + \sum \theta_i [f_i^-, f_i^+]$ is exact.
- Similarly, each variable $f_i$ appears only once in the expression $\sum \theta_{ii} f_i^2$; the resulting interval $\sum \theta_{ii} [f_i^-, f_i^+]^2$ is, therefore, also exact.
- For every $i$ and $j > i$, the interval $\theta_{ij} [f_i^-, f_i^+] \cdot [f_j^-, f_j^+]$ is also exact. There are $n(n-1)/2$ such terms, that correspond to all possible pairs $(i, j)$ with $1 \le i < j \le n$.

The resulting $n(n-1)/2 + 2$ intervals need to be added, and they are, in general, dependent (in the above-described sense). So, to get their sum, we must apply $n(n-1)/2 + 1$ operations of adding two dependent intervals.

To get better estimates for $j(\vec{x})$, we would like to reformulate the expression for $\eta(\vec{\theta}, \vec{f})$ in such a way that it would require fewer computations with dependent estimates and thus, hopefully, lead to better enclosures.

## A Simple Reformulation

One possibility is to combine, for each $i$, all the terms $\theta_{ij} f_i f_j$ that correspond to this $i$ and different $j = i+1, ..., n$, into a single group. Then, we can represent the enclosure $[\eta^-, \eta^+]$ as the sum of the following terms:

- The interval $\theta_0 + \sum \theta_i [f_i^-, f_i^+]$, that is (as we have already mentioned) the exact range of the linear part $\theta_0 + \sum \theta_i f_i$.
- The interval $\sum \theta_{ii} [f_i^-, f_i^+]^2$, that gives the exact range of the part $\sum \theta_{ii} f_i^2$.
- For $i < n$, the interval

  $$\sum_{j=i+1}^{n} \theta_{ij} [f_i^-, f_i^+] \cdot [f_j^-, f_j^+]$$

  is also an exact representation of the term $\sum_j \theta_{ij} f_i f_j$.

Totally, we have $2 + (n-1) = n+1$ terms, so, to get the final enclosure, we need $n$ operations (additions) with dependent intervals.

This representation, however, still leads to an overestimate, so we would like to come up with a new reformulation that would lead to even fewer operations with dependent intervals.

## The Idea of Gauss Decomposition

The possibility of such a reformulation stems from the fact that the function $\eta(\vec{\theta}, \vec{f})$ that we want to optimize is quadratic in $\vec{f}$, and therefore, we can use a special technique called *Gauss decomposition* (see, e.g., [4]) to bound the fluctuation of the performance characteristic.

This method is based on a theorem (proved by Gauss) that any quadratic polynomial $\eta(\vec{f})$ of $n$ variables can be presented as a linear combination of the squares of $q \le n$ (linearly independent) linear polynomials $\alpha_1(\vec{f}), ..., \alpha_q(\vec{f})$ of the same variables $f_i$: $\eta(\vec{f}) = C_1 \alpha_1^2 + ... + C_q \alpha_q^2$.

For example, a function $\eta(f_1, f_2) = f_1 f_2$ can be decomposed as

$$\eta = \frac{1}{4}\alpha_1^2 - \frac{1}{4}\alpha_2^2,$$

where

$$\alpha_1 = f_1 + f_2, \quad \alpha_2 = f_1 - f_2. \tag{5.7}$$

After we get a Gauss decomposition, we can use ***interval computations*** techniques to estimate the range of $\eta$, and thus, to get the bounds for $j(\vec{x})$. Namely:

- First, we use the intervals $[f_i^-, f_i^+] = [x_i - e_{fi}, x_i + e_{fi}]$ of possible values of $f_i$ to compute the intervals $[\alpha_k^-, \alpha_k^+]$ of possible values of their linear combinations $\alpha_k$.

  Since the functions $\alpha_k$ are linear, each variable $f_i$ appears only once in each $\alpha_k$, and therefore, the resulting intervals $[\alpha_k^-, \alpha_k^+]$ are exact.

- Second, by using formulas of interval computation, we find an enclosure $[\eta^-, \eta^+] = C_1[\alpha_1^-, \alpha_1^+]^2 + ... + C_q[\alpha_q^-, \alpha_q^+]^2$ for the interval of possible values of $\eta$ (in this formula, the operations with intervals are defined as above).

Since the interval that corresponds to each $\alpha_q$ is exact, each of $q \le n$ intervals $[\alpha_q^-, \alpha_q^+]$ is also exact. So, to compute the final enclosure, we only need $q - 1 \le n - 1$ operations with dependent intervals.

To illustrate this method, let us consider the above example of $\eta = f_1 f_2$, and assume that $f_1$ and $f_2$ belong to the interval $[f_1^-, f_1^+] = [f_2^-, f_2^+] = [-1, 1]$. Using interval computations, we find that $\alpha_1 \in [\alpha_1^-, \alpha_1^+] = [-2, 2]$ and $\alpha_2 \in [\alpha_2^-, \alpha_2^+] = [-2, 2]$. As we have already mentioned, these intervals $[\alpha_k^-, \alpha_k^+]$ are exact.

From interval computations, we can conclude that

$$(\alpha_1, \alpha_2) \in [\alpha_1^-, \alpha_1^+] \times [\alpha_2^-, \alpha_2^+] = [-2, 2] \times [-2, 2].$$

The exact uncertainty set for $\alpha_1$ and $\alpha_2$ resulting from (5.7) is represented by a normal (full) line in Figure 5, while the box obtained by interval analysis is represented by dashed lines. This figure illustrates the fact that interval methods often yield a pessimistic enclosure for the set of possible values of $(\alpha_1, ..., \alpha_q)$.

From the intervals $[\alpha_k^-, \alpha_k^+] = [-2, 2]$, we conclude that $[\alpha_k^-, \alpha_k^+]^2 = [0, 4]$ and therefore, the desired enclosure is

$$[\eta^-, \eta^+] = \frac{1}{4} \cdot [0, 4] - \frac{1}{4} \cdot [0, 4] = [0, 1] - [0, 1] = [-1, 1].$$

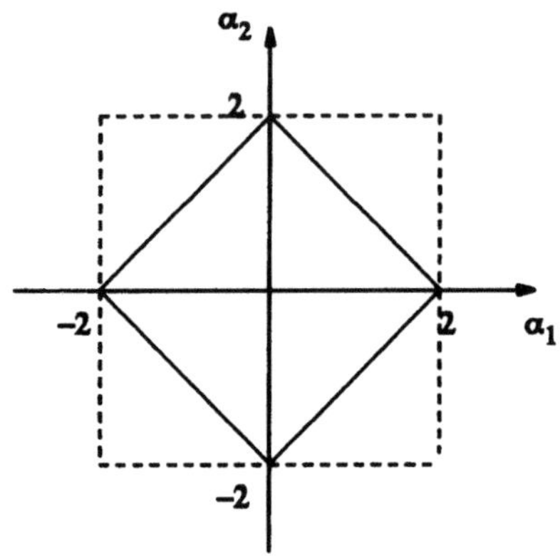


FIG 5: Uncertainty sets for $\vec{\alpha}$, as obtained by exact and interval computations.

In this particular example, this enclosure coincides with the exact range of the function $\eta$ (although in general, interval computations can lead to an overestimating enclosure).

## *Example 1 (continued)*

In our pedagogical example 1, with $h'(f) = y^* - 2 + 2f - f^2$, both "naive" interval computations, and the above-described simple reformulation lead to an overestimating enclosure. On the other hand, Gauss decomposition leads to an exact interval. Indeed, for this example, the modified criterion can be rewritten as

$$h'(f) = y^* - 2 + 2f - f^2 = y^* - 1 - \alpha^2$$

where $\alpha = f - 1$.

For $f$ belonging to $[x - e_f, x + e_f]$ the uncertainty interval for $\alpha$ is

$$[\alpha^-, \alpha^+] = [-1 + x - e_f, -1 + x + e_f].$$

Using interval computations, one can find the exact interval for $h'$ and thus the maximal value of $h(f)$; the results coincide with the exact bounds computed in Section 3.2.

# 4 MODEL PARAMETERS ESTIMATED IN CONTROLLED CONDITIONS

## 4.1 Estimation Stage

Assume now that the model parameters $\vec{\theta}$ are estimated from several ($N$) measurements of performance characteristic obtained in controlled conditions (where the error in the factors is negligible, so we may assume that we know the exact values of these factors).

Traditionally, in this case, *statistical* methods are used that are based on the knowledge of statistical characteristics of the measurement errors. However, as we have already mentioned, in many real-life situations, we do not know these statistical characteristics. The only information that we usually have consists of the upper bounds on the measurement error. (These bounds are provided by the manufacturer of the measuring instrument.) In this case, the bounded-error approach [9, 10, 11, 20, 24, 25, 40, 41] seems more appropriate than the usual statistical approach.

Let us describe the bounded-error approach in mathematical terms. For that, we will use the following notation:

- The values of the factors in the $k$−th measurement will be denoted by $\vec{x}^{(k)}$.
- The result of the $k$−th measurement will be denoted by $y^{(k)}$.
- The guaranteed upper bounds for absolute error on the result of the $k$−th measurement of the performance characteristic $y$ will be denoted by $e_y^{(k)}$.

The error $\varepsilon_y^{(k)} = y^{(k)} - \eta(\vec{\theta}, \vec{x}^{(k)})$ on the $k$−th measurement (i.e., the difference between the measured value $y^{(k)}$ and the predicted value $\eta(\vec{\theta}, \vec{x}^{(k)})$) cannot exceed the guaranteed error bound:

$$|\varepsilon_y^{(k)}| \le e_y^{(k)}.$$

In other words, from the result of the $k$−th measurement, we can conclude that the (unknown) actual values of the parameters $\vec{\theta}$ must satisfy the inequality

$$y^{(k)} - e_y^{(k)} \le \eta(\vec{\theta}, \vec{x}^{(k)}) \le y^{(k)} + e_y^{(k)}. \tag{5.8}$$

So, after we have performed all $N$ measurements, we conclude that the actual values $\vec{\theta}$ of the parameters must belong to the set

$$S_N = \{\vec{\theta} : y^{(k)} - e_y^{(k)} \le \eta(\vec{\theta}, \vec{x}^{(k)}) \le y^{(k)} + e_y^{(k)}, 1 \le k \le N\}. \tag{5.9}$$

Since $\eta(\vec{\theta}, \vec{f})$ is linear in $\vec{\theta}$, each of $N$ conditions (5.8) (that correspond to each of $N$ measurements) describes a strip of possible values of $\vec{\theta}$ that is bounded by two parallel hyperplanes:

$$H^+ = \{\vec{\theta} \in R^p : y^{(k)} - \eta(\vec{\theta}, \vec{x}^{(k)}) = e_y^{(k)}\},$$
$$H^- = \{\vec{\theta} \in R^p : y^{(k)} - \eta(\vec{\theta}, \vec{x}^{(k)}) = -e_y^{(k)}\}.$$

$S_N$ is the intersection of $N$ such strips (Figure 6) and is therefore a convex polyhedron (usually a polytope).

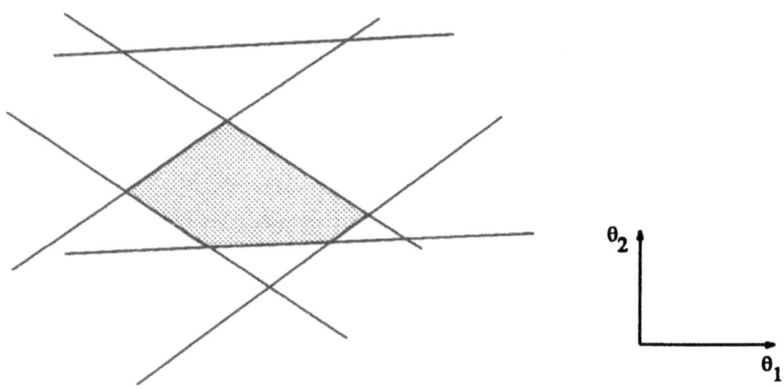


FIG 6: Posterior feasible parameter set (linear case).

Methods are available to characterize this set $S_N$ exactly [21, 39] or to compute simple enclosures guaranteed to contain $S_N$. As enclosures, one can use:

- *ellipsoids*; recursive enclosure methods are described in [1, 12, 28, 31], and non-recursive methods in [27];
- *rectangular parallelepipeds* (*orthotopes*); recursive enclosure methods are described in [2, 17, 30], and non-recursive methods in [18, 19];
- *simplexes* [26];
- (generic) *parallelepipeds* [6];
- *polyhedra with fixed maximal complexity* [3, 26]; etc.

## 4.2 Optimization Stage

### *The Choice of the Characteristic $j(\vec{x})$ that Describes the Quality of the Design $\vec{x}$*

On the preliminary stage of controlled experiments, we were able to measure the values of the factors precisely (i.e., with negligible error). However, at the manufacturing stage, we must take into consideration the fact that the actual values $f_i$ of the factors may differ from their nominal values: $f_i \neq x_i$, $\vec{f} \in F(\vec{x})$.

In addition to this uncertainty, we now have an additional uncertainty caused by the fact that we do not know the precise values $\vec{\theta}$ of the parameters. Instead, we only know the set $S_N$ that contains $\vec{\theta}$. In this case, for a given design $\vec{x}$, to find the worst-case loss $j(\vec{x})$, we have to consider not only all possible values of the factors $\vec{f} \in F(\vec{x})$, but also all possible values of the parameters $\vec{\theta} \in S_N$. As a result, we get the following worst-case characteristic of the quality of the design:

$$j(\vec{x}) = \max_{\vec{\theta} \in S_N, \vec{f} \in F(\vec{x})} (y^* - \eta(\vec{\theta}, \vec{f}))^2.$$

### *Estimating the Quality $j(\vec{x})$ of a Given Design $\vec{x}$*

Since $S_N$ is a convex set and $\eta$ is linear in $\vec{\theta}$, the maximum of $\eta(\vec{\theta}, \vec{f})$ with respect to $\vec{\theta}$ is attained at one of the vertices $\vec{\theta}^{(1)}, ..., \vec{\theta}^{(V)}$ of the set $S_N$. Therefore, when computing this maximum, it is sufficient to consider only these vertices $\vec{\theta} = \vec{\theta}^{(v)}$, and we have

$$j(\vec{x}) = \max(j^{(1)}(\vec{x}), ..., j^{(V)}(\vec{x})),$$

where $j^{(v)}(\vec{x})$ denotes the conditional maximum of $h(\vec{f}) = (y^* - \eta(\vec{\theta}, \vec{f}))^2$, under the condition that $\vec{\theta} = \vec{\theta}^{(v)}$, i.e.,

$$j^{(v)}(\vec{x}) = \max_{\vec{f} \in F(\vec{x})} (y^* - \eta(\vec{\theta}^{(v)}, \vec{f}))^2.$$

So, to estimate the value of $j(\vec{x})$, we can do the following:

- First, for every $v = 1, ..., V$, we estimate $j^{(v)}(\vec{x})$ using one of the methods described in Section 3.
- Second, we compute the maximum of these $V$ values. This maximum is the desired estimate for $j(\vec{x})$.

In this chapter, we use this method. However, the larger $N$ is, the more vertices we have to consider. To reduce the computation time, we may want to:

- approximate the set $S_N$ by a simpler set $S$: a simpler polyhedron, or an ellipsoid (using one of the above-mentioned techniques), and then
- for this set $S$, compute the value

$$\tilde{j}(\vec{x}) = \max_{\vec{\theta}\in S, \vec{f}\in F(\vec{x})} (y^* - \eta(\vec{\theta}, \vec{f}))^2,$$

that is an upper bound for $j(\vec{x})$.

## *Finding the Optimal Design*

To find the optimal design, we must find $\vec{x}$ that minimizes $j(\vec{x})$. For this purpose, we can use known global optimization techniques, e.g., adaptive random search.

In this paper we shall use an exact description of $S_N$, but if this exact description turns out to be too complicated then it can be replaced by any suitable limited-complexity set guaranteed to contain $S_N$.

# 4.3 Example 1 (continued)

## *Description of the Model*

Let us still assume that the actual model of the performance characteristic is:

$$\eta(f) = 2 - 2f + f^2.$$

In other words, we assume that $n = 1$, $\theta_0 = 2$, $\theta_1 = -2$, and $\theta_{11} = 1$.

We are simulating the case when the model is not known *a priori*. To determine the coefficients $\theta_0$, $\theta_1$, and $\theta_{11}$, we measure the value of the performance characteristic for $N = 3$ situations: for $x^{(1)} = 0$, $x^{(2)} = 1$, and for $x^{(3)} = 2$. We assume that the accuracy $e_y^{(k)}$ is equal to 0.01 for all three measurements. To actually simulate the error $\varepsilon_y^{(k)}$ from the interval $[-0.01, 0.01] = [-e_y^{(k)}, e_y^{(k)}]$, we use a (pseudo)random number generator that generates numbers uniformly distributed on this interval.

## *Estimation Stage: Graphing*

The resulting set $S_N = S_3$ is a 3-D figure, and thus difficult to present and interpret. This is why we will momentarily assume that the exact value of the parameter $\theta_0$ $(= 2)$ is known. The resulting set $S_N = S_3$ of possible values of $(\theta_1, \theta_{11})$ is shown on Figure 7 ($\theta_0$ is assumed to be known exactly).

This assumption that $\theta_0 = 2$ was done for graphing only. In the computations presented in this section, we do not make this assumption, and we determine all three parameters $\theta_0$, $\theta_1$, and $\theta_{11}$ from the (simulated) measurements.

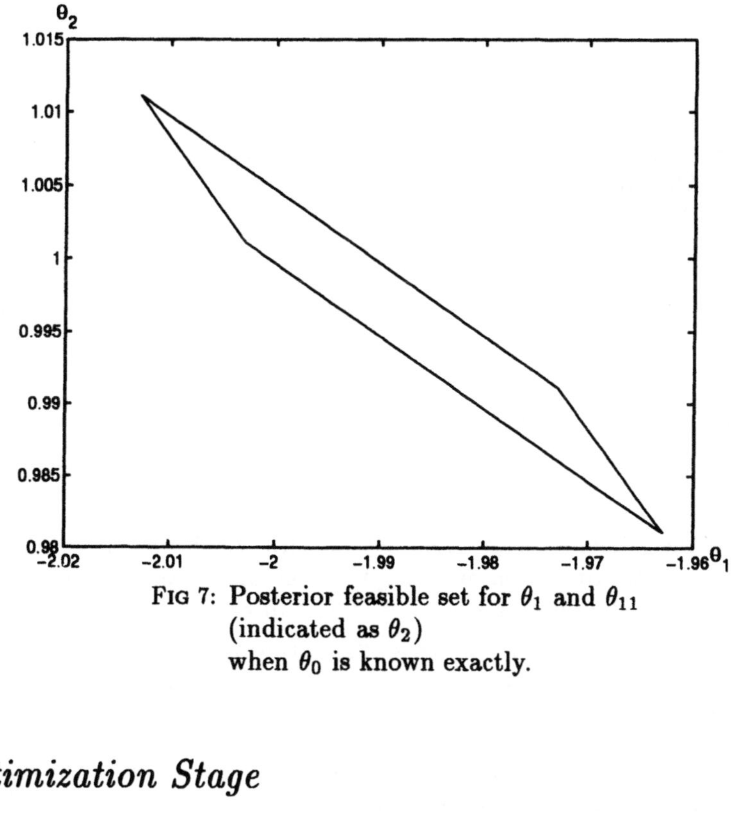


FIG 7: Posterior feasible set for $\theta_1$ and $\theta_{11}$ (indicated as $\theta_2$) when $\theta_0$ is known exactly.

## *Optimization Stage*

Worst case optimization was then performed, to select the best nominal value $x^*$ for the factor during mass production. The results are summarized in Table 2.

| Estimation stage | Optimization | stage | | |
|---|---|---|---|---|
| $e_y$ | $e_f$ | $y^*$ | $x^*$ | $j(x^*)$ |
| 0.1 | 0.0 | 1.25 | 0.4397 | 0.0156 |
| | | | 1.5379 | 0.0156 |
| 0.1 | 0.1 | 1.25 | 0.4478 | 0.0588 |
| | | | 1.5292 | 0.0588 |

Table 2. Optimization results (parameters are uncertain).

## *Analysis of the Results*

In this case, we have two sources of uncertainty:

- First, during mass production, we cannot maintain the exact values of the factors. This uncertainty is characterized by the accuracy $e_{fi}$ with which we control the factors.
- Second, we do not measure the performance characteristic precisely. The resulting uncertainty is characterized by the accuracy $e_y^{(k)}$ of the corresponding measurements.

Tables 1 and 2 show how these sources of uncertainty influence the optimal nominal design values $x^*$ and the quality $j(x^*)$ of the optimal design:

- To show the influence of the accuracy $e_f$ with which we can maintain the factors in mass production on the optimal nominal value $x^*$, we first give the results for the ideal case $e_f = 0$ (when all the factors are controlled). For the same accuracy $e_y^{(k)}$ in measuring performance, the larger $e_f$ is, the closer the optimal nominal values of the factor are to the stationary point; this behavior is similar to the one that we saw in Section 3.2.
- Note also that the optimal value $j(x^*)$ increases quickly with the errors in the estimation stage, as can be seen by comparing the results of Table 2 with those of Table 1.

# 5 CONTROLLED CONDITIONS ARE UNAVAILABLE

## 5.1 Estimation Stage

### *Formulation of the Problem in Mathematical Terms*

Consider, finally, the case of $N$ measurements performed with errors on both the factors and outputs. Let us describe this case in mathematical terms. For that, we will use the following notation:

- The nominal (set-up) values of the factors in the $k$−th measurement will be denoted by $\vec{x}^{(k)} = (x_1^{(k)}, ..., x_n^{(k)})^T$.
- The actual (unknown) values of the factors in the $k$−th measurement will be denoted by $\vec{f}^{(k)} = (f_1^{(k)}, ..., f_n^{(k)})^T$.
- The inaccuracy of our knowledge of the values of the factors in the $k$−th measurement will be denoted by $\vec{\varepsilon}_f^{(k)} = \vec{f}^{(k)} - \vec{x}^{(k)} = (\varepsilon_{f1}^{(k)}, ..., \varepsilon_{fn}^{(k)})^T$.
- The guaranteed accuracy with which we can control the values of the factors in the $k$−th measurement will be denoted by $\vec{e}_f^{(k)} = (e_{f1}^{(k)}, ..., e_{fn}^{(k)})^T$.
- The result of the $k$−th measurement will be denoted by $y^{(k)}$.
- The guaranteed upper bounds for the result of the $k$−th measurement of the performance characteristic $y$ will be denoted by $e_y^{(k)}$.

In these terms, from the $k$−th measurement, we can conclude:

- First, that for all $i$ from 1 to $n$, the error on the $i$−th factor

$$\varepsilon_{fi}^{(k)} = f_i^{(k)} - x_i^{(k)} \tag{5.10}$$

  cannot exceed the guaranteed error bound:

$$|\varepsilon_{fi}^{(k)}| \leq e_{fi}^{(k)}. \tag{5.11}$$

- Second, that the error

$$\varepsilon_y^{(k)} = y^{(k)} - \eta(\vec{\theta}, \vec{f}^{(k)}) \tag{5.12}$$

of the $k$–th measurement of the performance characteristic (i.e., the difference between the measured value $y^{(k)}$ and predicted value $\eta(\vec{\theta}, \vec{f}^{(k)})$) cannot exceed the guaranteed error bound:

$$|\varepsilon_y^{(k)}| \le e_y^{(k)}. \tag{5.13}$$

In other words, from the result of the $k$–th measurement, we can conclude that the (unknown) actual values of the parameters $\vec{\theta}$ must be such that for some $\varepsilon_{fi}^{(k)} \in [-e_{fi}^{(k)}, e_{fi}^{(k)}]$ and $\varepsilon_y^{(k)} \in [-e_y^{(k)}, e_y^{(k)}]$, the following equality holds:

$$y^{(k)} = \theta_0 + \sum_{i=1}^{n}(x_i^{(k)} + \varepsilon_{fi}^{(k)})\theta_i +$$

$$\sum_{i=1}^{n-1}\sum_{j=i+1}^{n} \theta_{ij}(x_i^{(k)} + \varepsilon_{fi}^{(k)})(x_j^{(k)} + \varepsilon_{fj}^{(k)}) + \sum_{i=1}^{n}\theta_{ii}(x_i^{(k)} + \varepsilon_{fi}^{(k)})^2 + \varepsilon_y^{(k)}. \tag{5.14}$$

As in Section 4, we will denote the set of all vectors $\vec{\theta}$ that satisfy these $N$ conditions (i.e., that is consistent with the results of all $N$ measurements) by $S_N$.

## *No Ready Method is Available to Solve this System of Interval Equations*

In interval computations, conditions of the type (5.14) are usually described as a system of ***interval linear equations*** with unknowns $\vec{\theta}$:

$$[y^{(k)} - e_y^{(k)}, y^{(k)} + e_y^{(k)}] = \theta_0 + \sum_{i=1}^{n}[x_i^{(k)} - e_{fi}^{(k)}, x_i^{(k)} + e_{fi}^{(k)}]\theta_i +$$

$$\sum_{i=1}^{n-1}\sum_{j=i+1}^{n} \theta_{ij}[x_i^{(k)} - e_{fi}^{(k)}, x_i^{(k)} + e_{fi}^{(k)}] \cdot [x_j^{(k)} - e_{fj}^{(k)}, x_j^{(k)} + e_{fj}^{(k)}] +$$

$$\sum_{i=1}^{n} \theta_{ii}[x_i^{(k)} - e_{fi}^{(k)}, x_i^{(k)} + e_{fi}^{(k)}]^2.$$

There exist efficient methods of solving interval linear equations for the case when the uncertainties in different coefficients are independent (see, e.g., [6, 7, 8, 16, 23, 24, 35] for a presentation in the context of parameter bounding). However, in our case, the coefficients are dependent, so these methods are not directly applicable.

*Comment.* This corresponds to an "errors-in-variables" estimation problem, known to be difficult and still a hot topic in statistical estimation.

## The Proposed Method

The following theorem shows that it is possible to include the set $S_N$ of all values of parameters that are consistent with the given measurements into a union of finitely many convex polyhedra (i.e., sets defined by linear inequalities).

**Theorem.** *A necessary condition for the vector $\vec{\theta} \in R^p$ to belong to the set $S_N$ (i.e., to be consistent with the results of $N$ measurements) is:*

$$|\varepsilon_y(\vec{\theta}, \vec{x}^{(k)})| \le \sum_{i=1}^{n} e_{fi}^{(k)} \theta_i \cdot \text{sign}(\theta_i) +$$

$$\sum_{i=1}^{n-1} \sum_{j=i+1}^{n} [e_{fi}^{(k)}|x_j^{(k)}| + e_{fj}^{(k)}|x_i^{(k)}| + e_{fi}^{(k)} e_{fj}^{(k)}] \cdot \theta_{ij} \cdot \text{sign}(\theta_{ij}) + \tag{5.15}$$

$$\sum_{i=1}^{n} [2e_{fi}^{(k)}|x_i^{(k)}| + (e_{fi}^{(k)})^2] \cdot \theta_{ii} \cdot \text{sign}(\theta_{ii}) + e_y^{(k)},\ k = 1, \ldots, N,$$

*where*

$$\varepsilon_y(\vec{\theta}, \vec{x}^{(k)}) = y^{(k)} - [\theta_0 + \sum_{i=1}^{n} \theta_i x_i^{(k)} + \sum_{i=1}^{n-1} \sum_{j=i+1}^{n} \theta_{ij} x_i^{(k)} x_j^{(k)} + \sum_{i=1}^{n} \theta_{ii} (x_i^{(k)})^2].$$

**Proof.** According to (5.14),

$$\varepsilon_y(\vec{\theta}, \vec{x}^{(k)}) = \sum_{i=1}^{n} \varepsilon_{fi}^{(k)} \theta_i + \sum_{i=1}^{n-1} \sum_{j=i+1}^{n} [\varepsilon_{fi}^{(k)} x_j^{(k)} + \varepsilon_{fj}^{(k)} x_i^{(k)} + \varepsilon_{fi}^{(k)} \varepsilon_{fj}^{(k)}] \cdot \theta_{ij}$$

$$+\sum_{i=1}^{n}[(\varepsilon_{fi}^{(k)})^2 + 2x_i^{(k)}\varepsilon_{fi}^{(k)}]\cdot\theta_{ii} + \varepsilon_y^{(k)}. \tag{5.16}$$

Equation (5.16) implies that

$$|\varepsilon_y(\vec{\theta}, \vec{x}^{(k)})| \le \sum_{i=1}^{n} |\varepsilon_{fi}^{(k)}\theta_i| + \sum_{i=1}^{n-1}\sum_{j=i+1}^{n} |[\varepsilon_{fi}^{(k)}x_j^{(k)} + \varepsilon_{fj}^{(k)}x_i^{(k)} + \varepsilon_{fi}^{(k)}\varepsilon_{fj}^{(k)}]\cdot\theta_{ij}|$$

$$+\sum_{i=1}^{n}|[(\varepsilon_{fi}^{(k)})^2 + 2x_i^{(k)}\varepsilon_{fi}^{(k)}]\cdot\theta_{ii}| + |\varepsilon_y^{(k)}|. \tag{5.17}$$

Taking (5.11) and (5.13) into account, we obtain:

$$|\varepsilon_y(\vec{\theta}, \vec{x}^{(k)})| \le \sum_{i=1}^{n} e_{fi}^{(k)}|\theta_i| +$$

$$\sum_{i=1}^{n-1}\sum_{j=i+1}^{n} [e_{fi}^{(k)}|x_j^{(k)}| + e_{fj}^{(k)}|x_i^{(k)}| + e_{fi}^{(k)}e_{fj}^{(k)}]\cdot|\theta_{ij}| + \tag{5.18}$$

$$\sum_{i=1}^{n}[(e_{fi}^{(k)})^2 + 2|x_i^{(k)}|e_{fi}^{(k)}]\cdot|\theta_{ii}| + e_y^{(k)}.$$

The absolute value $|a|$ of any variable $a$ can be replaced by $a \cdot \text{sign}(a)$, so that (5.18) is equivalent to:

$$|\varepsilon_y(\vec{\theta}, \vec{x}^{(k)})| \le \sum_{i=1}^{n} e_{fi}^{(k)}\theta_i \cdot \text{sign}(\theta_i) +$$

$$\sum_{i=1}^{n-1}\sum_{j=i+1}^{n} [e_{fi}^{(k)}|x_j^{(k)}| + e_{fj}^{(k)}|x_i^{(k)}| + e_{fi}^{(k)}e_{fj}^{(k)}]\cdot\theta_{ij}\cdot\text{sign}(\theta_{ij})$$

$$+\sum_{i=1}^{n}[(e_{fi}^{(k)})^2 + 2|x_i^{(k)}|e_{fi}^{(k)}]\theta_{ii}\cdot\text{sign}(\theta_{ii}) + e_y^{(k)}.$$

Q.E.D.

The set $T_N$ of all $\vec{\theta}$ that satisfy the inequality (5.15) is thus guaranteed to contain $S_N$. Since the values $|x_i^{(k)}|$ are known, and in each orthant $O_1$, $O_2$, ..., of the $p$–dimensional parameter space, all the signs of the parameters are constant, for each orthant of the parameter space, (5.15) amounts to $2N$ linear inequalities.

Thus, the set $S_N$ of all parameter vectors that are consistent with the results of all the measurements is contained in the union $T_N$ of the solutions obtained by linear techniques in all orthants of the parameter space, i.e., in the union of convex polyhedra.

*Comment.* This uncertainty region $T_N$ is generally pessimistic. Note that $T_N$ may be disconnected and unbounded. However, its intersection with each orthant is convex.

## 5.2 Optimization Stage

The optimization problem is very similar to that of Section 4, with the posterior feasible set $S_N$ replaced by a set $T_N$ guaranteed to contain it. We now want to solve the minimization problem $\tilde{\jmath}(\vec{x}) \rightarrow \min$, where

$$\tilde{\jmath}(\vec{x}) = \max_{\vec{\theta} \in T_N, \vec{f} \in F(\vec{x})} (y^* - \eta(\vec{f}, \vec{\theta}))^2. \tag{5.19}$$

To estimate this value, we can use the above-mentioned fact that the intersection of $T_N$ with each orthant $O_m$ is a convex polyhedron. Namely, we compute

$$\tilde{\jmath}(\vec{x}) = \max(\tilde{\jmath}_1(\vec{x}), \tilde{\jmath}_2(\vec{x}), \dots),$$

where

$$\tilde{\jmath}_m(\vec{x}) = \max_{\vec{\theta} \in T_N \cap O_m, \vec{f} \in F(\vec{x})} (y^* - \eta(\vec{f}, \vec{\theta}))^2$$

is a maximum over the intersection between $T_N$ and $m$–th orthant $O_m$. Since this intersection is a convex polyhedron, the value $\tilde{\jmath}_m(\vec{x})$ can be computed by methods described in Section 4.

## 5.3 Example 1 (continued)

### *Description of the Model*

Let us still assume that the actual model of the performance characteristic is:

$$\eta(f) = 2 - 2f + f^2.$$

In other words, we assume that $n = 1$, $\theta_0 = 2$, $\theta_1 = -2$, and $\theta_{11} = 1$.

To determine the coefficients $\theta_0$, $\theta_1$, and $\theta_{11}$, we measure the value of the performance characteristic for $N = 3$ situations: for $x^{(1)} = 0$, $x^{(2)} = 1$, and for $x^{(3)} = 2$. We assume that the accuracy $e_y^{(k)}$ is equal to 0.01 for all three measurements. We also assume that the accuracy $e_f^{(k)}$ with which we control the values of the factors is the same for all three measurements: $e_f^{(k)} = e_f$. We consider two cases: $e_f = 0.01$ and $e_f = 0.1$.

To actually simulate the errors $\varepsilon_y^{(k)}$ from the interval $[-0.01, 0.01] = [-e_y^{(k)}, e_y^{(k)}]$ and the errors $\varepsilon_f^{(k)} \in [-e_f, e_f]$, we use a (pseudo)random number generator that generates numbers uniformly distributed on these interval.

The estimation problem can be summarized as follows: we must find $\theta_0$, $\theta_1$, and $\theta_{11}$ from the simulated measurement results in which $y^{(k)} = 2 - 2f^{(k)} + (f^{(k)})^2 + \varepsilon_y^{(k)}$ and $f^{(k)} = x^{(k)} + \varepsilon_f^{(k)}$.

## *Results*

An exact description of the set $T_N$ has been obtained. The results of the optimization of the criterion (5.19) are given in Table 3.

| Estimation stage | | Optimization stage | | | |
|---|---|---|---|---|---|
| $e_y$ | $e_f$ | $e_f$ | $y^*$ | $x^*$ | $j(x^*)$ |
| 0.01 | 0.01 | 0.01 | 1.25 | 0.4908 | 0.0047 |
| 0.01 | 0.1 | 0.1 | 1.25 | 0.4409 | 0.0851 |

Table 3. Optimization results (with errors on the factors during the construction of the model).

## *Analysis of the Results*

Comparing the third line of Table 2 with the second line of Table 3, we see that errors in the factors at the estimation stage quickly deteriorate the value of $j(x^*)$, hence the interest of using controlled conditions during the construction of the model.

# 6 TEMPERATURE CONTROLLER

## 6.1 Description of the System

As a realistic application of our methods, let us consider the temperature controller presented on Figure 8. This example has been used as a test-case in

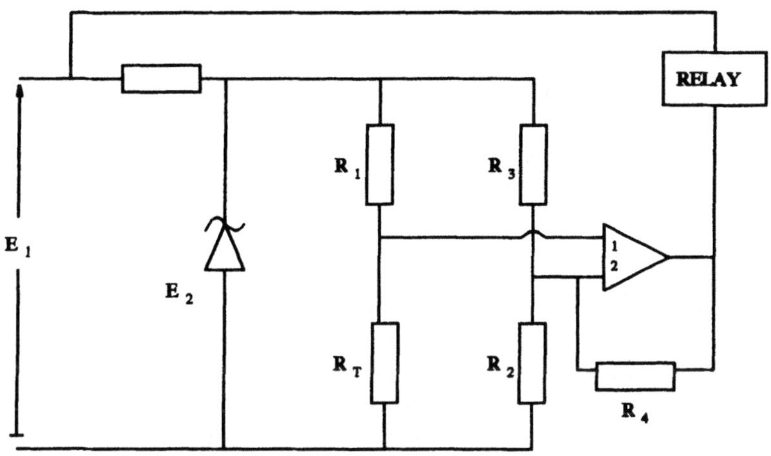


FIG 8: Temperature Controller Circuit.

[34, 42].

In this example, we have six factors $\tilde{f}_1 - \tilde{f}_6$ (a realistic number for industrial applications):

- Voltages $E_1$ ($\tilde{f}_1$) and $E_2$ ($\tilde{f}_6$), and
- Resistances $R_1$ ($\tilde{f}_2$), $R_2$ ($\tilde{f}_3$), $R_3$ ($\tilde{f}_4$), and $R_4$ ($\tilde{f}_5$).

We assume that all these factors can be controlled with a 5% tolerance, i.e., that for each nominal value $\tilde{x}_i$, the interval of possible actual values $\tilde{f}_i$ is equal to $[\tilde{x}_i - 0.05 \cdot \tilde{x}_i, \tilde{x}_i + 0.05 \cdot \tilde{x}_i]$ (i.e., in our denotations, $e_{fi}(\tilde{x}_i) = 0.05 \cdot \tilde{x}_i$).

As a performance characteristic, we take the value of the resistance $R_{T-on}$ at which a relay activates the heater. The target value $y^*$ for this characteristic is $3k\Omega$. The dependency of this performance characteristic on the factors is known:

$$R_{T-on} = \frac{R_1 R_2(E_2 R_4 + E_1 R_3)}{R_3(E_2 R_4 + E_2 R_2 - E_1 R_2)}. \tag{5.20}$$

We know the ranges $[x_{i\,\min}, x_{i\,\max}]$ of all the factors: namely, the voltages must be from the ranges $E_1 \in [7.5V, 9.5V]$ and $E_2 \in [4.5V, 7.5V]$, and the resistances must be from the ranges $R_1 \in [0.5k\Omega, 1.5k\Omega]$, $R_2 \in [6k\Omega, 10k\Omega]$, $R_3 \in [2k\Omega, 6k\Omega]$, and $R_4 \in [16k\Omega, 48k\Omega]$.

Our goal is to find the optimal design $\vec{x}^*$ of the temperature controller, i.e., the set of nominal values of the above-mentioned six factors for which the resulting value of the performance characteristic is *the most robust*, i.e., as close to the target value as possible (or, in mathematical terms: for which the worst-case fluctuation $j(\vec{x})$ around the target value is the smallest).

## 6.2 Preliminary Scaling

Before we start applying our technique, let us first re-scale the values of the factors (a classical procedure to limit numerical problems). Namely, we will re-scale each factor so that the re-scaled values will range from $-1$ to 1 (see Table 4).

| Original factors | $\tilde{x}_1$ | $\tilde{x}_2$ | $\tilde{x}_3$ | $\tilde{x}_4$ | $\tilde{x}_5$ | $\tilde{x}_6$ |
|---|---|---|---|---|---|---|
| Meaning | $E_1$ | $R_1$ | $R_2$ | $R_3$ | $R_4$ | $E_2$ |
| $-1$ | 7.5 | 0.5 | 6 | 2 | 16 | 4.5 |
| 0 | 8.5 | 1 | 8 | 4 | 32 | 6.0 |
| 1 | 9.5 | 1.5 | 10 | 6 | 48 | 7.5 |

Table 4. Experimental values for the temperature controller circuit.

The re-scaled values $x_i \in [-1, 1]$ are computed by the formulas

$$x_i = \frac{\tilde{x}_i - \tilde{x}_{i0}}{\Delta_i},$$

where

$$\tilde{x}_{i0} = \frac{\tilde{x}_{i\,\max} + \tilde{x}_{i\,\min}}{2} \quad \text{and} \quad \Delta_i = \frac{\tilde{x}_{i\,\max} - \tilde{x}_{i\,\min}}{2}, \quad i = 1, \ldots, 6.$$

## 6.3 Approximating the Original Model with the Quadratic Model

The given dependency of the performance characteristic on the factors is not quadratic. However, as we have mentioned in Section 2, we can get a pretty good quadratic approximation for the original dependency.

To find the approximation, we apply the least-squares method to the data computed by the analytical formula (5.20) according to an optimal composite design having a fraction of a full factorial design with $2^{6-1} = 32$ points, plus 12 star points (so that $N = 44$) [15]. As a result, we get the following quadratic model:

$$y = 2.6279 + 0.5015x_1 + 2.1091x_2 + 1.3961x_3 - 1.4635x_4 - 1.0313x_5$$
$$-1.0801x_6 + 0.2611x_1x_2 + 0.1839x_1x_3 + 0.3529x_1x_4 - 0.3116x_1x_5$$
$$-0.2848x_1x_6 + 0.7188x_2x_3 - 0.7310x_2x_4 - 0.5320x_2x_5 - 0.54x_2x_6$$
$$-0.1814x_3x_4 - 0.5238x_3x_5 - 0.5515x_3x_6 - 0.0479x_4x_5 - 0.0007x_4x_6$$
$$+0.7376x_5x_6 + 0.0079x_1^2 + 0.0001x_2^2$$
$$+0.0213x_3^2 + 0.7442x_4^2 + 0.2841x_5^2 + 0.6322x_6^2. \qquad (5.21)$$

## 6.4 Results

Since uncertainty on the value of parameters $\vec{\theta}$ of the model is not taken into account, we can apply methods from Section 3. As a result, we get the following optimal design:

$$\vec{x}^* = [8.3112,\ 0.6456,\ 9.2747,\ 2.0626,\ 34.5548,\ 7.1852].$$

The maximal fluctuation around the target value with this optimal design is equal to $j(\vec{x}^*) = 0.316$.

## 6.5 Advantages of our Approach

This example illustrates the advantages of our approach as compared to Taguchi's approach.

- The optimal design $\vec{x}^*$ has been selected in the complete admissible factor space and not among those tried during the experimental stage (as done by Taguchi and Phadke).
- We are not restricted to Taguchi's orthogonal arrays and can use the well developed experimental design methodology instead. In this example, we used $N = 44$ data points, while Taguchi and Phadke used an orthogonal array of size 18 for the design parameters and performed 18 experiments with different noise parameters for each choice of the design factors, so that $N = 324$.

It is difficult to push the comparison much further, since the criterion used by Taguchi and Phadke is quite different from the one used here.

# 7 CONCLUSIONS

## 7.1 Model-Based Approaches to Quality Improvement have some Advantages over Taguchi's Method

The statistical and deterministic model-based approaches to quality improvement during the design stage have the following advantages:

- the number of experiments can be much smaller than with Taguchi's method,
- the interactions between factors can be easily taken into account,
- the optimal values for the factors are searched for in the complete admissible factor space,
- the optimality criterion is clearly related to the problem considered.

## 7.2 When should we Choose Statistical Model-Based Methods, and when Deterministic Methods

The choice between the statistical and deterministic model-based approaches can be made on the following basis:

- First, the choice of the approach depends on our goal:
  - If we require an *acceptable probability of failure*, then the minimum variance policy based on the statistical approach seems more adequate.
  - If we are looking for *zero defect* (guaranteed) quality, then the worst case design is preferable, as made possible by the deterministic approach.
- Second, the choice depends on the available information: depending on whether the information available on the noise factors is under the form of probability distribution or of tolerance intervals, one or the other type of approach will be easier to implement.

One additional advantage of the deterministic approach proposed here is that it takes into account errors on the factors during the estimation stage.

## 7.3 Successes, Problems, and Future Plans

### *Successes*

As evidenced by Section 6, results are already available for realistic problems in the case where parameter uncertainty is neglected.

### *Problems*

The number of vertices of the parameter uncertainty sets ($S_N$ or $T_N$) increases quickly with the number of factors, so that the exact description of the set may become exceedingly complicated.

### *Future Plans*

Possible solutions to overcome this difficulty correspond to the use of outer ellipsoids, of polytopes with reduced complexity.

## Acknowledgements

The authors wish to thank I. Vuchkov and L. Boyadjieva for many helpful discussions on model-based quality control. The help of V. Kreinovich and anonymous referees to transform the original conference communication into a survey chapter is also gratefully appreciated.

## REFERENCES

[1] G. Belforte, B. Bona and V. Cerone, "Parameter estimation algorithms for a set-membership description of uncertainty", *Automatica*, 1990, Vol. 26, pp. 887–898.

[2] G. Belforte and T. T. Tay, "Recursive estimation for linear models with set membership measurement errors", *Prep. 9th IFAC/IFORS Symp. Identification and System Parameter Estimation*, Budapest, 8–12 July, 1991, pp. 872–877.

[3] V. Broman and M. J. Shensa, "A compact algorithm for the intersection and approximation of $N$–dimensional polytopes", *Math. and Comput. in Simulation*, 1990, Vol. 32, pp. 469–480.

[4] G. Cagnac, E. Ramis and J. Commeau, *Nouveau cours de mathématiques spéciales*, Masson, Paris, 1965, Vol. 3, pp. 17–19.

[5] V. Cerone, "Feasible parameter sets for linear models with bounded errors in all variables", *Automatica*, 1993, Vol. 29, pp. 1551–1555.

[6] L. Chisci, A. Garulli and G. Zappa, "Recursive set membership state estimation via parallelotopes", *Prep. 10th IFAC Symp. on System Identification*, Copenhagen, 4-6 July, 1994, Vol. 3, pp. 383–388.

[7] T. Clément and S. Gentil, "Reformulation of parameter identification with unknown but bounded errors", *Math. and Comput. in Simulation*, 1988, Vol. 30, pp. 257–270.

[8] T. Clément and S. Gentil, "Recursive membership set estimation for output-error models", *Math. Comput. in Simulation*, 1990, Vol. 32, pp. 505–513.

[9] P. Combettes, "The foundations of set theoretic estimation", *Proc. of the IEEE*, 1993, Vol. 81, pp. 182–208.

[10] J. R. Deller, "Set membership identification in digital signal processing", *IEEE ASSP Magazine*, 1989, Vol. 6, pp. 4–20.

[11] J. R. Deller, M. Nayeri and S. F. Odeh, "Least-square identification with error bounds for real-time signal processing and control", *Proc. of the IEEE*, 1993, Vol. 81, pp. 813–849.

[12] E. Fogel and Y. F. Huang, "On the value of information in system identification – bounded noise case", *Automatica*, 1982, Vol. 18, pp. 229–238.

[13] E. R. Hansen, "Global optimization using interval analysis – the multidimensional case", *Numer. Math.*, 1980, Vol. 34, pp. 247–270.

[14] K. Ishikawa, "Quality and standardization: progress for economic success", *Quality Progress*, 1984, Vol. 1, pp. 16–20.

[15] J. M. Lucas, "Optimum composite designs", *Technometrics*, 1974, Vol. 16, pp. 561–567.

[16] Y. A. Merkuryev, "Identification of objects with unknown bounded disturbances", *Int. J. Control*, 1989, Vol. 50, pp. 2333–2340.

[17] H. Messaoud and G. Favier, "Recursive determination of parameter uncertainty intervals for linear models with unknown but bounded errors", *Prep. 10th IFAC Symp. on System Identification*, Copenhagen, 4–6 July, 1994, Vol. 3, pp. 365–369.

[18] H. Messaoud, G. Favier and R. Santos Mendes, "Adaptative robust pole placement by connecting identification and control", *Prep. 4th IFAC Int. Symp. Adaptative Systems in Control and Signal Processing*, Grenoble, 1992, pp. 41–46.

[19] M. Milanese and G. Belforte, "Estimation theory and uncertainty intervals evaluation in presence of unknown but bounded errors: linear families of models and estimators", *IEEE Trans. Autom. Control*, 1982, Vol. 27, pp. 408–414.

[20] M. Milanese and A. Vicino, "Optimal estimation theory for dynamic systems with set membership uncertainty: an overview", *Automatica*, 1991, Vol. 27, pp. 997–1009.

[21] S. H. Mo and J. P. Norton, "Fast and robust algorithms to compute exact polytope parameter bounds", *Math. and Comput. in Simulation*, 1990, Vol. 32, pp. 481–493.

[22] R. E. Moore, *Methods and Applications of Interval Analysis*, SIAM Studies in Applied Mathematics, Philadelphia, 1979.

[23] J. P. Norton, "Identification of parameter bounds for ARMAX models from records with bounded noise", *Int. J. Control*, 1987, Vol. 45, pp. 375–390.

[24] J. P. Norton, "Identification and application of bounded-parameter models", *Automatica*, 1987, Vol. 23, pp. 497–507.

[25] J. P. Norton (Ed.), "Special issue on bounded-error estimation", Issues I and II, *Int. J. of Adapt. Contr. and Signal Proc.*, 1994, Vol. 8, and 1995, Vol. 9.

[26] H. Piet-Lahanier and E. Walter, "Polyhedric approximation and tracking for bounded-error models", *Proc. IEEE Int. Symp. Circuits and Systems*, Chicago, 1993, pp. 782–785.

[27] L. Pronzato and E. Walter, "Minimum-volume ellipsoids containing compact sets: application to parameter bounding", *Automatica*, 1994, Vol. 30, pp. 1731–1739.

[28] L. Pronzato, E. Walter and H. Piet-Lahanier, "Mathematical equivalence of two ellipsoidal algorithms for bounded-error estimation", *Proc. 28th IEEE Conference on Decision and Control*, Tampa, 1989, pp. 1952–1955.

[29] L. Pronzato, E. Walter, A. Venot and J. -F. Lebruchec, "A general purpose global optimizer: implementation and applications", *Math. and Computers in Simulation*, 1984, Vol. 26, pp. 412-422.

[30] B. Pschenichnyy and V. S. Pokotilo, "A minmax approach to the estimation of linear regression parameters", *Engrg. Cybernetics*, 1983, pp. 77–85.

[31] F. S. Schweppe, *Uncertain Dynamic Systems*, Prentice–Hall, Englewood Cliffs, 1973.

[32] K. Shimitzu and E. Aiyoshi, "Necessary conditions for min-max problems and algorithm by a relaxation procedure", *IEEE Trans. Autom. Control*, 1980, Vol. 25, pp. 62–66.

[33] G. Taguchi, *Introduction to Quality Engineering*, APO, Tokyo, 1986.

[34] G. Taguchi and M. S. Phadke, "Quality engineering through design optimization", *IEEE Global Telecommunications Conference*, Atlanta, GA, 1984, pp. 1106–1113

[35] S. M. Veres and J. P. Norton, "Parameter-bounding algorithms for linear errors in variables models", *Prep. 9th IFAC/IFORS Symp. on Identification and System Parameter Estimation*, Budapest, 1991, pp. 1038–1043.

[36] I. N. Vuchkov and L. N. Boyadjieva, "The robustness against tolerances of performance characteristics described by second order polynomials", in *Optimal design and analysis of experiments* (Eds. Y. Dodge, V. Fedorov and H. Wynn), North Holland, Amsterdam, 1988, pp. 293–309.

[37] I. N. Vuchkov and L. N. Boyadjieva, "A model-based approach to the robustness against tolerances", *Proc. 33rd EOQC Annual Conference*, Vienna, 1989, pp. 585–592.

[38] I. N. Vuchkov and L. N. Boyadjieva, "Quality improvement through design of experiments with both product parameters and external noise factors", in *Model Oriented Data Analysis. A Survey of Recent Methods* (Eds. V. Fedorov, W. G. Müller and I. N. Vuchkov), Physica Verlag, Heildelberg, 1992.

[39] E. Walter and H. Piet-Lahanier, "Exact recursive polyhedral description of the feasible parameter set for bounded-error models", *IEEE Trans. Autom. Control,* 1989, Vol.34, No. 8, pp. 911–915.

[40] E. Walter and H. Piet-Lahanier, "Estimation of parameter bounds from bounded - error data: a survey", *Math. Comput. in Simulation*, 1990, Vol. 32, pp. 449–468.

[41] E. Walter and L. Pronzato, "Characterizing sets defined by inequalities", *Prep. 10th IFAC Symp. on System Identification*, Copenhagen, 4–6 July, 1994, Vol. 2, pp. 15–26.

[42] H. P. Wynn and A. Winterbottom, *Lectures on Experimental Design and Off-line Quality Control (Taguchi methods)*, City University, London, 1986.

# 6

# APPLICATIONS OF INTERVAL COMPUTATIONS TO REGIONAL ECONOMIC INPUT-OUTPUT MODELS

**Max E. Jerrell**

*College of Business Administration, Northern Arizona University, Flagstaff, AZ 86011-5066, email jerrell@nauvax.ucc.nau.edu*

## ABSTRACT

Economic input-output models are empirical realizations of general equilibrium economic models. They are particularly useful in determining how various industrial sectors of the economy are interrelated and in predicting how these sectors respond to changes in economic activity. In the United States state and regional models are frequently used to help formulate local economic policy. Wassily Leontief is generally credited with the first useful implementation [5].

## 1 ECONOMIC INPUT-OUTPUT MODELS

### Leontief Models

Let $a_{ij}$ be the amount of the $i$–th good used to produce one unit of the $j$–th good. Let $x_j$ be the amount of the $j$–th good produced and $b_j$ be the amount of final (consumer) demand for the $j$–th good. Demand for the $i$–th good is $\sum a_{ij}x_j + b_i$ and in equilibrium (supply = demand)

$$x_i = \sum_{j=1}^{n} a_{ij}x_j + b_i.$$

Equilibrium for all goods in matrix form is

$$x = Ax + b$$



*R. B. Kearfott and V. Kreinovich (eds.), Applications of Interval Computations*, 133–143.

so

$$(I - A)x = Lx = b$$

where $L$ is called the *Leontief matrix* and $A$ is called the *technical coefficients matrix*. If final demand is known, then the amount of the goods needed to satisfy this demand can be found by solving the above equation.

In practice these models are implemented using the dollar value of output of industrial *sectors* rather than individual *goods* - there are just too many goods. The computer industry is an example of an industrial sector. Some of the output from this industry will be used as inputs used to produce other goods and some will be sold as final goods (vector $b$).

# Multipliers

The solution of the equation requires that final demand, $b$, be forecast - a formidable task in itself.

Much knowledge can be gained without knowing $b$, by considering how economic activity will change if there is a change in final demand. This does not require computing the level of economic activity, only the change.

One method of examining how a change in final demand will affect an economy is to construct so-called *multipliers* [10]. Some of these multipliers will be considered next.

## *The First Example of a Multiplier*

As an example of an elementary multiplier consider

$$O = (L^{-1})'(1, 1, \ldots, 1)'.$$

Element $O_i$ gives the dollar amount of output by all industries in the region needed to satisfy an additional dollars worth of final demand for the $i$–th good [13].

Income and employment multipliers can also be calculated.

## *Income Multipliers*

Let

$$h_i = \frac{\text{household income generated by industry } i}{\text{total outlays in industry } i}.$$

Then the *wage income multiplier* for the $i$–th industry is calculated by

$$I_i = \frac{[(L^{-1})'h]_i}{h_i}.$$

Replacing wage income by total income (wage income plus proprietor's income) results in a *total income multiplier.*

## *Employment multiplier*

An *employment multiplier* is calculated by defining

$$e_i = \frac{\text{employment industry } i}{\text{total outlays in industry } i}$$

and calculating

$$E_i = \frac{[(L^{-1})'e]_i}{e_i}.$$

## *To Compute Different Multipliers, It is Desirable to Have the Inverse of Leontief Matrix*

Each multiplier can be computed by solving an appropriate system of linear equations $Lx =$ known vector. The more multipliers we know, the more knowledge we have about the economic situation. However, solving many systems of large size is computationally complicated. To avoid solving several equations, we can compute the *inverse* $L^{-1}$ of the Leontief's matrix $L$. If we know this inverse, then all the multipliers can be easily computed (by multiplying $L^{-1}$ and the right-hand side vector that corresponds to this multiplier).

# 2 TECHNICAL COEFFICIENTS ARE ONLY KNOWN WITH UNCERTAINTY

The technical coefficients of the Leontief matrix are not known but must be estimated. The U.S. Census Bureau conducts a census of U.S. businesses every five years. The information from the census is transferred to the Bureau of Economic Analysis (BEA) which uses it to construct the national input-output technical coefficient matrix. There has been about a 5 to 7 year time delay from the initial collection of the data to the production of the final matrix.

The consequences of these time delays can be serious. The national input-output tables are used in estimating gross national product (GNP) and gross domestic product (GDP). When the Department of Commerce shifted from the 1972 census benchmark to the 1977 benchmark this raised the original 1983 GNP estimate by $100 billion, an error equivalent to a year's worth of strong economic growth [3]. Errors of this magnitude can lead policy makers to inappropriate action (e.g., creating inflation by fighting a non-existent recession).

Things are worse for regional analysts who rarely have the resources to conduct surveys of regional businesses. Usually they are forced to infer the regional technical coefficients by adjusting national tables and using other ad hoc procedures [13]. It is reasonable to believe that the technical coefficients for regional models are subject to some level of uncertainty. In practice, the effects of this uncertainty are largely ignored.

## 3 STATISTICAL METHODS ARE NOT DIRECTLY APPLICABLE, HENCE, INTERVAL COMPUTATIONS MAY BE USEFUL

### Statistical Methods Exist

West [17] has derived a formal expression for the probability density of the multipliers assuming that the technical coefficients are independent and that they can be characterized as probability distributions with small variances. This is certainly a significant improvement over assuming the coefficients are point values and that the multiplier results are known with certainty. If probability distributions are assumed, then West's method will generate confidence intervals.

## Statistical Methods Are Not Directly Applicable

Yet significant practical problems remain. We do not know what the probability distributions of the coefficients are in practice:

- On the *national* level: Since the national tables are at least 5 years old before publishing it is reasonable to believe that the distribution has changed.
- On the *regional* level: The regional analyst likely has no sample information at all. The regional analyst most likely will have no basis for assigning distributions, but will have to assign subjective levels of uncertainty.

## Interval Computations May Be Useful

Instead of probabilities, we usually know only the *estimates* $\tilde{l}_{ij}$ for the coefficients of the Leontief matrix, and estimates of the accuracy $\Delta_{ij}$ of these coefficients. In other words, the only thing we know about the actual (unknown) coefficients $l_{ij}$ is that they belong to a known *interval* $\mathbf{l}_{ij} = [\underline{l}_{ij}, \overline{l}_{ij}]$, where $\underline{l}_{ij} = \tilde{l}_{ij} - \Delta_{ij}$ and $\overline{l}_{ij} = \tilde{l}_{ij} + \Delta_{ij}$. These intervals form an *interval matrix* $\mathbf{L}$.

Similarly, the only thing we know about the demand $b_i$ for $i$–th good is the estimate $\tilde{b}_i$ of this demand and estimates of accuracy $\Delta_i$ of $\tilde{b}_i$ . In other words, we only know the *intervals* $\mathbf{b}_i = [\underline{b}_i, \overline{b}_i]$ (where $\underline{b}_i = \tilde{b}_i - \Delta_i$ and $\overline{b}_i = \tilde{b}_i + \Delta_i$) of possible values of $b_i$. These intervals form an *interval vector* $\mathbf{b}$.

As we have mentioned, we need the Leontief matrix for two reasons:

- First, we want to find the amount $x_i$ of $i$–th good that is necessary to satisfy the expected demand.

  If we knew the exact values of $L$ and $b$, we would be able to find $x$ by solving a linear system $Lx = b$. Since we do not know the exact values of the demand and the exact values of the coefficients, we can only describe the *possible values* of $x_i$, i.e., possible vectors $x$ for which $Lx = b$ for some $L \in \mathbf{L}$ and $b \in \mathbf{b}$. Finding a set of such vectors is a typical problem of *interval computations* (see, e.g., [11]): it is called *solving a system of interval linear equations.*

- Second, we want to compute the inverse Leontief matrix to be able to compute multipliers.

  Since we do not know the exact values of the coefficients $l_{ij}$ of the matrix $L$, we can only compute the ***intervals of possible values*** of the coefficients of the inverse matrix. This is also a problem of interval computations: ***inverting an interval matrix.***

*Historical comment.* The idea of using intervals to describe Leontief input-output models was first proposed by J. Rohn [14, 15]. Rohn has derived the conditions under which the input-output system will have economically feasible solutions when the technical coefficients are intervals. Lorenzen [6, 7], Lorenzen and Maas [8], and Maier [9] have also considered input-output models with interval data.

# 4 COMPUTATIONAL ALGORITHMS

## Interval Problems Originating From Economic Input-Output Models Are Feasible

### *General Properties of $M-$matrices*

In general, finding a solution of a system of linear interval equations can be computationally very complicated (see, e.g., [16]). There are, however, classes of interval matrices $\mathbf{L}$ for which feasible algorithms for solving the corresponding systems of linear equations $\mathbf{L}x = \mathbf{b}$ are available. One of these classes consists of the so-called *$M-$matrices.* We refer the reader to [11] for exact definitions and proofs. For our purposes, we will only need the following results:

- If $l_{ij} \leq 0$ for all $i \neq j$, and $Lu > 0$ for some positive vector $u$, then $L$ is an $M-$matrix.
- If the bounds $\underline{L}$ and $\overline{L}$ of an interval matrix $\mathbf{L} = [\underline{L}, \overline{L}]$ are both $M-$matrices, then $\mathbf{L}^{-1} = [\overline{L}^{-1}, \underline{L}^{-1}]$.
- If both bounds of an interval matrix $\mathbf{L}$ are $M-$matrices, and $\mathbf{b}$ is a non-negative vector (i.e., if $\underline{b}_i \geq 0$ for all $i$), then:

- the interval $\mathbf{x}$ of possible values of the solution $x = (x_1, ..., x_n)$ (i.e., of possible values of $L^{-1}b$) is equal to $[\overline{L}^{-1}\underline{b}, \underline{L}^{-1}\overline{b}]$(see [1, 11]); and
- Gauss elimination will produce this interval $\mathbf{x}$.

## *Leontief Matrix is an M−matrix*

Let us show that the Leontief matrix is an $M$−matrix. To prove that, we will show that $l_{ij} \leq 0$ for all $i \neq j$, and that $Lu > 0$ for some positive vector $u$.

Consider the first row of $L$.

$$(1 - a_{11}, -a_{12}, -a_{13}, \ldots, -a_{1n}).$$

All of the production coefficients $a_{ij}$ must be non-negative (a negative value would mean that a negative amount of good $i$ is used to produce $j$). Hence all of the off-diagonal terms of $L$ must be non-positive. This will hold for all rows of the Leontief matrix and so all off-diagonal terms will be non-positive.

To show the second condition let $u = (1, 1, \ldots, 1)'$. The condition will be met for the first element of $Lu$ if

$$1 - a_{11} - a_{12} - \ldots - a_{1n} > 0. \qquad (6.1)$$

This condition must be met for most industries because the terms $a_{1k}$ represent the proportion of industry 1's sales to other industries. A negative result for (1) would mean that more output is being sold than is being produced. A zero result could occur if the good is solely and intermediate good (one with no final demand). This would be rare in an input-output model and if it did occur a simple fix is available - aggregate this industry with a closely related one. These comments hold for all elements of $Lu$ and so the Leontief matrix should be an $M$−matrix.

## *The Interval Vector* b *is Non-Negative*

Only the condition $b \geq 0$ need be considered here because Consumer demand must be non-negative. Therefore, all possible values $b_i \in \mathbf{b}_i$ are non-negative. In particular, the value $\underline{b}_i \in \mathbf{b}_i$ must be non-negative.

## The Algorithm

Since both bounds for Leontief's matrix **L** are $M-$matrices, and **b** is a non-negative vector, the above-given properties of $M-$matrices enable us to use the following algorithm for solving the above-described interval problems (computing the inverse of **L** and solving the system of interval linear equations $\mathbf{L}x = \mathbf{b}$):

**Algorithm.**

- *For $i = 1, ..., n$, we use an interval version of Gaussian elimination method to solve the system $\mathbf{L}x = u_i$, where $u_i$ is a unit vector with the $i-th$ element equal to one and the other elements equal to zero. The resulting vectors $\mathbf{x}^{(i)}$ form the desired inverse matrix $\mathbf{L}^{-1}$.*
- *To compute the intervals $\mathbf{x}$ of possible values of production, we multiply $\mathbf{L}^{-1}$ by $\mathbf{b}$.*

# 5 AN EXAMPLE

The College of Business Administration at Northern Arizona University has been charged with analyzing the economic impact of the University on Coconino county, Arizona and on the state of Arizona. Because of the short length of time given for the project (4 months) and because sampling costs are prohibitive, we are using a commercially vended product, IMPLAN, to estimate the technical coefficients for both the county and state. IMPLAN adjusts the national coefficients to regional coefficients which provides a source of uncertainty [12]. IMPLAN produces a technical coefficient matrix for 171 industries. These were aggregated into 9 industries for purposes of display. The technical coefficient matrix for Coconino county is shown below.

$$\begin{pmatrix}
.0987 & .0000 & .0004 & .0036 & .0000 & .0010 & .0006 & .0002 & .0001 \\
.0000 & .0053 & .0003 & .0001 & .0013 & .0000 & .0000 & .0000 & .0001 \\
.0039 & .0013 & .0006 & .0018 & .0079 & .0015 & .0260 & .0046 & .0147 \\
.0089 & .0022 & .0440 & .1018 & .0141 & .0107 & .0018 & .0108 & .0013 \\
.0220 & .0119 & .0355 & .0444 & .0721 & .0315 & .0074 & .0220 & .0247 \\
.0075 & .0022 & .0537 & .0115 & .0040 & .0047 & .0003 & .0067 & .0005 \\
.0569 & .0150 & .0176 & .0129 & .0243 & .0315 & .0451 & .0379 & .0018 \\
.0151 & .0052 & .0790 & .0185 & .0271 & .0234 & .0111 & .0537 & .0017 \\
.0062 & .0026 & .0028 & .0051 & .0048 & .0064 & .0037 & .0078 & .0037
\end{pmatrix}$$

The multipliers for Coconino county and the state of Arizona are shown in Tables 1 and 2. The industries are aggregated using IMPLAN Industrial Numbers (IIN). The industry aggregates are agriculture, forestry and fisheries (IIN 1), mining (IIN 29), construction (IIN 48), manufacturing (IIN 58), transportation, communications and utilities (IIN 443), trade (IIN 447), finance, insurance and real estate (IIN 456), services (IIN 463), and government (IIN 512).

These multipliers assume that the technical coefficient matrix for both Coconino county and the state are subject to uncertainty of ±1%. Wage and total income are assumed to have a level of uncertainty of ±0.1%.

The calculations were coded in C-XSC [4], a library of C++ classes. This library provides an interval class and interval vector and matrix classes. This interval implementation provides directed rounding. The values from IMPLAN are input as interval values and all values, including the subjective uncertainty, were treated as interval values.

| Industry IIN | Output Multiplier | Wage Income Multiplier | Total Income Multiplier | Employment Multiplier |
|---|---|---|---|---|
| 1 | [1.259, 1.266] | [1.319, 1.328] | [1.265, 1.273] | [1.236, 1.242] |
| 29 | [1.051, 1.053] | [1.150, 1.155] | [1.039, 1.041] | [1.417, 1.428] |
| 48 | [1.270, 1.277] | [1.472, 1.484] | [1.552, 1.566] | [1.476, 1.488] |
| 58 | [1.236, 1.243] | [1.304, 1.312] | [1.304, 1.312] | [1.374, 1.384] |
| 433 | [1.180, 1.185] | [1.213, 1.219] | [1.202, 1.208] | [1.324, 1.333] |
| 447 | [1.126, 1.130] | [1.079, 1.081] | [1.125, 1.129] | [1.053, 1.055] |
| 456 | [1.110, 1.114] | [1.222, 1.229] | [1.090, 1.094] | [1.168, 1.173] |
| 463 | [1.164, 1.168] | [1.118, 1.121] | [1.155, 1.160] | [1.103, 1.106] |
| 512 | [1.057, 1.059] | [1.017, 1.018] | [1.027, 1.029] | [1.022, 1.024] |

Table 1. Multipliers for Coconino County

| Industry IIN | Output Multiplier | Wage Income Multiplier | Total Income Multiplier | Employment Multiplier |
|---|---|---|---|---|
| 1 | [1.337, 1.347] | [1.493, 1.507] | [1.375, 1.385] | [1.292, 1.301] |
| 29 | [1.483, 1.498] | [2.043, 2.073] | [1.794, 1.817] | [2.689, 2.739] |
| 48 | [1.335, 1.344] | [1.541, 1.555] | [1.613, 1.629] | [1.583, 1.598] |
| 58 | [1.286, 1.294] | [1.323, 1.333] | [1.329, 1.338] | [1.486, 1.500] |
| 433 | [1.224, 1.230] | [1.280, 1.288] | [1.230, 1.237] | [1.389, 1.400] |
| 447 | [1.159, 1.164] | [1.108, 1.111] | [1.140, 1.145] | [1.086, 1.089] |
| 456 | [1.165, 1.170] | [1.292, 1.301] | [1.170, 1.175] | [1.257, 1.265] |
| 463 | [1.251, 1.258] | [1.191, 1.197] | [1.221, 1.228] | [1.173, 1.178] |
| 512 | [1.086, 1.089] | [1.036, 1.038] | [1.050, 1.052] | [1.042, 1.044] |

Table 2. Multipliers for Arizona

# REFERENCES

[1] W. Barth and E. Nuding, "Optimale Lösung von Intervallgleichungssystemen", *Computing*, 1984, Vol. 24, pp. 117–125.

[2] G. J. D. Hewings, "The Role of Prior Information in Updating Regional Input-Output Models", *Socio-Economic Planning Sciences*, 1984, Vol. 18, pp. 319–336.

[3] H. Kelly and A. Wyckoff, "Distorted Image: How government statistics misrepresent the economy", *Technology Review*, February/March 1989, pp. 52-60.

[4] R. Klatte, U. Kulisch, A. Wiethoff, C. Lawo, and M. Rauch, *C-XSC: A C++ Class Library for Extended Scientific Computing*, Springer-Verlag, Berlin, 1993.

[5] W. Leontief, *The Structure of the American Economy, 1919-1939*, 2nd ed., Oxford University Press, N. Y., 1951.

[6] G. Lorenzen, *Adaptive Verfahren der Input-Output Analyse* (Studien zur angewandten Wirtschaftsforschung und Statistik, Heft 17), Vandenhoeck & Ruprecht, Göttingen, 1985.

[7] G. Lorenzen, "Input-output multipliers when data are incomplete or unreliable", *Environment and Planning A*, 1989, Vol. 21, pp. 1075–1092.

[8] G. Lorenzen and C. Maas, "Zur Input-Output Analyse mit Intervalldaten", *Jahrbücher für Nationalökonomie und Statistik*, 1989, Vol. 206/3, G. Fischer Verlag, Stuttgart, pp. 257–263

[9] T. Maier, *Intervall-Input-Output-Rechnung*, Verlag Anton Hain, Meisenheim, 1985.

[10] W. H. Miernyk, *Input-Output Analysis*, Random House, N.Y., 1965.

[11] A. Neumaier, *Interval Methods for Systems of Equations*, Cambridge University Press, Cambridge, 1990.

[12] D. Olson and S. Lindall, *Micro IMPLAN User's Guide, Version 91-F*, Minnesota IMPLAN Group, Inc., 1994.

[13] H. W. Richardson, *Input-Output and Regional Economics*, Halsted Press, N.Y., 1972.

[14] J. Rohn, "Input-Output Planning with Inexact Data", *Freiburger Intervall-Berichte*, 1978, Vol. 78/9.

[15] J. Rohn, "Input-Output Model with Interval Data", *Econometrica*, 1980, Vol. 48, No. 3, pp. 767–769.

[16] J. Rohn and V. Kreinovich, "Computing exact componentwise bounds on solutions of linear systems with interval data is NP-hard," *SIAM J. Matr. Anal. Appl.*, 1995, Vol. 16, pp. 415–420.

[17] G. R. West, "A Stochastic Analysis of an Input-Output Model", *Econometrica*, 1986, Vol. 54, No. 2, pp. 363–374.

# 7

# INTERVAL ARITHMETIC IN QUANTUM MECHANICS

**Charles L. Fefferman* and Luis A. Seco****

**Department of Mathematics, Princeton University, Princeton NJ 08544*

***Department of Mathematics, University of Toronto, Toronto, Canada M5S 1A4, email seco@math.toronto.edu*

Quantum mechanics is an area which, over the last ten years or so, has sparked a respectable amount of rigorous computer assisted work (see, for example [9, 13, 20, 39, 38], and their applications in [10, 11, 12, 3, 4, 5, 14, 15, 16, 17, 18, 19, 20]). The purpose of this review is to select a piece of that body of work, and to give a more or less detailed account, both of the quantum mechanical problem surrounding the computer work and of the computer assisted proof itself. We hope this will be enlightening since much of the other computer assisted work in quantum mechanics shares many of the main features presented below.

The underlying theme under discussion here is the proof presented in [14, 15, 16, 17, 18, 19, 20, 21] of a precise asymptotic formula for the ground–state energy of a non–relativistic atom. One of the ingredients of that proof, namely the content of [21], was to establish an elementary inequality that plays a crucial role in the rest of the proof of that formula. This inequality was settled as a computer-assisted proof using interval arithmetic.

The inequality concerns the Thomas–Fermi potential $V_{\rm TF}(r) = -y(r)/r$, where $y(r)$ is defined as the solution of

$$y''(x) = x^{-1/2} y^{3/2}(x), \quad y(0) = 1, \quad y(\infty) = 0. \tag{0.1}$$

Define

$$F(\Omega) = \int \left( \frac{y(x)}{x} - \frac{\Omega^2}{x^2} \right)_+^{1/2} dx, \qquad \Omega \in (0, \Omega_c),$$

where $a_+ = \max(a, 0)$, and a number $\Omega_c$ will be defined at the beginning of Section 2. The function $F(\Omega)$ depends smoothly on $\Omega$ [41], and the main result in [21] is as follows:



*R. B. Kearfott and V. Kreinovich (eds.), Applications of Interval Computations*, 145–167.

**Theorem 0.1.** *$F''(\Omega) \le c < 0$ for all $\Omega \in (0, \Omega_c)$.*

This is a quantitative form of the non-periodicity of almost all zero-energy orbits for the Hamiltonian $H = |\xi|^2 + V_{\rm TF}(|x|)$ on $\mathbf{R}^6 = \{(x,\xi) \mid x \in \mathbf{R}^3,\ \xi \in \mathbf{R}^3\}$. In fact, an easy computation shows that a zero-energy orbit with angular momentum $\Omega$ is periodic if and only if the derivative $F'(\Omega)$ is a rational multiple of $\pi$ (see [1].) Hence, Theorem 0.1 shows that closed zero-energy orbits arise for only countably many $\Omega$.

This review will be organized as follows: first, we will present the quantum mechanical picture relevant for this theorem: this gives a meaning to Theorem 0.1, but can be skipped if the reader is not interested in its applications. Afterwards, we will present the basic computer set-up for the proof of Theorem 0.1, which will involve two main things: the rigorous solution to the Thomas-Fermi equation, and the actual computation of the function $F$ above; the final sections of this review will be devoted to these issues.

Our review here will be just a brief sketch of the actual proof. We refer the reader to the original paper [21] for the complete details. The paper itself with the proof of Theorem 0.1, together with the complete programs used in our proof are publicly available by anonymous `ftp` from the machine `math.utexas.edu` (Internet number 128.83.133.215). Our programs are stored in the directory `/pub/papers/seco`.

# 1 QUANTUM MECHANICS

The ground-state energy $E(N, Z)$ for $N$ electrons and a nucleus of charge $Z$ is defined [1] as the infimum of the spectrum of the Hamiltonian

$$H_{NZ} = \sum_{k=1}^{N} (-\Delta_{x_k} - Z|x_k|^{-1}) + \sum_{1 \le j < k \le N} |x_j - x_k|^{-1},$$

acting on antisymmetric $\Psi(x_1, \ldots, x_N) \in L^2(\mathbf{R}^{3N})$. The ground-state energy of an atom is then defined as

$$E(Z) = \min_{N \ge 1} E(N, Z),$$

and our problem is to compute $E(Z)$ asymptotically for large $Z$. Building on the previous work of Thomas [43], Fermi [22], Dirac [6] and Scott [37] (see the

[1] In this survey we neglect electron spin to simplify notation.

survey article of Lieb [30]), Schwinger [36] proposed the refined formula

$$E(Z) \approx -c_0 Z^{7/3} + \tfrac{1}{8} Z^2 - c_1 Z^{5/3} \tag{1.1}$$

for some explicitly defined positive constants $c_0$, $c_1$. After the early work of Lieb and Simon [31] on molecules, Hughes and Siedentop-Weikard [24, 40, 41, 42] gave a rigorous proof of the "Scott conjecture", namely $E(Z) = -c_0 Z^{7/3} + \frac{1}{8} Z^2 + O(Z^\gamma)$, with $\gamma < 2$. Recently, Ivrii and Sigal [27] proved the analogue of the Scott conjecture for molecules. Our main result, announced in [14] and proved in [15, 16, 17, 18, 19, 20, 21], is as follows.

**Theorem 1.1.** $E(Z) = -c_0 Z^{7/3} + \frac{1}{8} Z^2 - c_1 Z^{5/3} + O(Z^{5/3-\varepsilon_0})$ *with* $\varepsilon_0 = \frac{1}{2835}$.

The starting point in discussing atoms is an elementary observation on free particles in a box. For $N$ free particles in a box $Q \subset \mathbf{R}^3$, the minimum possible kinetic energy $\mathrm{KE}(N, Q)$ is equal to the lowest eigenvalue of $-\Delta$ acting on antisymmetric $\Psi(x_1, \cdots, x_N) \in L^2(Q^N)$ with appropriate boundary conditions. One computes $\mathrm{KE}(N, Q)$ trivially, by separation of variables. For large $N$, the answer is

$$\mathrm{KE}(N, Q) \approx c_{\mathrm{TF}} \rho^{5/3} |Q| \tag{1.2}$$

where $\rho = N/|Q|$ is the density of particles in the box, and $c_{\mathrm{TF}}$ is a universal constant.

This suggests a way to approximate the energy $\langle H_{NZ}\Psi, \Psi\rangle$ of a wave function $\Psi(x_1, \ldots, x_N)$ in terms of the electron density

$$\rho(x_1) = N \int_{\mathbf{R}^{3(N-1)}} |\Psi(x_1, x_2, \cdots, x_N)|^2 dx_2 \ldots dx_N.$$

In fact, we set

$$\mathcal{E}_{\mathrm{TF}}(\rho) = c_{\mathrm{TF}} \int_{\mathbf{R}^3} \rho^{5/3}(x) dx - \int_{\mathbf{R}^3} \frac{Z}{|x|} \rho(x) dx + \frac{1}{2} \iint_{\mathbf{R}^3 \times \mathbf{R}^3} \frac{\rho(x)\rho(y)}{|x-y|} dx dy. \tag{1.3}$$

Here, the first term on the right is an approximation to the kinetic energy motivated by (1.2), and the remaining terms on the right are simply the classical electric potential energy for a charge density $\rho$ and a nucleus of charge $Z$. Thomas and Fermi independently proposed that the ground-state energy $E(Z)$ is approximately equal to the minimum of $\mathcal{E}_{\mathrm{TF}}(\rho)$ over all possible densities $\rho(x)$. Moreover, they proposed the minimizing density $\rho_{\mathrm{TF}}$ for (1.3) as an approximation to the electron density for an atom in its ground state. This is an immense simplification, since the original problem deals with $\Psi(x_1 \ldots x_N)$ for $N \gg 1$, while Thomas–Fermi theory deals merely with a function on $\mathbf{R}^3$. An elementary computation with the Euler-Lagrange equation for (1.3) leads to an ordinary differential equation for $\rho_{\mathrm{TF}}$, which is essentially (0.1). In particular, Thomas–Fermi theory predicts that

$$E(Z) \approx -c_0 Z^{7/3},$$

which is correct as far as it goes, but much too crude.

A more refined prediction for $E(Z)$ comes from the *Hartree-Fock approximation*[2]. The idea is that since the electron density is approximately $\rho_{\mathrm{TF}}$, each electron behaves as if it was moving in a potential

$$V_{\mathrm{TF}}(x) = -\frac{Z}{|x|} + \int_{\mathbf{R}^3} \frac{\rho_{\mathrm{TF}}(y)dy}{|x-y|}.$$

Therefore it is reasonable to approximate the ground-state of the true Hamiltonian $H_{NZ}$ by that of the much simpler Hamiltonian

$$H_{\mathrm{hf}} = \sum_{k=1}^{N} (-\Delta_{x_k} + V_{\mathrm{TF}}(x_k)) \text{, acting on antisymmetric } \Psi(x_1, \ldots, x_N). \quad (1.4)$$

Unlike the original Hamiltonian, (1.4) can be diagonalized using separation of variables, and the state of lowest energy can be written explicitly in terms of the eigenfunctions of $-\Delta + V_{\mathrm{TF}}$. So again, the problem is reduced from $3N$ to 3 variables. In fact, suppose $E_k$ are the (negative) eigenvalues of $-\Delta + V_{\mathrm{TF}}$, and let $\varphi_k(x)$ be the corresponding (normalized) eigenfunctions. Then the ground-state wave function for (1.4), which we call $\Psi_{\mathrm{hf}}$, is an antisymmetrized product of the $\varphi_k$. As an approximation to the ground-state energy of an atom, it is natural to use

$$E_{\mathrm{hf}}(Z) = \langle H_{NZ}\Psi_{\mathrm{hf}}, \Psi_{\mathrm{hf}}\rangle \,.$$

Note that we are using the exact Hamiltonian $H_{NZ}$, even though $\Psi_{\mathrm{hf}}$ arose from the simplified Hamiltonian (1.4). Elementary computation gives the formula

$$E_{\mathrm{hf}}(Z) = \operatorname{sneg}(-\Delta + V_{\mathrm{TF}}) - \tfrac{1}{2}\iint_{\mathbf{R}^3\times\mathbf{R}^3} \frac{\rho_{\mathrm{TF}}(x)\rho_{\mathrm{TF}}(y)}{|x-y|} dx dy -$$

$$\tfrac{1}{2}\iint_{\mathbf{R}^3\times\mathbf{R}^3} \frac{|S(x,y)|^2}{|x-y|} dx dy$$

$$+\tfrac{1}{2}\iint_{\mathbf{R}^3\times\mathbf{R}^3} [\rho_{\mathrm{hf}}(x) - \rho_{\mathrm{TF}}(x)][\rho_{\mathrm{hf}}(y) - \rho_{\mathrm{TF}}(y)] \frac{dx dy}{|x-y|}, \quad (1.5)$$

[2]This is not exactly the same as the usual Hartree-Fock approximation.

with

$$\text{sneg}(-\Delta+V_{\text{TF}}) = \sum_k E_k, \quad \rho_{\text{hf}}(x) = \sum_k |\varphi_k(x)|^2, \quad S(x,y) = \sum_k \varphi_k(x)\overline{\varphi_k(y)}.$$

To get more explicit information from (1.5), we approximate $\text{sneg}(-\Delta + V_{\text{TF}})$, $\rho_{\text{hf}}$ and $S$. The ***semi-classical approximations*** for these quantities are as follows.

$$\text{sneg}(-\Delta + V_{\text{TF}}) \approx -\tfrac{1}{15\pi^2}\int_{\mathbf{R}^3} |V_{\text{TF}}(x)|^{5/2}dx, \tag{1.6a}$$

$$\rho_{\text{hf}}(x) \approx \tfrac{1}{6\pi^2}|V_{\text{TF}}(x)|^{3/2}, \tag{1.6b}$$

$$\tfrac{1}{2}\iint_{\mathbf{R}^3\times\mathbf{R}^3} |S(x,y)|^2 \frac{dxdy}{|x-y|} \approx c_D \int_{\mathbf{R}^3} \rho_{\text{TF}}^{4/3}(x)dx, \tag{1.6c}$$

for a universal constant $c_D$.

We omit the motivation for (1.6a), (1.6b) and (1.6c) and content ourselves with the remark that they are closely related to Weyl's theorem on eigenvalues of the Laplacian. Formula (1.6c) and its application to atoms are due to Dirac.

Putting (1.6a), (1.6b) and (1.6c) into (1.5), we obtain the semi-classical approximation for $E_{\text{hf}}(Z)$. From the first two terms on the right in (1.5), we recover the Thomas–Fermi energy $-c_0Z^{7/3}$. The third term on the right of (1.5) takes the form $-c_1'Z^{5/3}$ for a universal constant $c_1'$. The final term in (1.5) vanishes in the semi-classical approximation. Altogether, we have

$$E_{\text{hf}}(Z) \approx -c_0Z^{7/3} - c_1'Z^{5/3}. \tag{1.7}$$

The last term in (1.7) is called the Dirac correction. Comparing (1.7) with the correct formula in Theorem 1.1, we see that the $Z^2$-term is missing from (1.7), and the $Z^{5/3}$ coefficient is wrong. The trouble is that (1.6a) is only a crude approximation.

A refined form of (1.6a) was proposed by Scott [37] and Schwinger [36]. For potentials $V$ with a Coulomb singularity $V(x) \approx -Z|x|^{-1}$ at the origin, their formula is

$$\text{sneg}(-\Delta + V) \approx -\tfrac{1}{15\pi^2}\int_{V<0} |V|^{5/2} + \tfrac{1}{8}Z^2 + \tfrac{1}{48\pi^2}\int_{V<0} |V|^{1/2}\Delta V. \tag{1.8}$$

Scott guessed the $Z^2$-term by working out the elementary example

$$V(x) = E_0 - Z|x|^{-1}.$$

Schwinger deduced the last term in (1.8) from the form of the heat kernel for $e^{-t(-\Delta+V)}$, which in turn he guessed from the known case of the harmonic oscillator. Using (1.8), (1.6b) and (1.6c) to approximate the right-hand side of (1.5), we obtain Schwinger's formula (1.1) for the ground-state energy.

The rigorous comparison of $E_{\text{hf}}$ with $E(Z)$ was originally done in [15], but that proof has been greatly simplified first by Bach [2] and then Graf–Solovej in [23]. Theorem 0.1 affects (1.8) as follows: Although (1.8) was guessed by Schwinger and Scott by comparison with the hydrogen atom and the harmonic oscillator, it is easily seen that (1.8) in fact fails for these two potentials; thus, that it continues to be true for the Thomas–Fermi potential must be a deep fact. This is the aperiodicity described at the beginning of this review, which is the essential content of Theorem 0.1.

Aperiodicity of zero–energy Hamiltonian paths is well-known to play a crucial role in the study of eigenvalues and eigenfunctions. In our setting, Theorem 0.1 enters because our formulas for the eigenvalue sum and density involve expressions of the form

$$S = \sum_{1 \le l < Z^{1/3}\Omega_c} \beta\left(\frac{Z^{1/3}}{\pi} F(Z^{-1/3} l)\right)$$

for elementary functions such as $\beta(t) = t - [t] - \frac{1}{2}$. (Here $[t]$ is the greatest integer in $t$.) Since $\beta$ is bounded, we obtain trivially the estimate $S = \mathcal{O}\left(Z^{1/3}\right)$. If $F(\Omega) = \pi\mu\Omega + \nu$ with $\mu$ rational, then the trivial estimate for $S$ is easily seen to be the best possible. On the other hand, if $d^2F/d\Omega^2 < c < 0$, then one can prove that the numbers

$$\phi_l = Z^{1/3} F(Z^{-1/3} l)$$

are equidistributed modulo $\pi$. (The argument is close to Hardy's estimates on the number of lattice points in a disc.) Since $\beta(t)$ is periodic and has average zero, it follows that $S = \mathcal{O}(Z^\gamma)$ with $\gamma < \frac{1}{3}$.

Thus, Theorem 0.1 allows us to improve on the trivial estimate for the sum $S$, which appears in the eigenvalue sum and density for $H_{\text{hf}}$.

The proof of Theorem 0.1 is necessarily rather delicate. For small perturbations of $V_{\text{TF}}$ in a natural topology, the analog of Theorem 0.1 fails. Therefore, we have to make strong use of the differential equation defining $y(r)$. We remark, however, that without a computer it can also be seen that $F''$ vanishes at most finitely many times (Proposition 4.8 in [21]; see also the independent proof in [25]), which also implies that zero–energy periodic orbits have measure zero, which in turn also implies the same results stated above for sums $S$, and therefore our result for atomic energies. Theorem 0.1, however, is better because it implies better error terms for all those formulas. Moreover, if one wants to understand ground–state energies to a greater accuracy, (see [3, 4, 5]), then Theorem 0.1, in its full strength, is unavoidable.

# 2 COMPUTER–ASSISTED SET–UP

The first issue at hand is a derivation for a formula for $F''(\Omega)$. This is not a trivial matter at first sight, since differentiation twice inside the integral sign gives rise to an unconditionally divergent integral.

To understand the formula for $F''$, consider the new function $u(x) = xy(x)$; this function has a single maximum $\Omega_c^2$ reached at $r_c$, and for $\Omega \in [0, \Omega_c]$ there are two solutions, denoted by $r_1(\Omega)$ and $r_2(\Omega)$, to the equation $u(r) = \Omega^2$. Then,

$$F''(\Omega) = -\lim_{\delta \to 0} \left( \int_{r_1(\Omega)+\delta}^{r_2(\Omega)-\delta} \left(u(x) - \Omega^2\right)^{-3/2} y(x) dx + c(\Omega)\delta^{-1/2} \right), \qquad (2.1)$$

where $c(\Omega)$ is uniquely specified by requiring the finiteness of the limit. Note here the delicate cancelations taking place in the limit in (2.1). The proof of (2.1) can be found in [21]. There it can also be seen that the function $F$ in fact extends as an analytic function to a complex neighborhood of $(0, \Omega_c]$, and that $F$ blows up like $\Omega^{-\gamma}$ at $\Omega = 0$, for $\gamma = \frac{9-\sqrt{73}}{2}$.

The proof of theorem 0.1 will be done by obtaining strictly negative uniform bounds for (2.1) on a finite number of small intervals $\Omega_i^*$ covering $(0, \Omega_c)$. This was done using *interval arithmetic*, which we now briefly review.

Let $\mathcal{R}$ be the set of "representable numbers" in a computer, that is those numbers that the computer can represent exactly. Since computers can only perform arithmetic in an approximate way, the idea to perform rigorous arithmetic is to instruct the computer on how to produce upper and lower bounds to the true results of arithmetic operations between representable numbers; in other words, we work with intervals with endpoints in $\mathcal{R}$, and we implement arithmetic operations on intervals in such a way that given two intervals, the computer will produce a third that is guaranteed to contain the result of all arithmetic operations between points in the initial intervals.

The next step is to perform a similar kind of arithmetic, where objects are functions in some Banach space, not numbers. A convenient Banach space to use in this theory is the space of piecewise analytic functions, with a lower bound on the size of the domains of analyticity. Occasionally, it will be convenient to switch to genuine real variable theory, for which we will do our work on $C^0[-1,1]$. The reason for this is that inversion of functions in $\mathbf{R}^1$ is a little easier than the complex counterpart, mainly because the domain of definition problem is trivial in the real case.

More precisely, consider the Banach algebras

$$H^1 = \left\{ f(z) = \sum_{n=0}^{\infty} a_n z^n \middle| \sum_{n=0}^{\infty} |a_n| < \infty \right\},$$

and

$$C^0 = \{f(x) | f \text{ is continuous on } [-1,1]\},$$

with norms

$$\|f\|_1 = \sum_{n=0}^{\infty} |a_n|, \qquad \|f\|_\infty = \sup |f(x)|,$$

respectively. $H^1$ is a subspace of the set of analytic functions in the unit disk.

Our substitute for intervals in $H^1$ are sets (sometimes called "neighborhoods", even if they are not actually neighborhoods) $\mathcal{U}^1(I_0, \ldots, I_N; C_h, C_g; k)$ of the form

$$\left\{ f(z) = \sum_{n=0}^{\infty} a_n z^n + z^k g(z) \middle| a_n \in I_n,\ 0 \leq n \leq N,\ \sum_{n=N+1}^{\infty} |a_n| \leq C_h,\ \|g\|_1 \leq C_g \right\} \tag{2.2}$$

where $C_h$ and $C_g$ are positive real numbers and the $I_n$ are intervals in the real line. For the computer implementation, $C_h$ and $C_g$ will run over the set of computer–representable numbers, and the intervals will be those with representable endpoints. We refer to $C_h$ and $C_g$ as high and general order error terms respectively, for obvious reasons. The number $k$ will be referred as **type** (set to infinity if $C_g = 0$), and $N$ will be referred to as **order** of $\mathcal{U}$. In our implementation, $N$ will not be fixed, but chosen adaptively during the execution of the programs.

By trivial scaling, we will be able to do analysis on

$$H^1(|z - z_0| \leq r) = \left\{ f(z) = \sum_{n=0}^{\infty} a_n \left(\frac{z - z_0}{r}\right)^n \middle| \sum_{n=0}^{\infty} |a_n| < \infty \right\}$$

a subspace of the set of analytic functions on the disk of center $z_0$ and radius $r$.

As for $C^0$, we will use sets – "neighborhoods" – of the kind

$$\mathcal{U}^0(I_0,\ldots,I_N;C_h,C_g;k;S)=\left\{f(z)=\sum_{n=0}^{N}a_nz^n+z^{N+1}h(z)+z^kg(z)\,\middle|\right.$$

$$\left.a_n\in I_n,\ 0\le n\le N,\ \sup_{z\in S}|h(z)|\le C_h\,,\ \sup_{z\in S}|g(z)|\le C_g\right\} \qquad (2.3)$$

where $S$ is a subset of $[-1,1]$, and $h$ and $g$ are continuous functions on $S$. We will use the superscript 0 or 1 whenever we want to emphasize in which topology we are taking these "neighborhoods".

The $\mathcal{U}^0$ will not allow us to perform as many operations –such as differentiation– as the smaller $\mathcal{U}^1$, but we can still add, multiply, raise to fractional powers and integrate (among others) in terms of them; furthermore, the formulas for these neighborhood operations are exactly the same as those for the $\mathcal{U}^1$. We refer the reader to [35, 39] and [21] for a description of some of these operations.

These $\mathcal{U}$ become more useful when they are rescaled appropriately and strung together: in this way, one can have complete local and global control of analytic functions defined over very large sets. This constitutes a mild novelty in this set–up, since a single variable, consisting of a pointer to an array of $\mathcal{U}$ will capture the global behavior of a piecewise analytic function.

Although computer-assisted analysis has become fairly standard, we refer the reader to [34] and [28] for a description of the basic ideas. The technique for solving ODE's is adapted from [39] and [38], and is tailored to handle our particular ODE. See [33] for a thorough discussion on ODE solving techniques, with very good general algorithms. Also, we refer the reader to [7, 8, 13, 29, 32] and [35] for a sample of computer–assisted proofs of a wide variety of problems. Main ideas in our approach go back to those proofs.

Our interval arithmetic package is an adaptation of the one used in [38] and [39], which in turn is an adaptation of the one developed by D. Rana. See [35] and [38] for details on the software.

# 3 THE THOMAS–FERMI EQUATION

In this section we will be concerned with the problem of getting good bounds for the solution of the Thomas–Fermi equation (0.1). The analysis will be broken into two main parts: the solution of initial value problems, and the solution of the boundary value problem. The first will be further broken into three main problems: initial value problem at 0, at infinity and elsewhere.

We will use the following well known results [26] about $y$:

$$-w_0 = \lim_{r\to 0} y'(r) < 0, \tag{3.1}$$

exists, and that $y$ admits a power series expansion

$$y(r) = 144r^{-3}\Big(\sum_{n=0}^{\infty} b_n r^{-n\alpha}\Big), \tag{3.2}$$

convergent for $r$ large enough, with $b_0 = 1$, $b_1 < 0$ and $\alpha = \frac{1}{2}(\sqrt{73} - 7)$. Also, $y$ is always positive, decreasing, and it is the only such solution of the ODE satisfying (3.1) and (3.2).

## The Initial Value Problem Outside the Singularities

In this section we will be concerned with the solution to the Initial Value Problem

$$u''(x) = x^{-1/2}u^{3/2}(x), \quad u(x_0) = u_0 \quad u'(x_0) = u_1 \tag{3.3}$$

for $x_0, u_0 > 0$, which can be viewed as the fixed point of

$$T(u) = u_0 + \int_{x_0}^{x}\left(u_1 + \int_{x_0}^{t} \frac{u^{3/2}(s)}{s^{1/2}}\, ds\right) dt.$$

If we regard $u$ as

$$u(x) = u_0 + u_1 \cdot r \cdot z + z^2 f(z)$$

with $z = (x - x_0)/r$, $f \in H^1$ and $r$ small, we see that $T$ induces in a trivial way an operator $\tilde{T}$ of which $f$ is its fixed point. Thus, the following algorithm is a simple consequence of the fixed point theorem on Banach spaces. The relevant norm – denoted by $\|\ \|$ – throughout this section we will be $\|\ \|_1$.

**Algorithm 3.1.** *We deduce conditions on $u_0$, $u_1$, $x_0$, $r$ and $\alpha$ under which $\tilde{T}$ is a well-defined contraction in $B(0,\alpha) \subset H^1$, and we compute an upper bound for $\|\tilde{T}\|_{\mathrm{Lip}}$.*

This algorithm is then applied to the following

**Algorithm 3.2.** *Given appropriate intervals $x^*$, $u_0^*$ and $u_1^*$, and representable $r$, we construct a neighborhood $\mathcal{U}(I_0, \dots, I_N; 0, C_g; 0)$ such that for any $x_0 \in x^*$, $u_0 \in u_0^*$, and $u_1 \in u_1^*$, and any solution $u$ of (3.3) with any of these initial conditions, we have*

$$u(x) = u_0 + u_1 \cdot (x - x_0) + z^2 f(z), \qquad z = \frac{x - x_0}{r},$$

*for some $f \in \mathcal{U}$.*

**Description.** First, we construct, in a heuristic way, a polynomial

$$p(z) = \sum_{i=0}^{N} p_i z^i,$$

which approximately solves $\tilde{T}p = p$, and we set $\alpha$ such that $\|p\| \leq \alpha$. Next, we look for $\alpha_0 \geq \alpha$ such that the conditions on $x_0$, $u_0$, $u_1$, $r$ and $\alpha_0$ given by Algorithm 3.1 hold uniformly for all $x_0 \in x^*$, $u_0 \in u_0^*$ and $u_1 \in u_1^*$. Next, since $f$ is the fixed point of $\tilde{T}$, we have

$$\|p - f\| \leq \frac{\|p - \tilde{T}p\|}{1 - \|\tilde{T}\|_{\mathrm{Lip}}}$$

We get the required $\mathcal{U}$, by putting $I_i = [p_i, p_i]$ for $i = 0, \dots, N$ and setting $C_g$ equal to the ratio above. Q.E.D. This has the following easy consequence :

**Algorithm 3.3.** *Given disjoint intervals $x_0^*$ and $x_1^*$, and intervals $u_0^*$ and $u_1^*$, we construct intervals $y_0^*$ and $y_1^*$ such that all solutions $u$ to (3.3) with initial values equal to any $u_0 \in u_0^*$ and any $u_1 \in u_1^*$, for any $x \in x_0^*$ are guaranteed to exist as positive solutions on $[x, x']$, and furthermore satisfy*

$$u(x') \in y_0^*, \qquad u'(x') \in y_1^*,$$

*for all $x' \in x_1^*$.*

# The Initial Value Problem at 0

It concerns the solution to the Initial Value Problem

$$u''(x) = x^{-1/2} u^{3/2}(x) \quad u(0) = 1 \quad u'(0) = -w \tag{3.4}$$

for $w > 0$, which can be viewed as the fixed point of

$$T(u) = 1 + \int_0^x \left( -w + \int_0^t \frac{u^{3/2}(s)}{s^{1/2}}\, ds \right) dt.$$

If we put

$$u(x) = 1 - w \cdot r \cdot z^2 + z^3 f(z), \tag{3.5}$$

with $z = (x/r)^{1/2}$, $f \in H^1$, and $r$ is small, $T$ induces again an operator $\tilde{T}$ of which $f$ is its fixed point. After this, we just proceed in a similar manner to the previous case.

# The Initial Value Problem at Infinity

This is

$$u''(x) = x^{-1/2} u^{3/2}(x) \quad u(\infty) = 0 \quad b_1 = b \tag{3.6}$$

where the last condition is interpreted in the sense of (3.2). The solution will be written as

$$u(x) = \frac{144}{x^3} \left( 1 + b x^{-\alpha} + z^2 f(z) \right), \tag{3.7}$$

where $f \in H^1$, $z = R^\alpha x^{-\alpha}$, for some $R$ large. If we set

$$T_1(f) = bR^{-\alpha} z + z^2 f(z)$$

$$T_2(g) = (1+g)^{3/2}$$

$$T_3(g) = 12 \sum_{n=2}^{\infty} \frac{a_n z^{n-2}}{(n\alpha + 3)(n\alpha + 4)}, \quad g(z) = \sum_{n \geq 0} a_n z^n.$$

A solution of (3.6) comes from an $f$ satisfying $\tilde{T} f = f$, with $\tilde{T} = T_3 \circ T_2 \circ T_1$. As before, we would like to go ahead and check that indeed $\tilde{T}$ is a contraction, but the problem here is considerably more subtle than in the previous cases, due to the fact that $T_3$ does not scale with $R$. As a consequence, contraction properties of $\tilde{T}$ either hold or don't, and taking large $R$ won't help much. We are lucky, however, that the norm of $T_2$ is essentially $\frac{3}{2}$, and that the norm of $T_3$ is essentially

$$\frac{12}{(2\alpha+3)(2\alpha+4)} < \frac{1}{2}$$

which says that the Lipschitz norm of $\tilde{T}$ will approximately be $\frac{3}{4}$. The following result is thus crucial.

**Lemma 3.4.** *Put $\beta = 0.3$. Assume that $|\bar{b}| = R^{-\alpha}|b| \leq 0.23$. Then $\tilde{T}$ is a contraction in $B(0,\beta)$, and $\|\tilde{T}\|_{\text{Lip}} \leq 0.8652$.*

Once this is settled, everything proceeds as before, and we can easily conclude the following:

**Algorithm 3.5.** *Given $b^*$ (interval) and $R$ (representable), we produce $\mathcal{U}_1$ such that, for any $b \in b^*$, the solution $u$ of (3.6) is given by*

$$y(x) = \frac{144}{x^3}\left(1 + bx^{-\alpha} + z^2 f(z)\right), \qquad z = R^{\alpha}x^{-\alpha},$$

*with $f \in \mathcal{U}_1$. Here, $\mathcal{U}_1$ depends only on $b^*$, i.e., it is independent of which particular $b$ in $b^*$ we are considering. In particular, we can produce two intervals $u_0^*$ and $u_1^*$ such that, if $u$ is the solution to (3.6), we have $u(R) \in u_0^*$ and $u'(R) \in u_1^*$.*

# The Boundary Value Problem

Next we discuss how to solve

$$u''(x) = x^{-1/2}u^{3/2}(x) \quad u(0) = 1 \quad u(\infty) = 0.$$

Unlike the previous section, we will avoid a serious formulation of the method (which becomes a little technical), and instead we concentrate on the main simple ideas.

What we will be using here is a rigorous variant of the *shooting* method: we will begin making a heuristic guess for, say, $y'(0)$, and we will then follow the solution forward until it violates one of the properties we know to be true for the true Thomas–Fermi function: this will determine whether our choice was an upper bound or a lower bound for the actual value for $y'(0)$. This yields bounds for $-w_0 = y'(0)$. We then carry out a similar procedure at $\infty$. Let's get into a little more detail:

Pick a value $w$ and, with the aid of the algorithms in the previous section, obtain bounds for the solution of

$$u''(x) = x^{-1/2} u^{3/2}(x) \quad u(0) = 1 \quad u'(0) = -w,$$

around small intervals covering, say, $[0, M]$. If, at some point, we can deduce using these bounds that $u$ is increasing, a trivial ODE analysis shows that this choice $w$ is a *lower* bound for $w_0 = -y'(0)$. If however, the solution vanished, the same trivial ODE analysis yields that $w_0 < w$. The problem here is that the computer alone won't ever be able to tell us that the solution vanishes, since it needs it to be positive to perform the different algorithms above: so, in practice, even if the solution would vanish at some point, the computer will just require, at each step, smaller and smaller intervals to propagate the solution forward, without ever giving negative bounds for $u(x)$. Hence, the following simple lemma is helpful:

**Lemma 3.6.** *Let $u$ be the solution of (3.3), with $u_1 < 0$. If*

$$\frac{2u_0^{5/2}}{x_0^{1/2}} \leq u_1^2, \tag{3.8}$$

*then, there exists a point $t > x_0$ such that $u$ can be extended as a well-defined positive solution of the ODE to $[x_0, t)$ and, furthermore, $\inf_{x \in (x_0, t)} u(x) = 0$.*

As a consequence, the word *vanishes* used above is simply replaced by the more feasible property (3.8), and we obtain the following

**Algorithm 3.7.** *Given a representable $w$, we construct an algorithm that, if successful, will indicate whether $w < w_0$ or $w > w_0$. Hence, given appropriate $x_i \in \mathcal{R}$, we can produce $y_i^*$ and $y'^*_i \in \mathcal{I}$, $i = 0, \ldots, m$, such that*

$$y(x_i) \in y_i^*, \qquad y'(x_i) \in y'^*_i, \qquad i = 0, \ldots, m.$$

We now do the same at infinity.

**Algorithm 3.8.** *Given a representable $b$, and assuming bounds for $w_0$, we construct an algorithm that, if successful, will indicate whether $b < b_1$ or $b > b_1$. Next, we use these bounds to produce $x_i \in \mathcal{R}$, and $y_i^*$, $y'^*_i \in \mathcal{I}$, $i = 1, \ldots, m$, such that*

$$y(x_i) \in y_i^*, \qquad y'(x_i) \in y'^*_i, \qquad i = 1, \ldots, m.$$

The details are easy. In fact we can use the extra information given by Algorithm 3.3 to detect easily when the test solution goes beyond bounds. Note that afterwards we can take the best of the two sources of information about the Thomas–Fermi function, coming from 0 or from infinity.

A delicate point here is the choice of the $x_i$ above, but we refer the interested reader to [21] for details.

Note also that the refined bounds for $u$ given by Algorithm 3.8, allow us to sharpen our bounds for $w_0$. These improved bounds for $w_0$ will in turn improve on the older bounds for $b_1$, and we are all set for a recursive algorithm, the result of which is quite precise bounds for the Thomas–Fermi function:

**Algorithm 3.9.** *We can produce $x_i$, $r_i \in \mathcal{R}$, and*

$$\mathcal{U}_i(I_0^i, \ldots, I_N^i; C_{h,i}, C_{g,i}, 2), \qquad i = 1, \ldots, m,$$

*such that*

$$y(x_i + z \cdot r_i) \in \mathcal{U}_i(I_0^i, \ldots, I_N^i; C_{h,i}, C_{g,i}, 2), \qquad i = 1, \ldots, m,$$

*and*

$$\bigcup_{i=1}^{m} (x_i - r_i, x_i + r_i) = (x_1 - r_1, x_m + r_m) \subset (0, \infty)$$

What follows is some rigorous numerical values we obtained for several Thomas–Fermi parameters.

**Lemma 3.10.** *The following inequalities hold:*

$$1.588071022611278 \leq w_0 \leq 1.588071022611471$$

$$-13.2709738479 \geq b_1 \geq -13.2709738482$$

$$0.486348538043594 \leq \Omega_c^2 \leq 0.486348538046869$$

$$2.104025280219 \leq r_c \leq 2.104025280274$$

# 4 THE APERIODICITY INEQUALITY

Once we have appropriate bounds for the Thomas–Fermi function, the next step is to use them to prove Theorem 0.1.

There is the straightforward approach of simply computing expression (2.1) for a finite collection of small intervals $\Omega_i^*$ which cover all of $(0,\Omega)_c$. As we will see here, this will work with more or less trivial modifications only for for intervals which are not near 0 or $\Omega_c$; intervals near 0 will require a more sophisticated approach to computing (2.1), but near $\Omega_c$ we will have to use a completely different approach. This section explains the analysis involved in the computation of $F''$ for the different $\Omega_i^*$.

## Case 1

We begin with the computation of $F''(\Omega^*)$ for $\Omega^*$ sufficiently far from 0 and $\Omega_c$. Use (2.1) to write $F(\Omega) = I_1(\Omega) + I_2(\Omega) + I_3(\Omega)$ with

$$I_1 = \int_a^b \left(u(r) - \Omega^2\right)^{-3/2} y(r)dr \tag{4.1a}$$

$$I_2 = \lim_{\delta\to 0}\left(\int_{r_1(\Omega)+\delta}^{a} \left(u(r) - \Omega^2\right)^{-3/2} y(r)dr - G_1(\Omega)\delta^{-1/2}\right) \tag{4.1b}$$

$$I_3 = \lim_{\delta\to 0}\left(\int_{b}^{r_2(\Omega)-\delta} \left(u(r) - \Omega^2\right)^{-3/2} y(r)dr - G_2(\Omega)\delta^{-1/2}\right) \tag{4.1c}$$

with the $G_i$ such that the limit is finite, and $a < b$ carefully chosen.

The computation of $I_1$ seems to be rather easy at first sight: we just break up

$$I_1 = \sum_{i=1}^{n} \int_{t_i}^{t_{i+1}} \left(u(r) - \Omega^2\right)^{-3/2} y(r)dr,$$

with $t_1 = a$ and $t_{n+1} = b$. This involves only basic operations with $u$ and thus feasible with the standard interval arithmetic packages. A deeper issue is that of speed: it turns out that we will be forced to take the intervals $\Omega_i^*$ very small, and thus need to compute $I_1$ many times; in fact, so many that the direct computation approach fails due to time considerations, and we are forced to use a certain amount of trickery which we will omit here.

The computation of $I_2$ and $I_3$ is much easier: we use Algorithm 3.2 to compute $\mathcal{U}_1$ such that

$$u(x) = \Omega^2 + zf(z), \qquad z = \frac{x - r_1(\Omega)}{r}, \qquad f \in \mathcal{U}_1, \tag{4.2}$$

where $r \geq |a - r_1(\Omega)|$ and $\mathcal{U}_1$ is uniform for all $\Omega \in \Omega^*$. Therefore,

$$\int_{r_1(\Omega)+\delta}^{a} \left(u(x) - \Omega^2\right)^{-3/2} y(x)dx = \int_{r_1(\Omega)+\delta}^{a} z^{-3/2}\tilde{f}(z)dx,$$

for a new function $\tilde{f}(z) = y(x)f^{-3/2}(z)$, that can also be enclosed in a computable $\mathcal{U}$. Thus, if

$$\tilde{f}(z) = \sum_{n\geq 0} a_n z^n \in \mathcal{U}(J_0, \ldots, J_N; C_h, C_g; 1),$$

we see that

$$\int_{r_1(\Omega)+\delta}^{a} \left(u(x) - \Omega^2\right)^{-3/2} y(x)dx = r\sum_{n\geq 0} \frac{a_n}{n - \frac{1}{2}} z^{n-\frac{1}{2}} \Bigg|_{z=\frac{\delta}{r}}^{z=\frac{a-r_1(\Omega)}{r}},$$

which implies

$$I_2 \in r \cdot \sum_{n=0}^{N} \frac{J_n}{n - \frac{1}{2}} \left(\frac{a - r_1(\Omega)}{r}\right)^{n-\frac{1}{2}} \pm \epsilon,$$

with $|\epsilon| \leq r\left(\frac{C_h}{N+\frac{1}{2}} + 2C_g\right)$. $I_3$ is done in a similar manner.

## Case 2

The problem now is that when $\Omega$ is small, (4.2) would have to be done either very far (for $I_3$) of very close to 0 (for $I_2$), and this is impractical. The answer to this degenerate case goes through the change of variables given by $r(t)$, the inverse of $u$ around 0 for $I_2$, and around infinity for $I_3$; we will concentrate here on $I_2$. This change of variables yields

$$I_2 = \frac{d}{d\Omega}\left(\Omega \int_{r_1(\Omega)}^{a} \left(u(r) - \Omega^2\right)^{-1/2} \frac{dr}{r}\right) = \frac{d}{d\Omega}\left(\Omega \int_{\Omega^2}^{u(a)} \left(t - \Omega^2\right)^{-1/2} w(t)dt\right),$$

for $w(t) = r'(t)/r(t)$. The first step is therefore to compute $w$, which hinges on the following inversion algorithm.

**Algorithm 4.1.** *Given $N \geq 0$, we produce intervals $a_2^*, \ldots, a_N^*$, and a constant $C_N$, such that*

$$\left| r(t) - t\left(1 + \sum_{n=2}^{N} a_n \bar{t}^{\,n}\right) \right| \leq C_N t \cdot \bar{t}^{\,N+1}, \quad \bar{t} = \left(\frac{t}{x_0}\right)^{1/2},$$

*for constants $a_i \in a_i^*$, $i = 2, \ldots, N$, and for $t \leq \eta$.*

The point to realize here is that the function $u(x)$ is not analytic either around 0 (where it is still is analytic in $\sqrt{x}$), and is even worse around infinity, where it is a combination of $x^{-2}$ and powers of $x^{(7-\sqrt{73})/2}$. In spite of the fact that this function is not analytic, we can in principle design an algorithm for this case by combining different manipulations with analytic functions. However, it turns out to be more effective to switch here to genuine real variable theory, and to use the Banach algebra $C^0$ instead of analytic functions. As a consequence, the calculation of the derivatives has to be dealt with in a different manner than before. Of course we omit all details, and simply state the main final result:

**Algorithm 4.2.** *We produce a neighborhood $\mathcal{U}^0(I_0, \ldots, I_N; C_h, 0; \infty)$ such that*

$$h(t) \stackrel{\text{def}}{\equiv} tw'(t) + w(t) = f(\bar{t}), \qquad f \in \mathcal{U}^0,$$

*for $t \leq \eta$.*

Say then that

$$f_h(\bar{t}) = \sum_{n=0}^{N} a_n \bar{t}^{\,n} + H(t), \quad |H(\bar{t})| \leq \varepsilon_h \, |\bar{t}|^{N+1}, \quad 0 < t \leq \eta,$$

for $a_n$ in computable intervals and $\varepsilon_n$ bounded by a computable number. Then

$$\frac{d}{d\Omega}\left(\Omega \int_{r_1(\Omega)}^{\delta} \left(u(r) - \Omega^2\right)^{-1/2} \frac{dr}{r}\right) = \frac{d}{d\Omega}\left(\Omega^2 \int_1^{\Omega^{-2}u(\delta)} (t-1)^{-1/2}\, w(t\Omega^2)dt\right)$$

$$= 2\Omega \int_1^{\Omega^{-2}u(\delta)} (t-1)^{-1/2}\, h(t\Omega^2)dt - 2\left(u(\delta) - \Omega^2\right)^{-1/2} w(u(\delta))u(\delta)$$

$$= 2\Omega \sum_{n=0}^{N} a_n x_0^{-n/2} \Omega^n \int_1^{\Omega^{-2}u(\delta)} (t-1)^{-1/2}\, t^{\frac{n}{2}} dt + \tilde{h}(\Omega) - \left(u(\delta) - \Omega^2\right)^{-1/2} \frac{2u(\delta)}{\delta u'(\delta)}, \tag{4.3a}$$

with

$$\left|\tilde{h}(\Omega)\right| \leq 2\Omega^{N+2}\epsilon_h x_0^{-\frac{(N+1)}{2}} \int_1^{\Omega^{-2}u(\delta)} (t-1)^{-1/2} t^{\frac{(N+1)}{2}} dt \qquad (4.3b)$$

This formula, and a similar formula for $I_3$, allow us to compute $I_2$ and $I_3$ for many small $\Omega^*$. There is a limit again, however, since as $\Omega \to 0$ the integrals above are extended over an intervals spreading to infinity. This calls for a final – but fortunately easy – analysis to control these integrals for very small $\Omega^*$. We simply state that this can be done and refer the interested reader to [21] for all the details.

## Case 3

We are finally left with the proof that $F''(\Omega) \leq c < 0$ for $\Omega$ sufficiently close to $\Omega_c$. The previous method fails for these $\Omega$ mainly because the derivatives of $u$ at the crossing points $r_i$ degenerate to 0. The idea here is to change the approach completely, and consider the change of variables given by $r(t)$, the inverse of $t(r) = \pm\sqrt{\Omega_c^2 - u(r)}$, which allows us to rewrite

$$-F'(\Omega) = \Omega \int_{-1}^{1} (1-t^2)^{-1/2} w(Dt)dt, \quad w(t) = \frac{r(t)}{\dot{r}(t)}, \quad D^2 = \Omega_c^2 - \Omega^2. \quad (4.4)$$

The main difficulty here is to gain control over the function $r(t)$ (or $w(t)$), which is done via the following technical lemma in complex variables.

**Lemma 4.3.** *Let $u \in H^1(|z - r_c| \leq R)$, smooth on the boundary of $B(r_c, R)$, of the form*

$$u(x) = \Omega_c^2 - u_2 R^2 z^2 + z^3 f(z), \qquad z = \frac{x - r_c}{R}, \qquad f(0) = u_3 R^3,$$

*satisfying*

1. $\|f\| \leq h$, $u_2 > 0$ *and* $u_2 R^2 > h$.
2. *For a constant $M$ we have* $\left|u^{(4)}(x)\right| \leq M$ *for* $|z| \leq 1$.

*Then, $t(x)$ can be extended analytically to $B(r_c, R)$, and there is an inverse $r(w)$ of $t(x)$, analytic in $|w| < T$, where $T \le \sqrt{u_2 R^2 - h}$ and*

$$\sup_{|w|\le T} |r'(w)| \le \frac{2\sqrt{u_2 + hR^{-2}}}{2u_2 - 3|u_3|R - \frac{1}{6}MR^2}$$

$$\left|\frac{d^{n+1}r}{dw^{n+1}}(0)\right| \le n!\,T^{-n}\,\frac{2\sqrt{u_2 + hR^{-2}}}{2u_2 - 3|u_3|R - \frac{1}{6}MR^2} \qquad n \ge 0$$

As a consequence, we obtain the following:

**Algorithm 4.4.** *Set $u(x) = \Omega_c^2 - z^2 f(z)$, with $z = (x - r_c)/R$. If we know that $f \in \mathcal{U}_0(I_0, \ldots, I_{2N+1}; C_h, 0; \infty)$ and that $A^0 \le |y(x)| \le A_0$ when $|x - r_c| \le R$, for bounds $A^0 < A_0$, we can then construct $T$ and $\mathcal{U}_1$, such that $w(t) = g(t/T)$, with $g \in \mathcal{U}_1$. We use this to compute a bound for $F''$ sufficiently close to $\Omega_c$.*

**Description.** The construction of $\mathcal{U}_1$ is rather easy from the lemma above. Therefore, we know that $w(t) = \sum w_n (t/T)^n$ and we have good control on the $w_n$. We insert this into (4.4) to obtain

$$-\tfrac{1}{\pi}F'(\Omega) = \Omega \sum_{n=0}^{N} w_{2n}\left(\frac{D}{T}\right)^{2n} \alpha_n + \Omega \sum_{n>N} w_{2n}\left(\frac{D}{T}\right)^{2n} \alpha_n,$$

with

$$\alpha_n = \frac{1}{\pi}\int_{-1}^{1} (1-t^2)^{-1/2} t^{2n}\,dt = \binom{2n}{n} 2^{-2n}. \tag{4.5}$$

The first term above is computed and the second is just an error term, which is easily estimated in absolute value from above. Q.E.D.

## Acknowledgements

Charles L. Fefferman was partially supported by an NSF grant at Princeton University.

## REFERENCES

[1] V. Arnold, *Mathematical Methods of Classical Mechanics*, Graduate Texts in Math. No. 60, Springer-Verlag, N.Y., 1978.

[2] V. Bach, "Accuracy of Mean Field Approximations for Atoms and Molecules", *Comm. Math. Phys.*, 1993, Vol. 155, No. 2, pp. 295–310.

[3] A. Córdoba, C. Fefferman, and L. Seco, "A Trigonometric Sum relevant to the Non–relativistic Theory of Atoms", *Proc. Nat. Acad. Sci. USA*, June 1994, Vol. 91, pp. 5776–5778.

[4] A. Córdoba, C. Fefferman, and L. Seco, "Weyl Sums and Atomic Energy Oscillations", *Revista Matemática Iberoamericana*, 1995, Vol. 11, No. 1, pp. 167–228.

[5] A. Córdoba, C. Fefferman, and L. Seco, "A Number–Theoretic Estimate for the Thomas–Fermi Density", 1995 (to appear).

[6] P. Dirac, "Note on Exchange Phenomena in the Thomas–Fermi Atom", *Proc. Cambridge Philos. Soc.*, 1930, Vol. 26, pp. 376—385.

[7] J. P. Eckmann, H. Koch, and P. Wittwer, *A computer Assisted Proof of Universality in Area Preserving Maps*, Memoirs A.M.S., Vol 289, American Mathematical Society, Providence, R.I., 1984.

[8] J. P. Eckmann and P. Wittwer, *Computer Methods and Borel Summability Applied to Feigenbaum's equation*, Lecture Notes in Mathematics, Vol. 227, Springer-Verlag, N.Y., 1985.

[9] C. Falcolini, C. Fefferman, and R. Llave, In preparation.

[10] C. Fefferman, "Atoms and Analytic Number Theory", In: *A.M.S. Centennial Publication, Vol. II*, American Math. Society, Providence, RI, 1992, p. 27–36.

[11] C. Fefferman, "The Atomic and Molecular Structure of Matter", *Revista Matemática Iberoamericana*, 1985, Vol. 1, No. 1, pp. 1–44.

[12] C. Fefferman, "The $N$–Body Problem in Quantum Mechanics", *Comm. Pure and Appl. Math.*, 1986, Vol. 39, No. S, pp. S67–S110.

[13] C. Fefferman and R. Llave, "Relativistic Stability of Matter, I", *Revista Matematica Iberoamericana*, 1986, Vol. 2, No. 1&2, pp. 119–213.

[14] C. Fefferman and L. Seco, "The Ground–State Energy of a Large Atom", *Bull. A.M.S.*, 1990, Vol. 23, No. 2, pp. 525—530.

[15] C. Fefferman and L. Seco, "Eigenvalues and Eigenfunctions of Ordinary Differential Operators", *Adv. Math.*, 1992, Vol. 95, No. 2, pp. 145–305.

[16] C. Fefferman and L. Seco, "The Eigenvalue Sum for a One–Dimensional Potential", *Advances in Math*, Oct. 1994, Vol. 108, No. 2, pp. 263–335.

[17] C. Fefferman and L. Seco, "The Density in a One-Dimensional Potential", *Advances in Math.*, Sept. 1994, Vol. 107, No. 2, pp. 187–364.

[18] C. Fefferman and L. Seco, "The Eigenvalue Sum for a Three–Dimensional Radial Potential", *Advances in Math.*, 1995 (to appear).

[19] C. Fefferman and L. Seco, "The Density in a Three–Dimensional Radial Potential", *Advances in Math.*, 1995 (to appear).

[20] C. Fefferman and L. Seco, "On the Dirac and Schwinger Corrections to the Ground-State Energy of an Atom", *Advances in Math.*, Aug. 1994, Vol. 107, No. 1, pp. 1–185.

[21] C. Fefferman and L. Seco, "Aperiodicity of the Hamiltonian Flow in the Thomas–Fermi Potential", *Revista Matemática Iberoamericana*, 1993, Vol. 9, No. 3, pp. 409–551.

[22] E. Fermi, "Un Metodo Statistico per la Determinazione di alcune Priorieta dell'Atome", *Rend. Accad. Naz. Lincei*, 1927, Vol. 6, pp. 602—607.

[23] G. M. Graf and J. P. Solovej, "A Correlation Estimate with Applications to Quantum Systems with Coulomb Interactions", *Reviews in Math. Phys.*, 1994, Vol. 6, No. 5a, pp. 977–997.

[24] W. Hughes, "An Atomic Energy Lower Bound that Agrees with Scott's Correction", *Advances in Mathematics*, 1990, Vol. 79, pp. 213–270.

[25] B. Helffer, A. Knauf, H. Siedentop, and R. Weikard, "On the Absence of a First Order Correction for the Number of Bound States of a Schrödinger Operator with Coulomb Singularity", *Comm. P.D.E.*, 1992, Vol. 17, No. 3&4, pp. 615–639.

[26] E. Hille, "On the Thomas–Fermi Equation", *Proc. Nat. Acad. Sci. USA*, 1969, Vol. 62, pp. 7—10.

[27] V. Ivrii and I. M. Sigal, "Asymptotics of the Ground State Energies of Large Coulomb Systems", *Annals of Math.*, 1993, Vol. 138, No. 2, pp. 243–335.

[28] E. W. Kaucher and W. L. Miranker, *Self-validating Numerics for Function Space Problems*, Academic Press, N.Y., 1984.

[29] O. Lanford, "A Computer–Assisted Proof of the Feigenbaum Conjecture", *Bull. AMS*, 1986, Vol. 6, pp. 427–434.

[30] E. Lieb, "Thomas–Fermi and Related Theories of Atoms and Molecules", *Reviews of Modern Physics*, 1981, Vol. 53, No. 4, pp. 603–641.

[31] E. Lieb and B. Simon, "Thomas–Fermi Theory of Atoms, Molecules and Solids", *Adv. Math.*, 1977, Vol. 23, pp. 22—116.

[32] R. Llave, "Computer Assisted Bounds in Stability of Matter", *Computer Aided Proofs in Analysis*, IMA Series in Math. and Appl., Vol 28, Springer, Cincinatti, 1989.

[33] R. Lohner, *Einschließung der Lösung gewöhnlicher Anfangs– und Randwertaufgaben und Anwendungen*, Dissertation, Universität Karlsruhe (TH), 1988.

[34] R. E. Moore, *Methods and Applications of Interval Analysis*, SIAM, Philadelphia, PA, 1979.

[35] D. Rana, *Proof of Accurate Upper and Lower Bounds for Stability Domains in Denominator Problems*, Thesis, Princeton University, 1987.

[36] J. Schwinger, "Thomas–Fermi Model: The Second Correction", *Physical Review*, 1981, Vol. A24, No. 5, pp. 2353—2361.

[37] J. M. C. Scott, "The Binding Energy of the Thomas–Fermi Atom", *Phil. Mag.*, 1952, Vol. 43, pp. 859—867.

[38] L. Seco, "Lower Bounds for the Ground State Energy of Atoms", Thesis, Princeton University, 1989.

[39] L. Seco, "Computer Assisted Lower Bounds for Atomic Energies", *Computer Aided Proofs in Analysis*, IMA Series in Math. and Appl., Vol 28, Springer, Cincinatti, 1989, pp. 241—251.

[40] H. Siedentop and R. Weikard, "On the Leading Energy Correction for the Statistical Model of the Atom: Interacting Case", *Communications in Mathematical Physics*, 1987, Vol. 112, pp. 471–490.

[41] H. Siedentop and R. Weikard, "On the Leading Correction of the Thomas–Fermi Model: Lower Bound", and an appendix by A. M. K. Müller, *Inv. Math.*, 1989, Vol. 97, pp. 159–193.

[42] H. Siedentop and R. Weikard, "A New Phase Space Localization Technique with Applications to the Sum of Negative Eigenvalues of Schrödinger Operators", *Ann. Scient. Ecole Normale Superieure*, 1991, Vol. 24, pp. 215–225.

[43] L. H. Thomas, "The Calculation of Atomic Fields", *Proc. Cambridge Philos. Soc.*, 1927, Vol. 23, pp. 542—548.

# 8

# INTERVAL COMPUTATIONS ON THE SPREADSHEET


**Eero Hyvönen and Stefano De Pascale**

*VTT Information Technology*
*P.O. Box 1201, 02044 VTT*
*Finland*

*{eero.hyvonen, stefano.depascale}@vtt.fi*


## ABSTRACT


This paper reviews work on using interval arithmetic as the basis for next generation spreadsheet programs capable of dealing with rounding errors, imprecise data, and numerical constraints. A series of ever more versatile computational models for spreadsheets are presented beginning with classical interval arithmetic and ending with interval constraint satisfaction. In order to demonstrate the ideas, an actual implementation of each model as a class library is presented and its integration with a commercial spreadsheet program is explained.


## 1 LIMITATIONS OF SPREADSHEET COMPUTING

Spreadsheet programs, such as MS Excel, Quattro Pro, Lotus 1–2–3, etc., are among the most widely used applications of computer science. Since the pioneering days of VisiCalc and others, spreadsheet programs have been enhanced immensely with new features. However, the underlying computational paradigm of evaluating arithmetical functions by using ordinary machine arithmetic has remained the same. The work presented in this paper shows that interval techniques provide a new and more versatile basis for spreadsheet computations in many ways. Since exact numbers are intervals of zero width, the generalizations proposed can be made without losing the possibility of using spreadsheets in the traditional way.




*R. B. Kearfott and V. Kreinovich (eds.), Applications of Interval Computations*, 169–209.

The fundamental computational task performed by spreadsheets is to evaluate a set of mutually dependent functions, i.e., a function system of the following kind.

**Definition** Let $S = \{Y_i = F_i(\ldots)\}$ be a set of arithmetical functions and $V$ the set of variables used in it. Variables $Y_i$ representing function values are called output variables $OUT$. $S$ is called function system ($FS$) if there is at most one function for each $Y_i \in OUT$. An output variable $Y_1$ that is related to itself via a circular chain of functions

$$Y_1 = F_1(\ldots, Y_2, \ldots), Y_2 = F_2(\ldots, Y_3, \ldots), \ldots, Y_n = F_n(\ldots, Y_1, \ldots),\ n \geq 1$$

is called ***recursion variable***. If $R$ is the set of recursion variables in a $FS$, then variables in $V - OUT \cup R$ are ***input*** variables $IN$. If $R = \{\}$, then the $FS$ is ***non–recursive***, otherwise ***recursive***. The task of ***cascaded function evaluation*** in a $FS$ is to evaluate the output variable values when the values of input variables are given.

For example, in function system

$$FS = \{x = y^2 + z + 3w,\ y = z^3 + e^z,\ w = \sin(xw) + z\} \tag{8.1}$$

the last function is recursive and it is also mutually recursive with the first function. The task of cascaded function evaluation is to compute output variable values $x$, $y$ and $w$ when input values $z$, $w$, and $x$ are given. Notice that recursion variables $x$ and $w$ are used simultaneously as input and output.

The current computational paradigm of spreadsheets is to evaluate non–recursive (non–circular) function systems by using ordinary machine arithmetic. From the user's view point, this simple scheme is insufficient in various practical situations. This paper focuses on three problem areas: limitations of machine arithmetic, exploitation of inexact data, and application to constraint solving.

## 1.1 Limitations of Machine Arithmetic

Consider the following harmless–looking spreadsheet situation:

$$\begin{aligned} A1 &= 10864 \\ B1 &= 18817 \\ C1 &= 9 * A1\hat{\ }4 - B1\hat{\ }4 + 2B1\hat{\ }2 \end{aligned} \tag{8.2}$$

When the system is evaluated by using the MS Excel spreadsheet program, value $C1 = 2$ is obtained although the correct value [44] should be $C1 = 1$ (!).

The problem is due to the rounding errors generated by finite precision machine arithmetic (double precision) used by the system. This example demonstrates that, in the general case, the spreadsheet user cannot be sure that even the first digit in the results is correct.

Another problem of machine arithmetic is how to deal with overflowed values. For example, the value of exp(710) overflows in ordinary double precision and generates an error in a spreadsheet program. However, in a tolerant system it should be possible to make computations with such values when reasonable, too. For example, consider the function system:

$$\begin{aligned} A1 &= 0 \\ A2 &= \exp(710) * A1 \\ A3 &= 2/\exp(710) \end{aligned}$$

MS Excel returns error values for $A2$ and $A3$ because the value of $e^{710}$ overflows. However, a more robust system could see that

$$\begin{aligned} A2 &= \exp(710) * 0 &= 0 \\ A3 &= 2/\exp(710) &= (0, 0+) \end{aligned}$$

where 0+ in the open interval $(0, 0+)$ denotes the smallest positive machine number available.

## 1.2 Exploiting Inexact Data

In spreadsheet applications all data cannot always be represented by exact numbers. For example, consider the electronic circuit of figure 1 below. The task is to compute voltage $U_1$ by function system

$$\begin{aligned} R_1 &= \frac{R_2 R_3}{R_2 + R_3} \\ U_1 &= R_1 I_1 \end{aligned} \tag{8.3}$$

when resistances $R_2$ and $R_3$ are given and current $I_1$ has been measured. In real life, resistances have manufacturing tolerances and $I_1$ can be measured only down to the accuracy of the measuring equipment used. A reliable value for $U_1$ cannot hence be computed by using ordinary arithmetic. However, by using interval analysis reliable bounds for $U_1$ can be determined.

Classical interval arithmetic (IA) [38] alone is not necessarily applicable to this kind of problem because it treats multiple variable instances in the functions

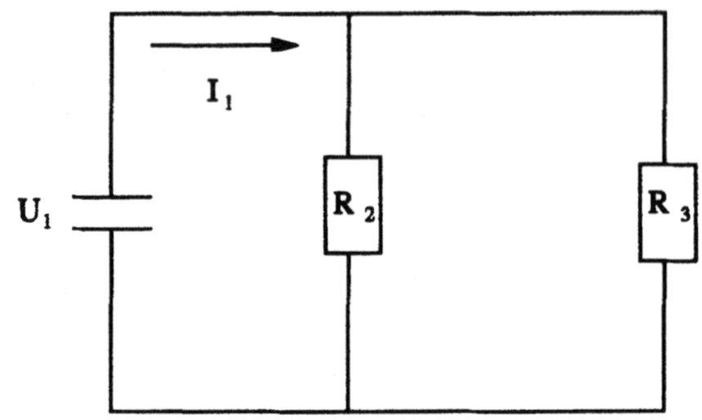


**Figure 1** An electronic circuit.

independently from each other (here variables $R_2$ and $R_3$ have two instances in the first function). As a result, there is usually too much width in the results. In spite of this difficulty, classical IA can be useful when intervals are narrow estimates for rounding errors [33] because then the extra width accumulated into the results often remains tolerable. However, in this paper we see interval arithmetic in the first place as a general formalism for representing uncertain, imprecise or missing numerical data. Intervals in such spreadsheet applications are usually wide, even infinite, and the problem of extra width is strongly emphasized. In order to evaluate the actual range of an interval function in the general case, global interval optimization techniques [43, 13] must be employed.

Another limitation of classical IA is its restricted notion of interval as a closed finite continuum of values. This may suffice when dealing with rounding error bounds. However, in spreadsheet computing, a notion of interval as general as possible is preferable including open ends, infinities, and discontinuities.

## 1.3 Constraint Solving

Spreadsheet functions are evaluated in one "direction" from input argument values to output function values. However, the functions could be used for more than that. For example, consider the temperature conversion formula from Celsius ($C$) to Fahrenheit degrees ($F$):

$$F = 1.8\,C + 32 \tag{8.4}$$

The formula computes $F$ in terms of $C$ but it could also be used for computing $C$ in terms of $F$ with a little extra algebraic manipulation.

In ordinary spreadsheets, the distinction between input and output variables is rigidly made although this is not necessary from the mathematical view point. It would often be quite useful to the user, if (s)he could set the value of a function and see what effect this constraint has on the arguments. Actually, in planning and designing applications the final goal of a computation (output) is typically given (e.g., the desired profit of a financial transaction) and the problem is to find input data by which the goal can be achieved. Using spreadsheets in such applications typically results in tedious trial–and–error sessions where the user tries to find a feasible solution (output values) by iteratively guessing argument values (input values).

Problems such as (8.4) can be approached by using interval constraint satisfaction techniques [8, 10, 15, 17] in which function sets (or more generally equation sets) are treated as symmetric relational constraints without committing to any input–output distinction. In addition to using the trial–and–error scheme, interval constraint techniques can also support top–down problem solving in which a feasible solution is searched for by stepwise refining the variable intervals of the problem until a solution emerges [16, 18].

## 1.4 Outline of Research

In this paper it is shown how interval techniques can be used for extending the usage of ordinary spreadsheets into the three directions discussed above. As a demonstration of the ideas, three extensions to a commercial spreadsheet system have been designed and implemented. MS Excel is used as the pilot environment but the ideas and techniques developed are general and can be applied to other similar systems as well. For reasons of portability, each interval extension is based on a general purpose interval arithmetic C++ class library of the InC++ library family [24] that has been integrated with the spreadsheet by an interface. The functionalities of the libraries can be incorporated into other spreadsheet systems by rewriting this interface only.

The following chapters review the three extensions one after another. In the first system, the spreadsheet program is provided with various interval types and operators for them. The main limitation of the system is that of classical IA: the results of function evaluations may have extra width in certain situations. The second extension overcomes this problem by using algebraic and numerical techniques for determining the actual bounds of interval functions. In both systems functions are evaluated in one direction only. This limitation is removed in the third extension in which spreadsheet functions are computed by solving the corresponding interval constraint satisfaction problem.

# 2 EXTENDED IA ON A SPREADSHEET

Interval arithmetic has usually been implemented by embedding it into programming tools that include:

1. *Programming languages.* Most interval arithmetic implementations such as Pascal–SC [6], Fortran–SC [5], PBasic [1], VPI [12] and BNR–Prolog [41] can be seen as extensions or additional libraries for various programming languages. The implementations are intended for programmers.
2. *Computer algebra packages.* Recently, interval extensions to various computer algebra packages such as Maple and Mathematica [29] have appeared. These implementations are intended also mainly for programmers but here in a conceptually higher level mathematical language than in (1).

We argue that spreadsheets provide one more natural platform for making interval arithmetic tools available. Here the intended users are not programmers but rather various spreadsheet end–users, a much larger audience. Interval computations can be implemented on the spreadsheets by using the macro languages and other programming devices provided by current spreadsheet products.

In this chapter, an interval extension called Extended Interval Arithmetic on MS Excel (EIA Excel) [11] is presented. The primary goal of EIA Excel is to provide the user with a notion of interval arithmetic in as versatile form as possible. For this reason, we first review properties of the extended IA used in the system. This arithmetic generalizes classical IA in many ways and is employed also in the other interval spreadsheet systems of this paper.

## 2.1 Extended Interval Arithmetic

### *Open–Ended Intervals*

In many fields of science and engineering, open and half open intervals are widely used in addition to the closed ones considered in classical IA. Hence, the extended IA should support all four interval types below:

$$[x,y] \qquad [x,y) \qquad (x,y] \qquad (x,y)$$

There are at least two approaches for implementing openness. One possibility is to approximate an open–ended interval by a closed one by rounding open ends inwards to the next smaller machine arithmetic number. After this classical closed IA can be applied. For instance:

$$(-2,3) \approx [-1,9999\ldots,2.9999\ldots]$$

In this approach open ends are only syntactic sugar for representing inward rounding. The problems of this approach are:

- Inward rounding discards possible values at the bounds. For instance, in the above example, ranges $(-2,-1.999\ldots)$ and $(2.999\ldots,3)$ are lost. Discarded ranges never contain any machine numbers but can still have effect on computations.
- Information concerning open bounds is lost and results are always closed.

A more proper approach is to generalize IA function evaluation rules for dealing with interval open ends (and with infinities $\pm\infty$) by considering different interval limit combinations separately for each arithmetic operation (see e.g. [8]). For example, two half open intervals $(a,b]$ can be added by rule:

$$(x,y]+(u,v]=(x+u,y+v]$$

However, in this approach rounding errors may cause problems. For example, when computing

$$(0.1,1]+(0.2,2]=(0.3\pm,3]$$

the open minimum, denoted by $0.3\pm = 0.1+0.2$ cannot be represented precisely by a machine number equal to 0.3, but must be represented by the next one smaller (0.3−) or larger (0.3+). When using normal programming languages, the rounding direction depends on the math library used. If the upward rounded value 0.3+ is selected, then the result $(0.3+,3]$ is incorrect because the correct minimum in range $(0.3,0.3+)$ is excluded. A better and safe minimum bound is 0.3−, but then the result $(0.3-,3]$ will contain extra values $(0.3-,0.3]$. Although the distance between neighboring machine numbers is tiny with small numbers (like with 0.3− and 0.3), it can be large (in absolute value) with large numbers. As a result, more or less excess width can be accumulated into the intervals during computations. This extra width as well as the extra width due to using classical IA (that assumes independent multiple variable instances) is much larger than the infinitesimal difference between an open and a closed interval bound.

There are still good reasons for introducing open ends. Firstly, the effect of rounding errors can usually in practice be filtered away by representing the interval bounds with less digits than are used for computations. Secondly, it is not possible to represent infinitely large and small or infinitely close bounds without the notion of open interval. For example, in extended interval arithmetic (to be presented later) interval division $\frac{1}{[-2,2]}$ can be performed by considering negative and positive values separately:

$$\frac{1}{[-2,2]} = \frac{1}{[-2,0)} \cup \frac{1}{(0,2]} = (-\infty,-1/2] \cup [1/2,+\infty)$$

If $[-2,0)$ had to be approximated by $[-2,0-]$, where $0-$ is the largest negative machine number, then the (non-representable) values $(0-,0)$ would have no effect on the value of division. However, in division the values of the divisor near zero are quite important because they generate the smallest and largest values of the function (interval limits). Thirdly, even in cases where a lot of extra width is accumulated in the result, openness never makes harm other than slightly complicates computations.

## *Complement Intervals*

An interval intuitively states a range of possible values. In some applications, it may also be useful to state ranges of impossible values. For this purpose, we introduce the notion of *complement* (*interval*). A closed complement is defined by

$$]x,y[ \;=\; \{a \mid a \leq x \text{ or } a \geq y\}$$

and intuitively contains values *not* in interval $(a,b)$. Open and half open complements can be accepted as well. The definitions of these variants are analogous to the interval case:

$$
\begin{aligned}
)x,y( &= \{a \mid a < x \text{ or } a > y\} \\
]x,y( &= \{a \mid a \leq x \text{ or } a > y\} \\
)x,y[ &= \{a \mid a < x \text{ or } a \geq y\}
\end{aligned}
$$

By using complements, IA rules can be extended for some discontinuous situations not defined in classical IA, such as division by an interval containing zero. For example:

$$\frac{[3,6]}{[-1,3]} = ]-3,1[$$

## *Discontinuous Intervals*

The complement is a simple case of the more general notion of ***discontinuous*** or ***multi–interval***. A discontinuous interval $X$ can be defined as the union of a set of intervals and complements $X_i$

$$X = \{X_1 \cup X_2 \cup \ldots \cup X_n\}$$

and it is represented in ordinary set notation. $X_i$ are called the ***constituents*** of $X$. For example:

$$X = \{]2,20[,[7,8),6,(15,25]\} \tag{8.5}$$

The order of constituents is free and the elements may overlap each other. Exact number 6 is implicitly considered interval $[6,6]$ here. The normalized representation of a discontinuous interval is the one in which

1. constituents are represented as ordinary intervals,
2. any two overlapping constituents $X \cap Y \neq \{\}$ are merged into $X \cup Y$ and
3. the constituents are sorted in increasing order.

It is easy to see that any discontinuous interval has a unique normalized representation. For instance, (8.5) in normalized form is:

$$X = \{(-\infty,2],6,[7,8),(15,+\infty)\}$$

## *Infinite Intervals*

A problem in implementing interval arithmetic by using finite precision machine arithmetic is how to deal with infinitely large intervals and value overflows. A solution approach to this problem is to use variable precision arithmetic [12], in which the user can select the number of bits to be used for number representation. By using more bits, less overflows are likely to happen but the problem still remains. The selection can also be made dynamically, like in C–XSC [30]. In many computer algebra systems [9], arbitrarily large integers and precise

| TYPE | COMMENT |
|---|---|
| | **Numbers** |
| $x$ | In outward rounding mode, numbers are changed automatically into intervals, if needed. |
| $-\infty, +\infty$ | Infinities |
| | **Intervals** |
| $[x, y]$ | Classical IA closed intervals |
| $(x, y)$ | Open intervals |
| $[x, y)$ | Half open intervals |
| $(x, y]$ | |
| $(-\infty, x)$ | Infinite intervals |
| $(-\infty, x]$ | |
| $(x, +\infty)$ | |
| $[x, +\infty)$ | |
| $(-\infty, +\infty)$ | |
| | **Complement intervals** |
| $]x, y[$ | $= \{a \mid a \leq x \text{ or } a \geq y\}$ |
| $)x, y($ | $= \{a \mid a < x \text{ or } a > y\}$ |
| $]x, y($ | $= \{a \mid a \leq x \text{ or } a > y\}$ |
| $)x, y[$ | $= \{a \mid a < x \text{ or } a \geq y\}$ |
| | **Discontinuous intervals** |
| $\{\}$ | Empty interval |
| $\{X_1, \ldots, X_n\}$ | Each $X_i$ is a number, an interval, or a complement |

**Table 1** Summary of interval types in extended IA.

rational numbers can be used. The price to be paid for such precision is of course increase in computational complexity.

When using computationally more efficient finite precision arithmetic, too large and too small values can be represented by special infinity values $+\infty$ and $-\infty$. Interval functions can then be modified by generalizing arithmetical operators for $\pm\infty$. For example, interval addition

$$(x, y) + (u, v) = (x + u, y + v)$$

can be generalized by the following rules for adding $\pm\infty$ [8]:

$$\begin{aligned} -\infty + x &= x + (-\infty) &= -\infty \\ +\infty + x &= x + (+\infty) &= +\infty \\ -\infty + (-\infty) &= -\infty \\ +\infty + (+\infty) &= +\infty \end{aligned}$$

Cleary argues that in his arithmetic system, cases $-\infty + (+\infty)$ and $+\infty + (-\infty)$ (and some other corresponding ill–defined cases with other operators) "can never occur and need not be specified" (p. 146). However, in practice, ill–defined situations may arise when subexpressions in combined expressions evaluate overflowed intervals (or if the user is allowed to use $\pm\infty$ as input for operators). For example, if ANSI C double precision is used, then the minimum and maximum of $e^{[710,720]}$ are larger than the largest representable machine number. Let us represent such an interval by $(\infty, \infty)$. This kind of degenerated values can be useful and cannot be discarded. For instance, if $(\infty, \infty)$ is multiplied by value 0, then the result should be 0. On the other hand, if $(\infty, \infty)$ is divided by $(\infty, \infty)$ then one can at least say that the possible values are positive:

$$y = \frac{e^{[710,720]}}{(\infty, \infty)} \subset (0, +\infty)$$

When using large or infinitely large intervals, the results of arithmetic operations often overflow and cannot be computed or represented precisely. However, in interval arithmetic it is possible to return as the result the closest obtainable outer approximation of the actual result (e.g., $(0, +\infty)$ in the above example). This is often useful, although one then typically loses information and gets extra width in the result. For example, in evaluating the function

$$y = \ln(e^{[690,710]})$$

it is not possible to get the precise result $[690, 710]$ but only $[690, +\infty)$ due to the maximum overflow in the inner exponent function call:

$$y = \ln(e^{[690,710]}) \approx \ln([4.60 \cdot 10^{299}, +\infty)) \approx [690, +\infty)$$

Fortunately, from the user's view point, the criterion for identifying an overflow is simple:

If there is a limit $\pm\infty$ in the result, then and only then a degenerating overflow has occurred.

Non–degenerating overflows compensated by other operations during the evaluation, such as multiplication of $[1, +\infty)$ by 0, make no harm as long as the final result is finite.

A summary of the interval types discussed above is presented in table 1.

## 2.2 Primitive Functions

Let us denote the interval and discontinuous interval functions corresponding to an exact value function $f$ by $f_i$ and $f_d$, respectively (e.g., $+_i$ means interval addition). The generalized discontinuous interval function $f_d(D_1, \ldots, D_n)$ corresponding to an interval function $f_i(I_1, \ldots, I_n)$ can be defined as follows:

$$f_d(D_1, \ldots, D_n) = \cup\{f_i(I_1, \ldots, I_n) \mid I_1 \in D_1, \ldots, I_n \in D_n\} \qquad (8.6)$$

The definition intuitively applies the interval function with respect to every possible combination of the constituent intervals used in the arguments. This guarantees that the resulting discontinuous interval is the actual set of all the possible values of the exact function. For illustration, a multiplication of two discontinuous intervals is performed below:

$$\begin{aligned} &\{[-2,-1],[3,4]\} \cdot_d \{[2,3],[4,5]\} \\ &= [-2,-1] \cdot_i [2,3] \cup [-2,-1] \cdot_i [4,5] \cup [3,4] \cdot_i [2,3] \cup [3,4] \cdot_i [4,5] \\ &= [-6,-2] \cup [-10,-4] \cup [6,12] \cup [12,20] \\ &= \{[-10,-2],[6,20]\} \end{aligned}$$

The number of constituent intervals in the resultant discontinuous interval in (8.6) varies between 1 and $k_1 k_2 \cdots k_n$ where $k_i$ is the number of intervals in discontinuous interval $D_i$, $i = 1 \ldots n$. In the example above, the maximum number $2 \cdot 2 = 4$ was reduced to 2 because two pairs of intervals were overlapping and could be merged by the union operation.

A major motivation for discontinuous interval arithmetic is the possibility of representing precisely function values even when the function has discontinuity points. For example, the definition of interval division with closed intervals can be generalized as follows:

$$\frac{[x,y]}{[u,v]} = \begin{cases} [x,y]\cdot[1/v,1/u], & \text{if } 0 \notin [u,v] \\ \{\}\ \text{ empty interval}, & \text{if } u=v=0 \\ \dfrac{[x,y]}{(0,v]}, & \text{if } u=0 \\ \dfrac{[x,y]}{[u,0)}, & \text{if } v=0 \\ \dfrac{[x,y]}{[u,0)} \cup \dfrac{[x,y]}{(0,v]}, & \text{otherwise} \end{cases}$$

By defining modified rules for open–ended and infinite intervals, the general interval (and discontinuous interval) division rule can be compiled. Examples:

$$\frac{6}{[-1,3]} = \frac{6}{[-1,0)} \cup \frac{6}{(0,3]} = (-\infty,-6] \cup [2,+\infty) = ]-6,2[$$

$$\frac{[3,6]}{[-1,3]} = (-\infty,-3] \cup [1,+\infty) = ]-3,1[$$

$$\frac{[-2,6]}{[-1,3]} = \frac{[-2,6]}{[-1,0)} \cup \frac{[-2,6]}{(0,3]} = (-\infty,+\infty)$$

In addition to interval division, there are also other important discontinuous interval operators, such as $\tan(x)$ and $\cot(x)$.

Classical IA evaluations are computationally more demanding than exact value computations because of the need to deal with both lower and higher bounds. The extensions to open and infinite intervals increase the complexity further because of the need to identify bound types and overflows. Introduction of discontinuity increases the complexity further. For example, applying an $m$:ary function with discontinuous intervals having $n$ interval constituents, needs $n^m$ ordinary interval evaluations. In cascaded evaluation of discontinuous functions, the number of constituents in the worst case grows exponentially. However, by normalizing intermediate results the growth of constituent number in practise usually gets smaller.

The idea of extending IA rules for the discontinuous case has been suggested probably first [13] by Hanson [14] and Kahan [27]. Our extended IA is a full–scale development of the idea. All interval types of table 1 can be used with interval functions. The scheme has been implemented as C++ class library LIA InC++ [20]. The library contains interval classes and overloads arithmetic

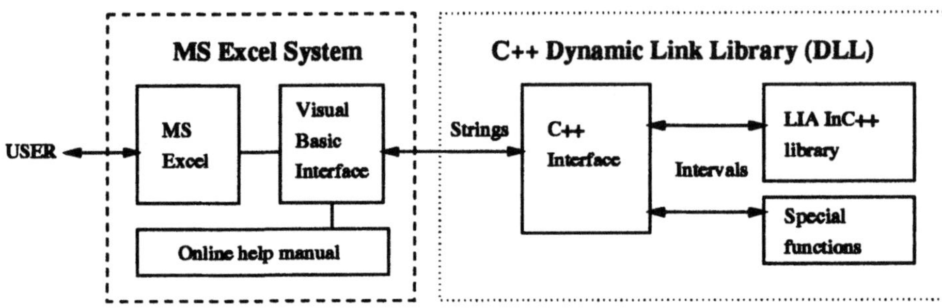


**Figure 2** EIA Excel Architecture.

operators and functions for them. The library covers ordinary rational, trigonometric, exponential, logarithmic and many other functions. To our knowledge, it is the only extended IA package discussed in the literature and currently available.

## 2.3 Interfacing MS Excel with Extended IA

The extended IA described above has been embedded in MS Excel. From the user's viewpoint, the system [11] is an Add–In library of new interval functions (such as $TSUM$, $TSIN$, $TLOG$, etc.) corresponding to the ordinary mathematical functions of MS Excel (such as $SUM$, $SIN$, $LOG$, respectively). The functions can be loaded into the system either automatically (from the MS Excel start–up directory) or by demand from the menu. This idea of providing the user with loadable additional functions is widely used in MS Excel, e.g., in its implementations of complex arithmetic, statistical functions, etc. The functions are used and evaluated like ordinary MS Excel functions by Excel's own control mechanism.

Figure 2 depicts the general architecture of EIA Excel. Arrows indicate data flows and lines internal connections between the modules. The user inputs intervals and functions into the spreadsheet in the usual way. Intervals are written as strings (because MS Excel does not support interval data types). Interval functions are written as normal MS Excel functions. They accept both numbers and interval strings as arguments; the function value is also an interval string or a number. The functions are defined in Visual Basic, the programming language provided for the user in MS Excel 5.0. When an interval function call is evaluated, the Visual Basic definition passes control to the underlying DLL–module that actually performs the computation. In the implementation, the

Object Linking and Embedding (OLE) scheme of Microsoft is used.

The Visual Basic interface defines new interval functions ($TSUM$, $TSIN$, etc.) in terms of intermediate C++ functions in the C++ interface module. Only MS Excel specific conventions for the interval functions are implemented in the Visual Basic interface. These include the following:

- The effect of errors encountered in earlier computations is propagated further. For example, if $A2 = 0$ in function $= TSUM(A1/A2, B1)$, the ordinary /–function of MS Excel returns a special error value $\#DIV/0$ that must be returned also as the value of the call to $TSUM$.
- Missing or empty arguments are skipped according to MS Excel conventions. For example, the value of interval addition $TSUM(A1, 2)$ is 2 if $A1$ is empty.
- Type checks and dynamic screening on arguments are performed. For example, the interval sine function $TSIN$ does not accept MS Excel's array arguments that represent actually argument sets (e.g., array $A1 : B2$ refers to cells $\{A1, A2, B1, B2\}$) but $TSUM$ does (in the same way as $SUM$). $TSUM$ also automatically skips arguments of wrong data type, but not error values (e.g., error value $\#DIV/0$ above).
- The interface is also used for implementing some miscellaneous extensions to MS Excel functionality. For example, the hypertext–based on–line help for EIA Excel (see figure 2) can be activated from a new entry in MS Excel's help menu.

The task of evaluating the interval function is passed to the C++ Interface. This layer is needed mainly because MS Excel does not support interval data types. The C++ interface transforms the strings and numbers of the MS Excel cells into interval objects, applies the actual interval operation to them by the functions of the interval library LIA InC++ and the Special Functions module, and transforms the result back into a string or a number settable in a cell by the Visual Basic Interface. Outward rounding is performed for primitive functions when needed.

The Special Functions -module contains definitions of some interval operations not defined in LIA InC++. For example, the "intelligent" formatting function supported by EIA Excel

$$TF(interval, n)$$

prints an *interval* in a format in which numbers and bounds are represented down to $n$ digits. Intervals (and constituent intervals) are merged automatically into numbers if the interval limits become equal at the given precision level. For example:

$$\begin{array}{lcl} TF(3.1234567, 6) & \Longrightarrow & 3.12346 \\ TF(\text{"}[2.1999999999, 2.2000000001]\text{"}, 10) & \Longrightarrow & 2.2 \\ TF(\text{"}\{[0, 1269], [1300, 2000]\}\text{"}, 2) & \Longrightarrow & [0, 2E+03] \end{array}$$

The second example shows that the function is quite necessary in interval computations in order to get rid of the irrelevant digits generated by outward rounding arithmetic.

# 3 GLOBAL IA ON A SPREADSHEET

The main limitation of classical IA and the interval techniques discussed in the previous section is that the function values evaluated often have more or less extra width. For example, sheet

$$\begin{array}{lcl} A1 & = & C1 \quad = \quad 1 \\ B1 & = & [1,2] \\ B3 & = & (A1 \cdot B1) - (B1 \cdot C1) \end{array} \tag{8.7}$$

evaluates $B3 = [-1, 1]$ although the actual range of B3 is $B3 = 1 \cdot B1 - B1 \cdot 1 = 0$. In the general case, the actual value range of a function $F$ can be evaluated only if $F$ does not have multiple variable instances, or if such variables have precise numeric values, which actually makes them non–variables [39]. The excess width diminishes as the values of the multiple instance variables get stricter and stricter.

This dependency problem has its incarnation also at the function set level. For example, consider the following function set version of (8.7):

$$\begin{array}{lcl} A1 & = & C1 \quad = \quad 1 \\ B1 & = & [1,2] \\ A2 & = & A1 \cdot B1 \\ C2 & = & B1 \cdot C1 \\ B3 & = & A2 - C2 \end{array} \tag{8.8}$$

Here the actual value of each function can be computed in separation (there are no multiple variable instances) but we still get $B3 = [-1, 1]$ (as in (8.7))

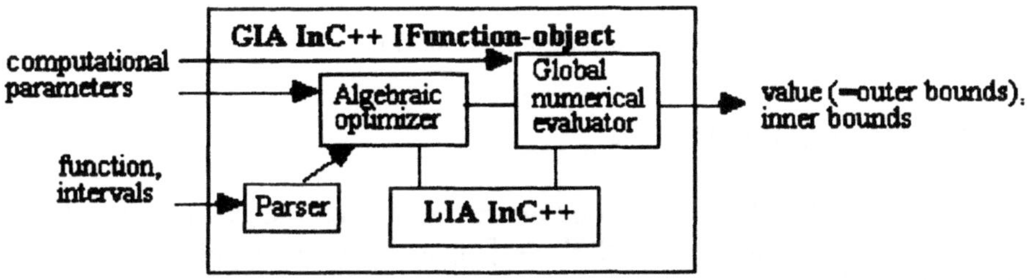


**Figure 3** Architecture of function objects (IFunction) in GIA InC++ library.

and not $B3 = 1 \cdot B1 - B1 \cdot 1 = 0$.

Obviously, the dependency problem must be solved both at the function and at the function sets levels in order to implement an interval spreadsheet program capable of evaluating the actual cell values. In the following, it is first shown how these tasks can be performed. After this, an implementation of global interval arithmetic on top of MS Excel is presented.

## 3.1 Global Evaluation of an Interval Function

Various techniques have been developed for evaluating the actual value range of an interval function, i.e., for its *global* evaluation. Best-known approaches are the various numerical branch-and-bound algorithms discussed in depth in [42, 43, 13]. This paper considers these techniques mainly from the practical user's view point. We proceed by presenting a global function evaluator library GIA InC++ [21, 24] in whose design practical issues encountered in spreadsheet-like computations have been the main concern.

GIA provides the user with a class "IFunction" for defining, parametrizing and evaluating an interval function. Figure 3 depicts the general architecture of IFunction-objects. The parser transforms user given functions and interval expressions (strings in ordinary mathematical notation) into an internal representation. An algebraic extension optimizer then transforms the parsed representation into a form optimal for global numerical interval arithmetic evaluation. Various computational parameters can be used for controlling both numerical evaluation and algebraic optimization. The LIA InC++ library discussed earlier is used as the underlying basis for evaluating interval function primitives. In the following, numerical evaluation and algebraic optimization in GIA InC++ are discussed in some more detail.

## *Numerical Evaluation*

GIA's numerical evaluation is based on the numerical branch–and–bound (BB) search scheme introduced by Skelboe [46] and developed further by others [2, 42, 43, 40]. Roughly speaking, the idea is to consider argument intervals in ever smaller and more precise parts. Efficiency is gained by identifying and pruning irrelevant subintervals from consideration as early as possible.

The BB algorithms above are intended for finding an arbitrarily close lower bound for the minimum of an interval function $F$. In GIA, the algorithm is applied also to $-F$ for getting the upper bound for the maximum. Hence, ***outer bounds*** $[min_{out}, max_{out}]$ of the actual function $F$ values $[min, max]$ can be computed. In addition, as a side effect of the computation, it is possible to obtain also ***inner bounds*** $[min_{in}, max_{in}]$ for the actual $F$ values $[min, max]$ (e.g., by using the so–called midpoint–test accelerator):

$$[min_{in}, max_{in}] \subseteq [min, max] \subseteq [min_{out}, max_{out}]$$

Computation is terminated based on either absolute or relative precision levels set by the user:

$$\begin{aligned} min_{in} - min_{out} &\leq \text{absolute precision} \\ max_{out} - max_{in} &\leq \text{absolute precision} \\ \frac{min_{in} - min_{out}}{min_{in}} &\leq \text{relative precision}, \quad min_{in} \neq 0 \\ \frac{max_{out} - max_{in}}{max_{in}} &\leq \text{relative precision}, \quad max_{in} \neq 0 \end{aligned} \tag{8.9}$$

Relative criteria are useful especially when the absolute values of the interval limits are large; absolute criteria are useful when they are around zero.

## *Recursive Functions*

In interval arithmetic spreadsheets it makes sense to accept recursive function systems in addition to the traditional non–recursive ones. For this purpose, GIA InC++ extends the BB scheme also to recursive functions of form:

$$X_i = F(X_1, \ldots, X_i, \ldots, X_n)$$

Mathematically, the actual range of feasible values of a recursive function is the range of its exact value ***fixed points***:

$$X_i = \{x_i \mid x_i = f(x_1, \ldots, x_i, \ldots, x_n), x_j \in X_j, j = 1, \ldots, n\} \tag{8.10}$$

The goal of GIA InC++ is to compute safe outer bounds $[min_{out}, max_{out}]$ for this set within user given precision levels.

When evaluating a recursive function, simple iteration by using the intersected value $F(X_1, \ldots, X_i, \ldots, X_n) \cap X_i$ as the next value for $X_i$ is not enough. The problem is that computation easily gets stuck in fixed point interval solutions. For example, in evaluating $X = \frac{1}{X}$ such an iterative computation could terminate at

$$X = \frac{1}{[0.1, 10]} = [0.1, 10]$$

and the actual range $[1, 1]$ within $[0.1, 10]$ is not found.

Another problem in evaluating a recursive function is that it is usually difficult to find (by the midpoint test) exact fixed points, i.e., feasible values of the function, because $f(x_1, \ldots, x_i, \ldots, x_n)$ seldom happens to evaluate $x_i$ precisely. As a result, inner bounds of the function are difficult to determine. Without finding such points, the BB algorithms cannot be terminated easily (cf. the termination conditions (8.9) above).

In order to solve this difficulty, GIA InC++ slightly loosens the notion of fixed point based on user given precision levels. Instead of bounding set (8.10) GIA determines bounds for:

$$X_i = \{x_i \mid x_i \approx f(x_1, \ldots, x_i, \ldots, x_n),\ x_j \in X_j,\ j = 1, \ldots, n\} \tag{8.11}$$

Situation $x_i \approx f(x_1, \ldots, x_i, \ldots, x_n)$ is considered a ***precision fixed point*** using the user given absolute or relative precision criteria. GIA determines the range of precision fixed points (as an interval) instead of the range of ordinary precise fixed points. Since every ordinary fixed point is a precision fixed point, the range (outer bound) of values determined by GIA InC++ is a superset of the actual fixed point values (8.10). The extra width in the obtained range approaches zero as desired absolute and relative precision approach to zero. With zero precision the sets of ordinary and precision fixed points are identical.

By using precision fixed points, it is easier to obtain also inner bounds for a recursive function. However, one should remember that the inner bounds are now determined for precision fixed points and not for ordinary fixed points. There are not necessarily any ordinary fixed points between inner and outer bounds, only precision fixed points. Although outer bounds in the recursive case safely bound all possible actual fixed points (if any), the inner bounds do not necessarily bound actual feasible values from the inside in the same sense as in the non–recursive case. For example, consider recursive function

$$X = \frac{Y - (X^5 + 15X^4 + 85X^3 + 225X^2 + 120)}{274}, \quad Y = 0, X = [-100, -1]$$

for bounding the zeros of the polynomial

$$Y = X^5 + 15X^4 + 85X^3 + 225X^2 + 274X + 120$$

There are five ordinary fixed points (zeros) $X \in \{-5, -4, -3, -2, -1\}$ and the bounds are hence $X = [-5, -1]$. If absolute and relative precision is $10^{-6}$, the minimums of outer and inner bounds are after some evaluation time $-5.0002\ldots$ and $-5.00005\ldots$ with no ordinary fixed points (but only precision fixed points) in between. Unlike in the non–recursive case, existence of ordinary feasible (fixed) points of a recursive function cannot in the general case be proved unless zero precision is used. Existence of ordinary fixed points implies existence of precision fixed points but not vice versa.

The task of evaluating a recursive function $X_i = F(X_1, \ldots, X_i, \ldots, X_n)$ is equivalent to the problem of finding the zeros of the function:

$$0 = F(X_1, \ldots, X_i, \ldots, X_n) - X_i$$

An efficient technique for the latter problem is the *interval Newton method* [13]. In GIA this technique can be employed by a special macro "newton" that automatically transforms a function into the corresponding (recursive) *interval Newton operator*. For example, the zeros of the above polynomial can be found by using macro:

$$\text{newton}(X^5 + 15X^4 + 85X^3 + 225X^2 + 274X + 120, X)$$

## *Algebraic Optimization*

Consider a real valued function $y = f(x_1, \ldots, x_n)$ and an interval valued function $Y = F(X_1, \ldots, X_n)$. $F$ is called an (interval) *extension* of $f$ iff:

$$f(x_1, \ldots, x_n) = F(x_1, \ldots, x_n)$$

The simplest extensional form is the *natural extension* that is obtained from $f(x_1, \ldots, x_n)$ by simply replacing function operations by the corresponding

| MODE | MEANING |
| --- | --- |
| Natural | Use natural extension (no algebraic optimization). |
| Taylor | Use nth order multivariable Taylor form. |
| Monotonicity | Use monotonicity test form [2]. |
| Optimize | Use monotonicity test form with derivatives optimized by nth order Taylor expansions. |

**Table 2** Some algebraic modes used in GIA InC++.

| Mode | Algebraic preproc. | Numerical evaluation | Monotonicity test used | Taylor order |
| --- | --- | --- | --- | --- |
| Natural | 0 | $> 6 \cdot 10^5$ | no | - |
| Taylor | $< 10$ | 220 | no | 2 |
| Monotonicity | 20 | 440 | yes | - |
| Optimize | 50 | 90 | yes | 2 |

**Table 3** Algebraic preprocessing and numerical evaluation times (msecs) of (8.12) illustrating the need for algebraic optimization.

interval operations. The problem with the natural extension is that the extra width accumulated is usually large. Various alternative ways for constructing extensional forms have been proposed for evaluating stricter results [42]. GIA uses several algebraic modes settable by the programmer. Some of them are listed in table 2.

There is no known general solution to the problem of finding the best extensional form for a function, only some heuristic guidelines. The excess width depends both on the properties of the function and on the argument interval values. Extension selection is, however, quite important from the practical computational view point. For example, some test evaluations of the polynomial

$$Y = X^4 - 10X^3 + 35X^2 - 50X + 24 \qquad (8.12)$$

with $X = [0, 4]$ are listed in table 3. The last mode involving most algebraic preprocessing is here clearly the best choice.

Since the average spreadsheet user is hardly interested in the peculiarities of extensional forms, GIA has also an automatic algebraic mode (used by default). In this mode, GIA heuristically determines for each given function an appropriate extensional form. As a general strategy used there, multiple instances of variables in the expression are tried to be eliminated by using computer algebraic techniques [9] such as polynomial canonization, symbolic rational division and a rule base for simplifying irrational expressions. Taylor expansions are also used when they do not generate more multiple instances of variables.

## *Computational Parameters*

A major pragmatic difference between a local interval system such as EIA Excel and a global one is computational complexity. Global numerical evaluation of an interval function can take an unpredictable long time, and the spreadsheet user must be provided with a versatile set of computational parameters for controlling the computations. In GIA, the following parameters can be used for controlling numerical evaluations:

- *Global vs. local evaluation*. The function objects can be evaluated either globally (default mode) or locally, i.e., in ordinary (extended) IA. Local mode is useful in situations where very fast response time is needed even at the price of extra excess width in the result.
- *Absolute and relative precision levels*. Less demand on desired precision levels often significantly shortens evaluation time.
- *Time limit*. In many applications, response times within predefined limits must be guaranteed (e.g., in real time systems). GIA supports such applications by providing a settable time limit. If the resultant interval has not been obtained within the limit, computation is forcibly terminated. The difference between inner and outer bounds tells how precise the result is.

## *Reusing Earlier Computations*

Computations in spreadsheets are often reactivated after a small change in only one argument. Given the complexity of global interval evaluations, it becomes necessary to maintain results of earlier computations and to try to reuse them as much as possible. It is not feasible to start each evaluation from scratch after any small modification in some argument value.

In GIA the function objects constructed maintain computational partial results obtained during the last evaluation. When evaluated again, the object dynamically checks changed input values with respect to the previous evaluation, makes only needed updates and changes in the previous computational results, and continues computation from this internal state on. Such reuse of earlier work often dramatically shortens needed computational time.

## 3.2 Global Evaluation of an Interval Function System

In [19] a technique is developed for evaluating a function system globally. The hearth of the technique is the simple algorithm **Local–Function–Evaluation** below that implements cascaded function evaluation in traditional spreadsheet computing. Function args($Y = F(\ldots)$) denotes the set of arguments in function $Y = F(\ldots)$.

**Algorithm: Local–Function–Evaluation**

1. *Initially, given is a function system* $S$ $\{Y_i = F_i(\ldots)\}$ *and values for all input (and recursion) variables.*
2. *Sort* $S$ *by the following criterion: if* $Y_i \in$ args($Y_{i+1} = F_{i+1}(\ldots)$)*, put* $Y_i = F_i(\ldots)$ *before* $Y_{i+1} = F_{i+1}(\ldots)$.
3. *If* $S$ *is empty, go to 5. Otherwise select and remove next function* $Y = F(\ldots)$ *from* $S$.
4. *Evaluate* $Y = F(\ldots)$. *Go to 3.*
5. *Terminate.*

However, this algorithm alone does not solve our problem in the interval case: The functions are evaluated independently from each other (locally), which means that common argument variables used in different functions are treated as if they were different variables having the same value. Furthermore, problems of circular function chains are not considered. Proper function system evaluation in interval arithmetic would require that the dependencies between different variable instances in different functions are respected globally. This dependency problem can be captured by the notions of local and global cascaded evaluation:

**Definition** Consider a function system $FS$ with input variables $X_1, \ldots, X_n$ and output variables $Y_1 = F_1(\ldots), \ldots, Y_k = F_k(\ldots)$. Cascaded interval evaluation is *local* iff each output (and recursion) variable $Y_i$ gets interval value:

$$Y_i = \{y \mid \exists\, x_1 \in X_1, \ldots, \exists\, x_n \in X_n \;:\; y = F_i(x_1, \ldots, x_n)\}.$$

Evaluation is *global*, if the value of each output (and recursion) variable $Y_i$ is the actual range of the feasible values w.r.t. all local functions:

$$Y_i = \{y_i \mid \exists\, x_1 \in X_1, \ldots, \exists\, x_n \in X_n \;:\; y_j = F_j(x_1, \ldots, x_n),\ j = 1 \ldots k\}.$$

The local propagation algorithm **Local–Function–Evaluation** can be applied for global cascaded interval evaluation if the function system is first algebraically globalized and then evaluated by using a global interval function evaluator. This scheme is summarized by algorithm **Global–Function–System–Evaluation**. The algorithm evaluates values for the output variables of the function system $FS$.

**Algorithm: Global–Function–System–Evaluation**

1. *Globalize. For each (local) function $f \in FS$ derive a global function $g$ by expanding $f$ by recursively replacing function arguments by the corresponding functions into a minimally large expression. The stopping criterion for the minimal expansion is that no variable used in $g$ should occur as an argument in the remaining functions of $FS$ not used for deriving $g$.*

2. *Optimize algebraically. For each global function, derive algebraically an efficient extensional form for numerical evaluation.*

3. *Evaluate numerically. Evaluate the global function system derived in steps 1–2 by the local propagation algorithm* **Local–Function–Evaluation** *(by using global interval function evaluation techniques discussed in section 3.1).*

For example, consider the function system:

$$\begin{aligned} B2 &= A1 \cdot A2 \\ C1 &= B1 + B2 \\ C2 &= B2 + A2 \\ D1 &= C1 + C2 \\ E1 &= D1 - D2 \end{aligned} \tag{8.13}$$

| Cell | Minimal Expansion | In Simplified Form |
|---|---|---|
| $B2$ | $= A1 \cdot A2$ | $= A1 \cdot A2$ |
| $C1$ | $= B1 + B2$ | $= B1 + B2$ |
| $C2$ | $= (A1 \cdot A2) + A2$ | $= A2 \cdot (A1 + 1)$ |
| $D1$ | $= (B1 + (A1 \cdot A2)) + ((A1 \cdot A2) + A2)$ | $= A1 \cdot A2 \cdot (B1 + A2)$ |
| $E1$ | $= D1 - D2$ | $= D1 - D2$ |

**Table 4** Globalized functions for the output variables of (8.13).

For simplicity, the functions here are simple primitives, but could in principle be functions of arbitrary complexity. The globalization phase (step 1) gives the following globalized functions for the output variables shown in table 4. The minimally expanded global functions for the output variables are given on table 4. The rightmost column gives the expansion in algebraically simplified form.

Application of the minimality criterion is important because it makes the global functions smaller and easier to evaluate. For example, there is no need to expand the function $E1 = D1 - D2$ further because neither $D1$ nor $D2$ is used as an argument in any other function. Intuitively, the global function for $D1$ already takes care of the problem of the multiple variable instances of $A1$ and $A2$, and $D1$ can be used as it is as the argument.

In step (2) further optimization is needed only for the function of $D1$ whose simplified form contains multiple variable instances. In step (3) this function must be evaluated by a global evaluator like GIA. For the other global functions simple local evaluation is sufficient.

## 3.3 Implementation on a Spreadsheet

The global cascaded evaluation scheme has been implemented as a C++ library called CIF InC++ (Cascaded Interval Functions In C++) [22]. The library provides the user with a class "Cif" for defining, maintaining, parametrizing and evaluating a function system. The main input of Cif–objects is functions and argument interval values in the form of strings and numbers. A separate parser is used for extracting the algebraic form of the functions for globalization. After inserting a function or removing one, the global definitions are updated automatically by a definition maintenance unit. Both algebraic globalization

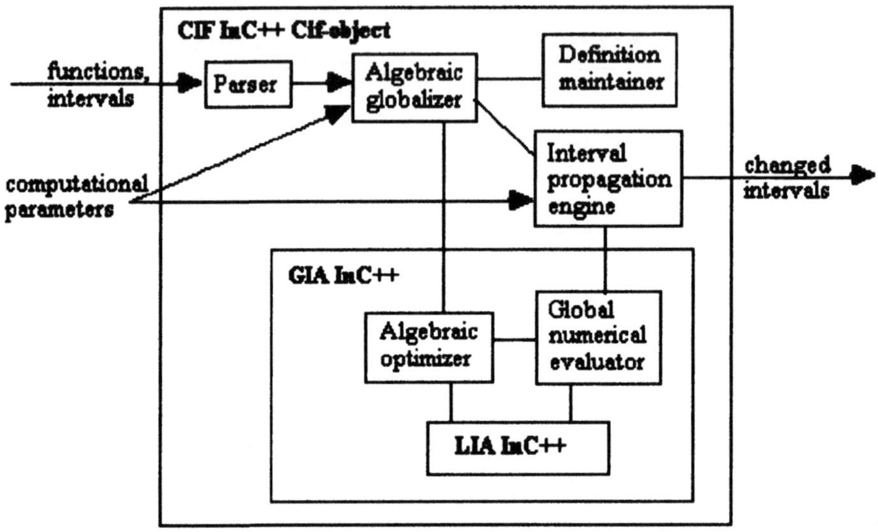


**Figure 4** The architecture of function system objects (Cif) in CIF InC++.

and interval propagation can be tuned by a set of (optional) computational parameters that are similar to those in GIA discussed earlier.

After its construction, the function system can be evaluated and changed variable values can be read from the Cif–object.

Integrating an external global IA module such as CIF with a spreadsheet program is more difficult than integrating LIA in EIA Excel. The main problem is that control of computations needed for propagating values globally is different from the local propagation scheme provided by spreadsheets. As a result, the whole computation procedure must be run in isolation from the host spreadsheet system. Furthermore, all changes made by the user to the local definitions should be reported to the underlying computational mechanism in order to make the corresponding changes at the global level.

In our implementation, each MS Excel workbook sheet is associated with a Cif–object that contains and maintains the definitions of the sheet consistent and performs cascaded function evaluations. The user inputs interval functions for global evaluation by typing into a cell a call of a special Add–In function $F$

$$= F(function)$$

implemented in the MS Excel macro language Visual Basic. Function F implements the interface between MS Excel and the underlying computational Cif–objects. When MS Excel evaluates a call to $F$, say

$$= F("A1 + A2 * C2")$$

in cell $B1$, then the following happens:

1. The function $B1 = A1 + A2 * C2$ is inserted into the corresponding Cif–object.
2. The current function system (Cif–object) is updated and evaluated.
3. Changed values are read from the Cif–object and output on the sheet.

In the same manner, if the user inputs or modifies an interval value in a cell, then the modification is made in the Cif–object and after evaluation, changed values are updated on the sheet.

Computational parameters for global interval evaluation can be set by a dialog box activated from an additional MS Excel menu.

# 4 INTERVAL CONSTRAINT SPREADSHEETS

Reconsider the function system (8.3):

$$\begin{aligned} R_1 &= \frac{R_2 R_3}{R_2 + R_3} \\ U_1 &= R_1 I_1 \end{aligned}$$

If arguments $R_2$, $R_3$, and $I_1$ are given, the voltage $U_1$ can be computed. However, from the mathematical view point, any unknown variable in (8.3) could be computed as a function of the others. This kind of "symmetric" computations where any variable can be used as input or output, can be performed by using constraint propagation techniques [47, 36].

The first spreadsheet programs using constraint propagation were released already in the early 80's. TK!Solver [31] was a well–known commercial implementation. In [15, 17] four major problems of numerical constraint propagation based on ordinary arithmetic were discussed:

- The systems cannot deal with imprecise data.
- The actual solutions cannot always be found even if this were mathematically possible. This may happen if the functions to be evaluated are locally but not globally underconstrained.
- Techniques for solving overconstrained problems were ad hoc.
- Although computations could be performed in any direction, input and output variables had to be selected by the user before propagation. This is often difficult.

As a solution approach, interval constraint satisfaction techniques [8, 10] based on interval arithmetic were proposed, and the approach was developed into a new generalized computational paradigm for spreadsheet computing in [16, 18]. Other approaches to interval constraint satisfaction based on related techniques are discussed in [28, 3, 32]. Application of interval constraints in evaluating actual function value ranges is discussed in [7].

Interval constraint spreadsheets have several attractions when compared with ordinary spreadsheet:

- Interval arithmetic makes it possible to deal with inexact data and rounding errors.
- Infinite solution sets of underconstrained problems can be abstracted as interval solutions.
- Consistency techniques make it possible to compute (or at least bound) values for any variable in a set of related functions or more generally constraint equations. No distinction between input variables (cells) and output variables is needed; the type of variables is determined dynamically.
- Top–down refinement can be used as a new problem solving paradigm. In this scheme, a solution is found by constraining cell intervals stepwise until exact solutions emerge. This is often more useful than the trial–and–error scheme used in ordinary spreadsheets.
- Interval computations are safe and sound — possible solutions are never lost.
- Ordinary numerical computation in spreadsheets is a special case of interval constraint satisfaction. Interval techniques offer new possibilities

without giving up the possibility of using traditional exact spreadsheet computing.

In the following, the notion of interval constraint satisfaction problem (ICSP) is first presented and techniques for solving it are discussed. After this, problems of incorporating interval constraint computations with a spreadsheet program are discussed. As a test and demonstration, MS Excel has been extended for interval constraint satisfaction.

## 4.1 Interval CSPs

The general interval constraint satisfaction problem (ICSP) [17] can be formulated as:

$$\begin{array}{l} E_1, \ldots, E_n \\ P_1 := X_1, \ldots, P_m := X_m \end{array} \tag{8.14}$$

Here $E_i$ are functions, equations (or inequalities, that can be transformed into interval equations), $P_j$ are variables used in them, and $X_j$ are real intervals (or discontinuous real intervals). The equations constitute a ***constraint net*** that consists of a set of variables and equational constraints connecting them. For example:

$$\begin{array}{l} \sin(Z) = X + Y + 0.4, \quad e^X + Z = -0.2, \quad X \cdot Y = Z, \quad X \leq 0 \\ Y = [0, 1000], \quad Z = [-100, -0.1] \end{array} \tag{8.15}$$

The ***interval situation*** (called also ***tolerance situation***) $\{P_1 := X_1, \ldots, P_m := X_m\}$ in an ICSP (8.14) refers to a set of exact value situations

$$\{P_1 := x_1, \ldots, P_m := x_m \mid x_i \in X_i,\ i = 1 \ldots m\}.$$

An ICSP is ***feasible*** iff its interval situation has an exact subsituation satisfying all constraints, i.e., the equations have a solution within the intervals:

$$\exists\, \{P_1 := x_1 \in X_1, \ldots, P_m := x_m \in X_m\} \;:\; E_1, \ldots, E_n \text{ satisfied.}$$

A variable $P_i \in X_i$ is ***consistent*** iff each interpretation $P_i := x$, $x \in X_i$, can be satisfied with respect to all constraints by some exact subsituation:

$$\begin{array}{l} \forall x \in X_i\ \exists\, \{P_1 := x_1 \in X_1, \ldots, P_i := x, \ldots, P_m := x_m \in X_m\} \;: \\ \qquad E_1, \ldots, E_n \text{ satisfied.} \end{array}$$

We say that an interval situation is (***globally***) ***consistent***, i.e., it is a (***global***) ***solution***, iff its every variable is consistent. Intuitively, an interval in a solution should not contain "extra" values that cannot be satisfied within the given intervals. The global solution is the tightest interval set that bounds the solutions for the ICSP. For example, it turns out that equations (8.15) have a single (global) solution

$$X \approx -1.25, \quad Y \approx 0.387, \quad Z \approx -0.485 \tag{8.16}$$

within the initial situation. Hence (8.15) is feasible.

A weaker but computationally more easily obtainable and hence useful notion of consistency is *local consistency*. A variable is *locally* consistent iff it is consistent with respect to all directly connected constraints in the constraint net. Local consistency of an interval situation (solution) means that all variables are locally consistent. For example, the local solution of

$$\begin{array}{l} X + T = Y, \quad Y + T = Z \\ X = 1, \; T \in [0, +\infty), \; Y \in [0, +\infty), \; Z = 11 \end{array} \tag{8.17}$$

is $T \in [0, 10]$, $Y \in [1, 11]$. Only at the global level one can determine that actually $Y$ must be the mean value of $X$ and $Z$, i.e., that the global solution is $Y = 6$, $T = 5$. Intervals in the local solution may have extra width but always correctly bound the global solutions.

In the general case, the solution of an ICSP may not be represented by a single interval solution, but as a set $S$ of them. For reasons of convenience, these solutions can be generalized by the situation $G = \{P_1 \in X_1, \ldots, P_n \in X_n\}$ such that for $i = 1 \ldots n$, $\min(X_i)$ is the minimum and $\max(X_i)$ is the maximum of all values $X_i$ of $P_i$ in the solutions $S$. Hence, the solution $G$ is the most constrained situation that can be obtained by refining the tolerances from both ends without loosing exact value solutions. If discontinuous intervals are employed, the generalized solution can be represented more accurately by a single discontinuous interval situation.

## 4.2 Solving ICSPs

A solution technique for solving interval CSPs is to use Waltz filtering for removing certainly infeasible regions from the variable intervals [10]. By this technique local solutions can be found. This technique is also employed in interval constraint implementations for logic programming such as [41, 45]. Another possibility is to generalize classical value propagation techniques [47] for interval arithmetic, i.e., to use tolerance propagation [15, 17]. In this technique, the structure of the constraint net underlying the constraint equations is exploited. A benefit of this approach is its extendibility from local techniques to global problem solving by algebraic techniques. In global tolerance propagation, better than local solutions can be obtained by propagating values over several looping constraint equations at the same time.

The techniques above aim at bounding the solution set of the equations within an initial interval box. If the actual solutions are to be generated, Interval Newton method [13] provides a promising technique [4]. Here the number of solutions is assumed to be finite, i.e., the problem should not be underconstrained.

In our work, local and some global tolerance propagation algorithms have been implemented as a C++ library ICE InC++ [23]. By default, this system uses local propagation but individual constraints are solved globally. If desired, additional global solution functions can be generated. ICE is designed to be used as a general purpose ICSP library in C++ programming [26]. From the programmer's view point, it is implemented as a C++ class Ice. Objects of this class correspond to ICSPs. Communication with Ice–objects is based on simple member functions that define ICSPs, set computational parameters, propagate tolerances, and read changed variable values. Using Ice–object is in principle quite similar to using IFunction–objects of GIA and Cif–objects of CIF.

The main technical contribution of ICE is to apply a combination of interval analysis, computer algebra and tolerance (interval) propagation techniques for determining local and, especially, stricter than local interval solutions in an efficient way. Other implementations for bounding solutions, such as [41, 34, 35, 45, 3, 37] do not make use of algebraic or global interval techniques but rely on local numerical ones.

## 4.3 Interval Constraint Excel

As a demonstration, a version of ICE has been integrated with MS Excel [25]. This chapter discusses modifications that need to be made in a spreadsheet program such as MS Excel when extending it with interval constraints. Important issues are reviewed mainly from the user's point of view and solution approaches taken in our demonstrational implementation IC Excel (Interval Constraint Excel) are described.

## *Inserting Equations*

Each spreadsheet is associated with an Ice–object (an instance of class Ice) that maintains internally the ICSP corresponding to the spreadsheet definitions. Interval values and constraints are defined by an Excel–function *I(equation)* that sends string *equation* to the underlying object updating its internal state accordingly. For example, when the user types $I("[2,3]")$ into cell $A13$, the value of variable $A13$ is updated in the Ice–object and the status of related constraints and solution functions is set undetermined. These constraints will later be processed during tolerance propagation. In the same way, $I("A1 * SIN(B2) = A1 + C2")$ would insert the equation as a constraint and update the internal constraint net. After inserting/modifying constraints and values and setting computational parameters (like desired precision levels), tolerance propagation can be executed, and changed variable values are updated to corresponding cells on the sheet.

In ordinary MS Excel, output cells have function definitions such as "$= A2 + B3 + 7$" and the function value is used and displayed as the value of the cell. A problem here is that when an interval constraint equation is inserted, such as "$A1 + B3 = C3 * B2$", there is no unique cell where to display the value, but all related variable cells must be considered. In contrast to the function case, the position of an equation in the sheet is immaterial but should be respected as a decision of the user. IC Excel works exactly like MS Excel when interval functions are inserted. When a constraint equation is inserted, IC Excel shows the equation itself in the cell, unless the cell has a value, in which case the value is shown.

## *Setting Output Cell Values*

In spreadsheet programs, it is not possible for the user to set the value of an output cell, because output cell values are defined by function expressions. In interval constraint satisfaction, it must be possible to set the value of any cell. IC Excel extends MS Excel in the following way: when double clicking a cell, a special DLL function gets activated and shows the value of the cell in a dialogue box for editing. The possible formula attached to the cell remains as it was.

## *Control of Computation*

Control of tolerance propagation is quite different from the evaluation mechanism used in MS Excel. As a result, all interval computations have to be done by shifting control to the underlying Ice–object. In IC Excel, the special function *I* described above makes this shift possible. Function *I* does not activate MS Excel's own value propagation algorithm, but only communicates with the underlying Ice–object.

## *Interrupts*

Since interval computations can take an unpredictable time to be completed, the user must be supported with an interrupt facility. In GIA InC++ library, computation can be terminated interactively and the resulting semi–global result used as the solution. IC Excel makes use of the facility: a lengthy computation can terminated by hitting a control key.

## *Relaxation*

Since all cell values are user settable, the user can easily end up in an infeasible situation. In such situations it is not always easy to see what cells should be modified in order to make the situation feasible again. In IC Excel, a local relaxation facility is available. In infeasible situations, computation is carried on until all constraints have been evaluated. After this, cell values can be enlarged locally in order to satisfy all related constraints, if demanded by the user. Often, but not always, this suffices to make an infeasible situation feasible.

## *Special Computational Parameters*

Interval computations can be controlled by a set of parameters. IC Excel makes use of those available in GIA, such as desired absolute/relative precision and response time limit.

## *Compatibility*

An important motivation behind the idea of interval constraint spreadsheets is compatibility with traditional spreadsheet computing: cascaded exact value evaluation can be seen as a special case of constraint satisfaction.

In IC Excel, the same spreadsheet can be used simultaneously for both ordinary computations and interval constraint satisfaction. When intervals change into numbers, traditional MS Excel features like advanced graphics become automatically available.

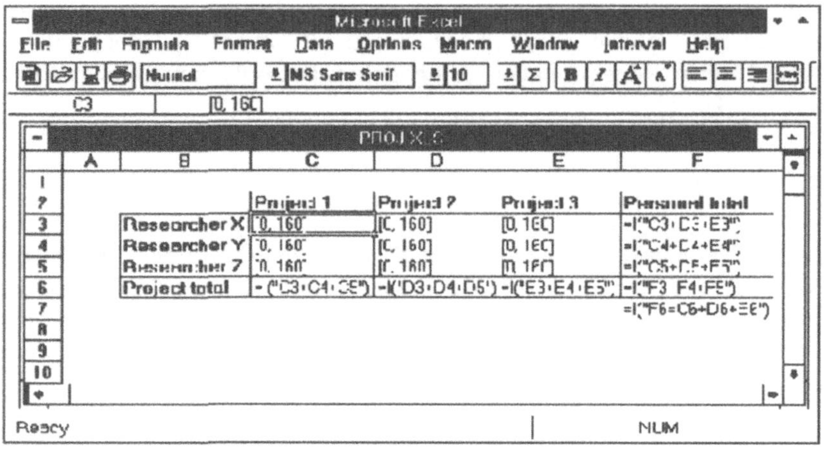


**Figure 5** The resource allocation problem. Initial values and formulas.

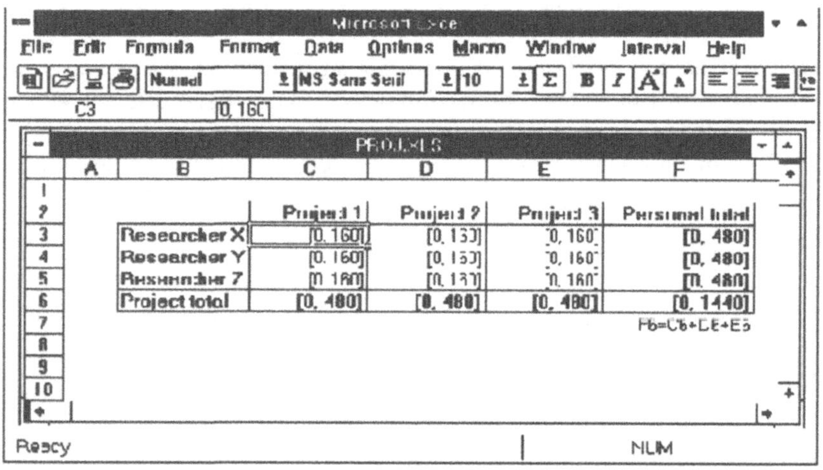


**Figure 6** The system's response to the initial state of figure 4.1. Modified cells are in bold font.

## 4.4 An Example

Consider the resource allocation problem where the task is to allocate the working time of $n$ researchers in $m$ projects during some time period. Let us assume that three researchers $X$, $Y$, and $Z$ are allocated to three projects, as shown

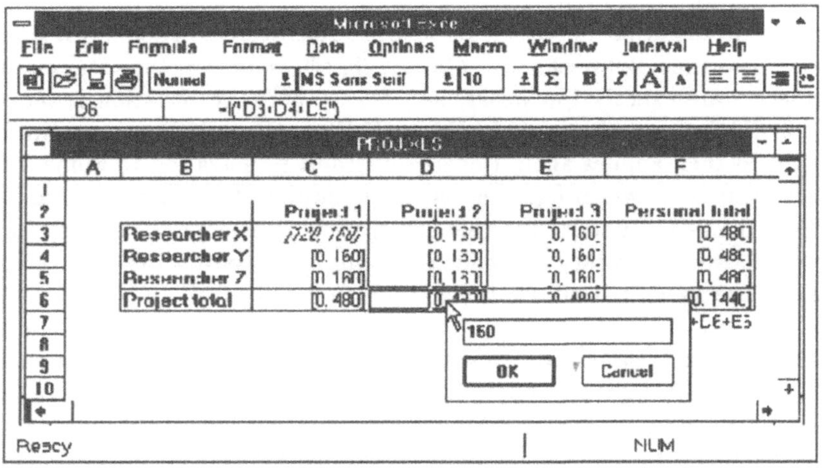


**Figure 7** Changing the total time for the resources of Project 2 to 150 weeks.

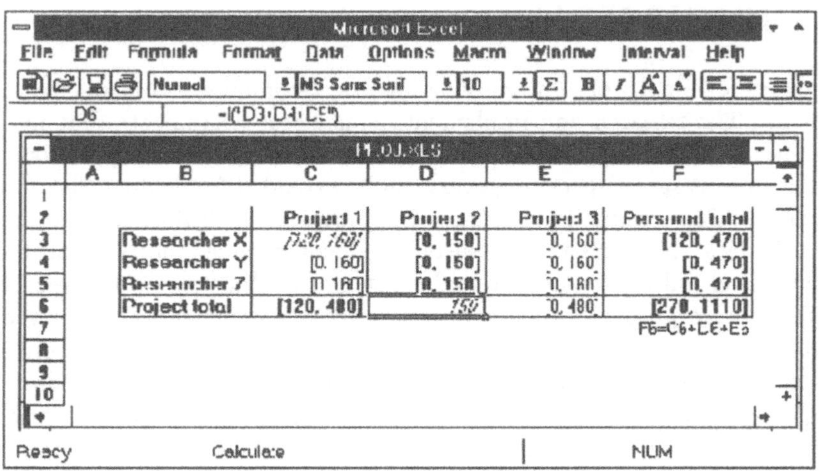


**Figure 8** The user has made two changes (italic font). The system's modifications in response are shown in bold font.

in figure 5. By applying (local) constraint propagation the system gives the response shown in figure 6. The user has now the possibility to constrain the intervals. Assume that he/she demands that researcher $X$ should not work on project 1 less than 120 weeks, i.e., $C3 = [120, 160]$, and that the total resources of project 2 are exactly 150 weeks, i.e., $D6 = 150$. The situation is shown in figure 7. Notice that the formula $D6 = I("D3 + D4 + D5")$ remains in force. The system then applies again (local) constraint propagation to refine values. The response is shown in figure 8, where the modified values are in bold font. Based on the new situation, the user can then further refine the tolerances, the system will infer necessary consequences, and so on until some exact solution satisfying the user's criteria for the solution is found.

In figure 7 the user simultaneously changed one input and one output variable (italic font). Although only two cell values were altered in this arithmetically quite simple system, it was possible to infer eight value modifications. For the user it is difficult to see every consequence of his/her changes immediately. In the general case, it is possible to change any combination of cell values simultaneously. Furthermore, more complicated non–linear algebraic constraints are often present in real–world problems of this type.

# 5 DISCUSSION

This paper reviewed work on applying interval techniques in spreadsheet computations. The research and implementations discussed indicate that interval arithmetic is a promising formalism by which the computational paradigm underlying current spreadsheet programs can be extended for dealing with limitations of machine arithmetic and, more importantly, for dealing with imprecise real world data. Three natural extensions — local extended interval arithmetic, global interval arithmetic and interval constraint arithmetic — for spreadsheet computing were proposed and their experimental implementations embedded in MS Excel were presented (EIA Excel, IF Excel and IC Excel, respectively). Each extension includes the traditional spreadsheet computation mode as its special case. The extensions can be implemented up to a reasonable level by using the ordinary programming features provided by current spreadsheet programs. However, a more seamless and efficient implementation would require that interval data types and techniques were embedded in the kernel of the spreadsheet system.

Local interval arithmetic functions can be embedded fairly easily in a spreadsheet program (EIA Excel). However, two difficult problems are encountered when extending spreadsheets for global interval computations (IF Excel) or for constraint solving (IC Excel). Firstly, from the programmer's view point, control of advanced interval computations cannot be supported by ordinary spreadsheet control mechanisms. The propagation engine must be implemented as a separate module, which brings in problems of maintaining the spreadsheet and the underlying computational system mutually consistent. Secondly, from the end-user's viewpoint, interval spreadsheet computations are time (and memory) consuming. The response times are to a large extent unpredictable depending on the functions to be evaluated and argument values used. Whatever algorithms are used, it is always easy to come up with functions or equations that cannot be evaluated in a reasonable time. Such complexity problems are unknown to ordinary spreadsheets. Hence, it is of vital importance that the user is provided with a versatile set of computational parameters, such as precision levels and the time limit feature, for controlling computations. It is also important that (s)he can get estimates for not only function values (outer bounds) but also for the precision of the results (inner bounds).

More information concerning the InC++ library family and the MS Excel extensions discussed can be found at WWW site

`http://www.vtt.fi/tte/projects/interval/.`

# Acknowledgements

This work has been supported financially by Technology Development Centre of Finland (TEKES). Thanks are also due to Trema Inc. and ViSolutions Inc. for co-operation.

# REFERENCES

[1] O. Aberth, *Precise Numerical Analysis*, Wm. C. Brown, Dubuque, Iowa, 1988.

[2] N. Asaithambi, S. Zuhe, and R. E. Moore, "On computing the range of values", *Computing*, 1982, Vol. 28, pp. 225–237.

[3] A. B. Babichev, O. B. Kadyrowa, T. P. Kashevarova, A. S. Leshchenko, and A. L. Semenov, "UniCalc, A Novel Approach to Solving Systems of Algebraic Equations", *Interval Computations*, 1993, No. 2, pp. 29–47.

[4] F. Benhamou, D. McAllester, and P. Van Hentenryck, *CLP (Intervals)*, Research report, LIM University of Marseille, France, 1994 (Revised).

[5] J. H. Bleher, S. M. Rump, U. Kulisch, M. Metzger, Ch. Ullrich, and W. Walter, "FORTRAN–SC: A Study of a FORTRAN Extension for Engineering / Scientific Computation with Access to ACRITH", *Computing*, 1987, Vol. 39, November 1987, pp. 93–110.

[6] G. Bohlender, L. B. Rall, Ch. Ullrich, and J. Wolff v. Gudenberg, "PASCAL–SC: a Computer Language for Scientific Computation", In: *Perspectives in Computing*, Vol. 17, Academic Press, Orlando, 1987.

[7] H. M. Chen and M. H. van Emden, "Adding Interval Constraints to the Moore–Skelboe Global Optimization Algorithm", In: *Reliable Computing*, Supplement, Proceedings of Application of Interval Computations, El Paso, Texas, 1995, pp. 54–57.

[8] J. C. Cleary, "Logical Arithmetic", *Future Computing Systems*, 1987, Vol. 2, No. 2, pp. 125–149.

[9] J. H. Davenport, Y. Siret, and E. Tournier, *Computer Algebra — Systems and Algorithms for Algebraic Computation*, Academic Press, New York, NY, 1993.

[10] E. Davis, "Constraint Propagation with Interval Labels", *Artificial Intelligence*, 1987, Vol. 32, pp. 99–118.

[11] S. De Pascale, S. and E. Hyvönen, *An Extended Interval Arithmetic Library for MS Excel*, Research report, VTT, Technical Research Centre of Finland, Information Technology, Espoo, 1994.

[12] J. S. Ely, *Prospects for Using Variable Precision Interval Software in C++ for Solving some Contemporary Scientific Problems*, Ph. D. dissertation, The Ohio State University, Ohio, 1990.

[13] E. R. Hansen, *Global Optimization Using Interval Analysis*, Marcel Dekker, New York, 1992.

[14] R. J. Hanson, *Interval Arithmetic as a Closed Arithmetic System on a Computer*, Report 197, Jet Propulsion Laboratory, 1968.

[15] E. Hyvönen, "Constraint Reasoning Based on Interval Arithmetic", In: *Proceedings of IJCAI-89*, Morgan Kaufmann, Los Altos, USA, 1989, pp. 1193–1198.

[16] E. Hyvönen, "Interval Constraint Spreadsheets for Financial Planning", In: *Proceedings of the First International Conference on Artificial Intelligence Applications on Wall Street*, IEEE Press, New York, 1991.

[17] E. Hyvönen, "Constraint Reasoning Based on Interval Arithmetic — The Tolerance Propagation Approach", *Artificial Intelligence*, 1992, Vol. 58, pp. 71–112.

[18] E. Hyvönen, "Spreadsheets Based on Interval Constraint Satisfaction", *Artificial Intelligence for Engineering Design, Analysis and Manufacturing*, 1994, Vol. 8, pp. 27–34.

[19] E. Hyvönen, "Evaluation of Cascaded Interval Function Constraints", In: D. E. Leasure (editor), *Proceedings of International Workshop on Constraint-Based Reasoning at FLAIRS-95,* Texas A&M University-Corpus Christi, Melbourne Beach, Florida, April 1995, pp. 69–75.

[20] E. Hyvönen and S. De Pascale, *LIA InC++: A Local Interval Arithmetic Library*, Technical Report, VTT, Technical Research Centre of Finland, Information Technology, Espoo, 1994a.

[21] E. Hyvönen and S. De Pascale, *GIA InC++: A Global Interval Arithmetic Library*, Technical Report, VTT, Technical Research Centre of Finland, Information Technology, Espoo, 1994b.

[22] E. Hyvönen and S. De Pascale, *CIF InC++: A Library for Cascaded Interval Function Evaluation*, Technical Report, VTT, Technical Research Centre of Finland, Information Technology, Espoo, 1994c.

[23] E. Hyvönen and S. De Pascale, *ICE InC++: A Library for Interval Constraint Equations*, Technical Report, VTT, Technical Research Centre of Finland, Information Technology, Espoo, 1994d.

[24] E. Hyvönen and S. De Pascale, "InC++ Library Family for Interval Computations", In: *Reliable Computing*, Supplement, Proceedings of Application of Interval Computations, El Paso, Texas, 1995, pp. 85–90.

[25] E. Hyvönen and S. De Pascale, "Interval Constraint Spreadsheets: An Implementation for Microsoft Excel", In: *Reliable Computing*, Supplement, Proceedings of Application of Interval Computations, El Paso, Texas, 1995, pp. 91–101.

[26] E. Hyvönen, S. De Pascale, and A. Lehtola, "Interval Constraint Programming in C++", In: B. Mayoh, E. Tyugu, and J. Penham (editors), *Constraint Programming*, Springer–Verlag, Berlin (NATO Advanced Science institute edition, 1994.

[27] W. M. Kahan, *A More Complete Interval Arithmetic*, Lecture Notes for a Summer Course at the University of Michigan, 1968.

[28] R. B. Kearfott, "Decomposition of Arithmetic Expressions to Improve the Behavior of Interval Iteration for Nonlinear Systems", *Computing*, 1991, Vol. 47, No. 2, pp. 169–191.

[29] J. Keiper, "Interval Arithmetic in Mathematica", *Interval Computations*, 1993, No. 3.

[30] R. Klatte, U. Kulisch, C. Lawo, M. Rauch, and A. Wiethoff, *C-XSC, A C++ Class Library for Extending Scientific Computing*, Springer–Verlag, Berlin/Heidelberg/New York, 1993.

[31] M. Konopasek and S. Jayaraman, *The TK!Solver Book*, McGraw–Hill, Berkley, CA, 1984.

[32] V. Kornienko, *Spreadsheets with Subdefinite Data*, Student Diploma Thesis, Novosibirsk State University, 1994 (in Russian).

[33] U. Kulisch and W. Miranker, "The Arithmetic of the Digital Computer: A New Approach", *SIAM Review*, 1986, Vol. 28, No. 1 (March), pp. 1–40.

[34] J. Lee and M. van Emden, *Numerical Computation can be Deduction in CHIP*, Paper, Department of Computer Science, University of Victoria, Canada, 1991.

[35] J. Lee and M. van Emden, *Adapting CLP(R) to Floating Point Arithmetic*, Paper, Department of Computer Science, University of Victoria, Canada, 1991.

[36] Wm. Leler, *Constraint Programming Languages — Their Specification and Generation*, Addison–Wesley, 1988.

[37] O. Lhomme, "Consistency Techniques for Numeric CSPs", In: *Proceedings of IJCAI-93*, Morgan Kaufmann, San Mateo, CA, 1993, pp. 232–238.

[38] R. E. Moore, *Interval Analysis*, Prentice Hall Inc., Englewood Cliffs, N. J., 1966.

[39] R. E. Moore, *Methods and Applications of Interval Analysis*, SIAM Studies in Applied Mathematics, Philadelphia, Pennsylvania, 1979.

[40] R. E. Moore, E. R. Hansen, and A. Leclerc, "Rigorous Methods for Global Optimization", In: *Recent Advances in Global Optimization*, Princeton University Press, Princeton, 1992, pp. 321–342.

[41] W. Older and A. Vellino, "Constraint Arithmetic on Real Intervals", In: F. Benhamou and A. Colmareur (editors), *Constraint Logic Programming*, MIT Press, Cambridge, MA, 1993.

[42] H. Ratschek and J. Rokne, *Computer Methods for the Range of Functions*, Ellis Horwood Limited, Chichester, England, 1984.

[43] H. Ratschek and J. Rokne, *New Computer Methods for Global Optimization*, Ellis Horwood Limited, Chichester, England, 1988.

[44] S. M. Rump, "How Reliable are Results of Computers? / Wie Zuverlässig sind die Ergebnisse Unserer Rechenanlagen?", In: *Jahrbuch Überlicke Mathematik*, Bibliographisches Institut, Mannheim, 1983, pp. 163–168.

[45] G. Sidebottom and W. Havens, "Hierarchical Arc–Consistency for Disjoint Real Intervals in Constraint Logic Programming", *Computational Intelligence*, 1992, Vol. 8, No. 4.

[46] S. Skelboe, "Computation of Rational Interval Functions", *BIT*, 1974, Vol. 14, No. 1, pp. 87–95.

[47] G. Steele, *The definition and Implementation of a Computer Programming Language Based on Constraints*, Dissertation, Massachusetts Institute of Technology, Department of Electrical Engineering and Computer Science, Cambridge, Massachusetts, 1980.

# 9

# SOLVING OPTIMIZATION PROBLEMS WITH HELP OF THE UNICALC SOLVER

**Alexander L. Semenov**

*Novosibirsk Division of the Russian Research Institute of Artificial Intelligence, pr. Lavrent'eva 6, Novosibirsk, Russia, 630090, email semenov@iis.nsk.su*

## 1 INTRODUCTION

The UniCalc solver [3] was designed at the Russian Institute of Artificial Intelligence to solve systems of nonlinear equations and inequalities with possibly inexact data. UniCalc uses the computation techniques based on ***the subdefinite computations method*** [13], which can be regarded as an analogue of constraint propagation with interval labels [4]. To implement this method, algorithms of interval mathematics are used, so UniCalc can also be used to solve interval problems. Another feature of UniCalc is its ability to solve different integer problems and problems with mixed data types (integers and reals). The efficiency of UniCalc is confirmed by the results of its successful applications to many problems, from benchmark (test) problems (see, e.g., [8, 10]) to real-life problems. Some optimization problems [11] have also been successfully solved by these techniques.

Optimization problems are very important in practice, so solving these problems attracts considerable interest, and many different methods exist to solve them [14]. To solve problems of global optimization, methods of interval analysis are often used [5]. UniCalc doesn't contain special methods for solving optimization problems, but the use of interval mathematics along with the original computational techniques of UniCalc enables the user to solve a wide set of different optimization problems: integer and real, local and global, with constraints and unconstrained. In the following text, we will consider some methods that we have successfully used to solve optimization problems.

The general plan of the paper is as follows. Section 2 describes the main features of the UniCalc solver. A brief overview of the method of subdefinite



*R. B. Kearfott and V. Kreinovich (eds.), Applications of Interval Computations*, 211–225.

computations and of some implementation issues is given in Section 3. In the next two sections we consider solving optimization problems: Section 4 describes integer programming problems and Section 5 is devoted to real-valued optimization. The directions for future research are given in Section 6.

# 2 THE UNICALC SOLVER

The UniCalc solver [3] was designed to solve arbitrary systems of algebraic and algebraic-differential relations. Here, by a *relation*, we mean an equation, inequality or a logical expression. According to the algorithm used, the system to be solved can be overdetermined or underdetermined, and the system's parameters (coefficients, variables, initial conditions for the Cauchy problem) can be imprecise and expressed as intervals. Such a system may contain only integer and real variables, or it may combine both integer and real variables.

As a result of solving algebraic systems, we either find a parallelepiped that contains all solutions of the system, or a message about the system's incompatibility is issued. If the system has a unique solution, then the parallelepiped will be reduced to a point (with a given accuracy). If the system has several solutions, then in order to locate all of them, it is necessary either to add the appropriate auxiliary relations, or to use the built-in tool for automatic root locating.

The solver is an integrated environment that enables the user to input the system to be solved, to modify this system (if necessary), to specify the desired accuracy of the solution, to actually perform the computations, and to view the results. To input and modify the system, UniCalc has a built-in text editor. The user can describe the input problems in a special source language that is very close to the commonly used mathematical notations. All problems are processed with the above algorithm. To translate the source input from the user-understandable source language into the language understandable for the computer, the solver includes a compiler and two pre-processors: a pre-processor for symbolic transformations [17], and a pre-processor for solving the systems of algebraic-differential relations [7].

The user interface of UniCalc offers a number of services to support problem solving. If the solution of a problem requires large computation time, it is possible to suspend calculations to see the intermediate results, and, depending on convergence rate, either to continue computation or to stop it.

The built-in root locating tool is used to separate roots and to narrow a box containing exactly one solution. This tool uses the following "bisection" process: it splits the resulting box in one of its dimensions (for the variable selected by the user — this variable may be either an independent variable or a function name), and repeat the basic algorithm for the sub-boxes. This process is continued recursively until either the width of the interval for the separated variable is smaller than the computation accuracy, or no more roots are found in this manner. After that, it splits the box in other dimensions as well until the width of the interval for each variable is smaller than the computation accuracy.

At present, UniCalc is implemented for PC-compatible computers running MS Windows or MS DOS with at least 640K of RAM, and for Sun workstations.

# 3 THE ALGORITHM OF SUBDEFINITE CALCULATIONS

In this section, we will give a brief description of the algorithm of subdefinite calculations and consider an example of the computations according to the algorithm. A more general and detailed description of the algorithm can be found in [13, 16].

## 3.1 Brief Description and Definitions

We consider a system of $m$ algebraic equations

$$F(\vec{x}) = 0, \tag{9.1}$$

where $\vec{x} = (x_1, \ldots, x_n)$ is a vector of real variables.

Let us consider the $i$-th equation of the system. In the general case it doesn't contain all $n$ variables. For notational convenience, we assume that the variables that occur in $i$–th equations are are the first $n_i, n_i \leq n$. In this case, we can write this equation in the following form:

$$f_i(x_1, \ldots, x_{n_i}) = 0.$$

We replace this equation by $n_i$ auxiliary ones: Namely, for each of the $n_i$ variables, we define an equation where this variable is expressed as a function of the other $n_i - 1$ variables. Thus we will have the following $n_i$ equations (their right hand sides will be called ***interpretation functions***):

$$\begin{aligned} x_1 &= f_1^{(i)}(x_2, x_3, \ldots, x_{n_i}) \\ \ldots \\ x_j &= f_j^{(i)}(x_1, x_2, \ldots, x_{j-1}, x_{j+1}, \ldots, x_{n_i}) \qquad (9.2) \\ \ldots \\ x_{n_i} &= f_{n_i}^{(i)}(x_1, x_2 \ldots, x_{n_i - 1}) \end{aligned}$$

Thus, if we know the values of any $n_i - 1$ variables from the set $x_1, x_2, \ldots, x_{n_i}$, we are able to find the value of the remaining variable. After we represent all $m$ equations of our system in such a form, the system (9.1) is replaced by the system of $k = n_1 + n_2 + \ldots + n_m$ equations.

In the beginning, we do not know the values of the variables, so, we start with assuming that these variables can take all possible values, and then, remove the values which don't satisfy the system. To define such computation process, we will give some definitions and denotations.

Let $A$ denote the domain of each of the variables $x_1, \ldots, x_n$. Here, $A$ is either $\mathbb{R}$ (the field of real numbers), or $\mathbf{Z}$ (the ring of integer numbers), or a subset of one these sets. Let us denote the set of all nonempty subsets of $A$ by $^*A$ and consider variables $^*x_1, \ldots, {}^*x_n$ whose domain is $^*A$. In other words, the values of these variables are subsets of the set $A$. We will call values from $^*A$ ***subdefinite values***, and the variables $^*x_1, \ldots, {}^*x_n$ ***subdefinite variables.***

On each step of our algorithm, the value of the variable $^*x_i$ will consist of all values $a \in A$ that are still considered possible values of $x_i$ for a solution $x$. In accordance with this idea, as the initial value $^*x_j^{(0)}$ of each variable, we take $^*x_j^{(0)} = A$. To describe how we change these values, we need to extend functions defined for variables from $X$ (in particular, arithmetic operations) to functions and operations with subdefinite variables:

The subdefinite extension $^*f$ of a function $f$ of $t$ variables is:

$$^*f(^*x_1, \ldots, {}^*x_t) = \{r = f(x_1, \ldots, x_t) \mid x_1 \in {}^*x_1, \ldots, x_t \in {}^*x_t\}.$$

In particular, if $\odot$ be an arithmetic operation, then its extension $^*\odot$ is:

$$^*a\,^*\!\odot\,^*b = \{r = a \odot b \mid a \in {}^*a, b \in {}^*b; {}^*a, {}^*b \in {}^*A\} \qquad (9.3)$$

Let us assume that on a certain ($k$-th) stage of the algorithm, the values of the subdefinite variables $^*x_1, \ldots, {}^*x_n$ are $^*x_1^{(k)}, \ldots, {}^*x_n^{(k)}$. This means that we know that for every solution $x = (x_1, ..., x_n)$, we have $x_i \in {}^*x_i^{(k)}$ for all $i$. ¿From $j$-th equation (9.2), we can now conclude that

$$x_j \in {}^*f_j^{(i)}({}^*x_1{}^{(k)}, {}^*x_2{}^{(k)}, \ldots, {}^*x_{j-1}{}^{(k)}, {}^*x_{j+1}{}^{(k)}, \ldots, {}^*x_{n_i}{}^{(k)}).$$

Since we also know that $x_j \in {}^*x_j^{(k)}$, we can conclude that $x_j$ belongs to the intersection:

$$x_j \in {}^*f_j^{(i)}({}^*x_1{}^{(k)}, {}^*x_2{}^{(k)}, \ldots, {}^*x_{j-1}{}^{(k)}, {}^*x_{j+1}{}^{(k)}, \ldots, {}^*x_{n_i}{}^{(k)}) \cap {}^*x_j{}^{(k)}.$$

Therefore, we can take this intersection as a new value ${}^*x_j^{(k+1)}$ of the subdefinite variable ${}^*x_j$:

$${}^*x_j^{(k+1)} := {}^*f_j^{(i)}({}^*x_1{}^{(k)}, {}^*x_2{}^{(k)}, \ldots, {}^*x_{j-1}{}^{(k)}, {}^*x_{j+1}{}^{(k)}, \ldots, {}^*x_{n_i}{}^{(k)}) \cap {}^*x_j{}^{(k)}.$$

The algorithm is as follows:

1. First, we start with $k = 0$ (no iterations done yet), and with the initial value ${}^*x_j^{(0)} = A$ of the subdefinite variable ${}^*x_j$. We apply apply all $n_1 + n_2 + \ldots + n_m$ functions ${}^*f_j^{(i)}$ to compute the new values ${}^*x_j^{(1)}$ of the variables ${}^*x_j$.
2. If the value of any variable ${}^*x_j$ has changed after the execution of one of the functions (i.e., if ${}^*x_j^{(k+1)} \neq {}^*x_j^{(k)}$), then we mark all functions $f_j^{(i)}$ that have this variable as an argument.
3. If the set of marked functions is empty then stop.
4. Execute an arbitrary function from the set of marked functions (thus creating the new value ${}^*x_j^{(k+1)}$ of the variable ${}^*x_j$), and go to Step 2.

The set of marked functions is called an ***active function set*** and a function which is currently being executed is called a ***working function***.

## 3.2 An Example with Integer Variables

Let the system consist of the following two two equations:

$$8a - b = 15, \; 6b + 4c = 230,$$

and suppose that it is known that the three unknowns $a$, $b$, and $c$ ($x_1$, $x_2$, and $x_3$ in the previous notation) are integers in the range from 1 to 100. Hence, the initial subdefinite value of each of these variables is

$$^*a^{(0)} = {}^*b^{(0)} = {}^*c^{(0)} = A = \{1, ..., 100\}.$$

The system (9.2) has in this case the following form:

$$\begin{aligned} f_1^{(1)} &: \; a := (b+15)/8; \\ f_2^{(1)} &: \; b := 8a - 15; \\ f_2^{(2)} &: \; b := (230 - 4c)/6; \\ f_3^{(2)} &: \; c := (230 - 6b)/4. \end{aligned}$$

Introducing the subdefinite description, we have:

$$\begin{aligned} {}^*f_1^{(1)} &: \; {}^*a := \{a = (b+15)/8 \mid b \in {}^*b\} \cap {}^*a; \\ {}^*f_2^{(1)} &: \; {}^*b := \{b = 8a - 15 \mid a \in {}^*a\} \cap {}^*b; \\ {}^*f_2^{(2)} &: \; {}^*b := \{b = (230 - 4c)/6 \mid c \in {}^*c\} \cap {}^*b; \\ {}^*f_3^{(2)} &: \; {}^*c := \{c = (230 - 6b)/4 \mid b \in {}^*b\} \cap {}^*c. \end{aligned}$$

Table 1 illustrates the algorithm's execution.

## 3.3 Some Implementation Issues

The implementation of the algorithm just described assumes the choice of a set $A$ corresponding to the class of models; in addition, acceptable operations and functions over values in $A$ and variables in $X$ are to be extended for objects in $^*A$ and $^*X$. A special technique called ***active data type technique*** has been proposed [12] and used to implement various types of subdefinite data (integers, reals, boolean values, sets, planar geometric metric objects, etc.) [19].

We consider only models over real numbers; therefore, the set $A$ is either the field of real numbers $\mathbb{R}$, or the ring of integer numbers $\mathbf{Z}$, or a subset of

| Subdefinite values | Working function | Active functions |
|---|---|---|
| $^*a = \{2,3,\ldots,14\}$ | $^*f_1^{(1)}$ | $^*f_1^{(1)}, {}^*f_2^{(1)}, {}^*f_2^{(2)}, {}^*f_3^{(2)}$ |
| $^*b = \{1,9,\ldots,97\}$ | $^*f_2^{(1)}$ | $^*f_2^{(1)}, {}^*f_2^{(2)}, {}^*f_3^{(2)}$ |
| $^*b = \{1,9,17,25,33\}$ | $^*f_2^{(2)}$ | $^*f_2^{(2)}, {}^*f_3^{(2)}, {}^*f_1^{(1)}$ |
| $^*c = \{8,20,32,44,56\}$ | $^*f_3^{(2)}$ | $^*f_3^{(2)}, {}^*f_1^{(1)}, {}^*f_2^{(2)}$ |
| $^*a = \{2,3,4,5,6\}$ | $^*f_1^{(1)}$ | $^*f_1^{(1)}, {}^*f_2^{(1)}, {}^*f_2^{(2)}$ |
| $^*b = \{1,9,17,25,33\}$ | $^*f_2^{(2)}$ | $^*f_2^{(2)}, {}^*f_2^{(1)}$ |
| $^*b = \{1,9,17,25,33\}$ | $^*f_2^{(1)}$ | $^*f_2^{(1)}$ |

**Table 1** The execution of the integer variable example.

one of these sets. We assume that all the arithmetic operations as well as exponentiation and root extraction are defined over values in the set $A$. The set of acceptable functions includes exponential, logarithmic and all trigonometric functions. Since in any given computer, only finitely many numbers can be represented, inside the computer, both sets $A = \mathbb{R}$ and $A = \mathbb{Z}$ will be actually represented by finite intervals which are bounded by some numbers $MinA$ and $MaxA$ playing the role of $-\infty$ and $+\infty$, respectively (these numbers differ for the real and integer sets $A$). With these bounds in mind, the computer-implemented operations on elements of $A$ are defined as follows: if $a, b \in A$ and $\odot$ is a operator, then

$$\begin{aligned} a \odot b &= \max(a \odot b, MinA) \text{ if } a \odot b < 0 \\ a \odot b &= \min(a \odot b, MaxA) \text{ if } a \odot b \geq 0 \end{aligned}$$

Here, $a \odot b$ is the right-hand side means the (ideal) result of applying a mathematical operation $\odot$ to the numbers $a$ and $b$, and $a \odot b$ in the left-hand side means the actual number produced by the computer-implemented operation $\odot$ for inputs $a$ and $b$.

For $A = \mathbb{R}$ and $A = \mathbb{Z}$, it is natural to take as $^*A$ the set of all intervals from the set $A$; using formula (9.3), and applying the appropriate operations of interval mathematics [1], we can extend all operations on elements of $A$ to operations on $^*A$.

Due to the possibility of using intervals with "infinite" limits, we can extend interval expansion to discontinuous functions, assuming, for example, that $1/0 = [0, MaxA]$ and $1/[-1,1] = [MinA, MaxA]$. Such an approach makes it possible to avoid the divide-by-zero situation that is fatal for other methods.

Besides the above-mentioned divide-by-zero problem, some other implementation problems have been solved. They are mostly due to violation of uniqueness in computing some functions of interval parameters (for instance, for the inverse trigonometric functions or for even power roots) which may require using multi-intervals (they are under implementation now [18]). The violation of uniqueness in the course of calculations is a rather common problem, since for each relation we have to calculate the interpretation functions for each of its variables. To overcome this difficulty, the intervals of function monotonicity are extracted to calculate the function values, and the general solution is constructed as the union of all solution intervals obtained. Suppose, for example, we have the following model:

$$x^2 - a = 0$$
$$4 \leq a \leq 25$$

The interpretation function $x := \sqrt{a}$ yields two intervals $x_1 = [-5, -2]$ and $x_2 = [2, 5]$, and the interval $x = [-5, 5]$ will be returned as the result (because it is the smallest interval that contains th union of the two solution intervals). If the relation $x > 0$ is added to the system, the resulting interval will be $x = [2, 5]$.

Another problem of interval calculations worth mentioning is the problem of undefined values. These values appear, for instance, when we evaluate expressions like $x/y$ or $x^y$, and $x$ and $y$ are intervals containing zero. In our solver, the method for handling such situations is chosen by the user, who can specify whether the presence of such arguments in these expressions is considered contradictory or not. The latter case requires the user to specify what should be considered to bc the result of the expression: the fully undefined value $[MinA, MaxA]$, or the value defined by continuity.

## 3.4 Representation of Models

The most appropriate data structure for the algorithm described above is the network representation of systems. A network is represented as a bipartite oriented graph with two types of vertices: objects and operators. Objects represent variables, and operators represent interpretation functions. The outgoing edges from an object point to those operators which have this object as an argument, and the incoming edges specify the operators which change this object. The network is associated with a discipline of its execution, which substantiates the inference algorithm. This substantiation depends on the computer's architecture, on its computational capabilities, as well as on a number of other factors. For instance, interpretation functions may be selected at step 4 according to certain priorities, from a queue or randomly, and they may be computed sequentially or in parallel. In the current implementation of the processor, the interpretation functions are selected depending on their ordinal number and are computed sequentially. If one uses computers designed for parallel processing (e.g., transputers), the strategy of network execution may be changed.

# 4 SOLVING INTEGER PROGRAMMING PROBLEMS

The possibility to solve problems of this class is the result of UniCalc's ability to solve integer equations and inequalities, and the existence of the root locating tool.

UniCalc can solve both linear and nonlinear integer problems. We considered only linear problems because, first, there are many different benchmark linear integer programming problems, and, second, our experience shows that many linear problems appear to be more difficult for UniCalc than many nonlinear problems.

We consider integer problems in the following form. We want to find extrema of the function

$$f(x) = cx$$

under the following constraints:

$$Ax \leq b, \quad x \geq 0, \tag{9.4}$$

where $A$ is $m \times n$ matrix, $c = (c_1, \ldots, c_n)$, $b = (b_1, \ldots, b_m)^T$, $x = (x_1, \ldots, x_n)^T$, $x \in \mathbf{Z}^n$.

To solve such a problem, we, first, use UniCalc to find the interval $[f_-, f_+]$ that contains the possible values of the objective function. This interval is thus guaranteed to contain the desired extremum. Then, we apply bisection to narrow this interval. Let us show how we can do that for the case when we are interested in maximizing $f(x)$ (the case of min is treated similarly). If at some step $k$, we have an interval $[f_-^{(k)}, f_+^{(k)}]$ that contains the desired maximum, then we divide this interval into two halves $[f_i^{(k)}, (f_i^{(k)} + f_+^{(k)})/2]$ and $[(f_i^{(k)} + f_+^{(k)})/2, f_+^{(k)}]$. We then apply UniCalc to find out if there exists an integer vector $x$ that satisfies the condition (9.4), and for which, $f(x)$ belongs to the upper half-interval (i.e., for which $(f_-^{(k)} + f_+^{(k)})/2 \le f(x) \le f_+^{(k)}$).

- We already know that the conditional maximum of $f(x)$ belongs to the interval $[f_-^{(k)}, f_+^{(k)}]$. This means, in particular, that if $x$ satisfies (9.4), then $f(x) \le f_+^{(k)}$. Therefore, if this auxiliary system does not have a solution, then no $x$ that satisfy (9.4) can belong to the upper half-interval, and therefore, the maximum of the function belongs to the lower half-interval. We can thus take the lower half-interval as a new enclosure $[f_-^{(k+1)}, f_+^{(k+1)}]$ for the maximum.

- If the auxiliary system has a solution, this means that there is a vector $x$ that satifies (9.4), and for which $f(x) \ge (f_-^{(k)} + f_+^{(k)})/2$, and hence, the conditional maximum of $f(x)$ is also $\ge (f_-^{(k)} + f_+^{(k)})/2$. Therefore, in this case, the upper half-interval contains the maximum and can be thus taken as a new enclosure.

In both cases, we have a new interval of half size. After we repeat this bisection process sufficiently many times, we will get a pretty good estimate for the optimum. At each step, UniCalc, applied to the corresponding auxiliary problem, also generates the intervals of possible values of $x_i$ for which the value of $f(x)$ is close to the maximum. These intervals thus contain the vector in which the conditional maximum is actually attained.

Our experiments showed that UniCalc can successfully solve problems of this class, including problems with Boolean variables, for example, the "knapsack" problem. (To test UniCalc's ability to solve integer programming problems, we applied UniCalc to benchmark problems from [15].) However, increasing the size of the systems causes the increase in UniCalc's computation time, and sometimes, this time becomes very large. To solve problems of a large size $n$, we intend to combine our algorithm with special methods of solving integer programming problems, for example, the simplex method or cutting methods.

Let us consider an example of solving the multidimensional knapsack problem [15] with help of UniCalc. Suppose we are to find maxima of the function:

$$f(x) = 100x_1 + 600x_2 + 1200x_3 + 2400x_4 + 400x_5 + 2000x_6$$

with the following constraints:

$$\begin{aligned}
8x_1 + 12x_2 + 13x_3 + 64x_4 + 22x_5 + 41x_6 &\leq 80; \\
8x_1 + 12x_2 + 13x_3 + 75x_4 + 22x_5 + 41x_6 &\leq 96; \\
3x_1 + 6x_2 + 4x_3 + 18x_4 + 6x_5 + 4x_6 &\leq 20; \\
5x_1 + 10x_2 + 8x_3 + 32x_4 + 6x_5 + 12x_6 &\leq 36; \\
5x_1 + 13x_2 + 8x_3 + 42x_4 + 6x_5 + 20x_6 &\leq 44; \\
5x_1 + 13x_2 + 8x_3 + 48x_4 + 6x_5 + 20x_6 &\leq 48; \\
8x_5 &\leq 10; \\
3x_1 + 4x_3 + 8x_5 &\leq 18; \\
3x_1 + 2x_2 + 4x_3 + 8x_5 + 4x_6 &\leq 22; \\
3x_1 + 2x_2 + 4x_3 + 8x_4 + 8x_5 + 4x_6 &\leq 24;
\end{aligned}$$

Writing this system in the UniCalc language and adding the conditions for the variables to be Boolean (since we consider intervals as $^*A$, these conditions are $x_i := [0, 1]$), we will find an interval for the function values: $[0, 6700]$. After running the above-described bisection (root locating) process over function values from right-to-left (we are seeking a maximum) we get the solution:

$$f_{max} = 3800, \quad x = (0, 1, 1, 0, 0, 1).$$

If some constraints from (9.4) are equations, then we can use the tool for symbolic solution of systems of linear equations (that is a built-in part of UniCalc) to reduce computation time. Using this tool, we can express some variables in terms of the others, substitute the resulting expressions into the original system and into the objective function, and, as a result, simplify the problem by reducing its number of variables. Our experiments have shown that for systems of large size $n$, this approach can significantly decrease computation time.

# 5 REAL-VALUED OPTIMIZATION

## 5.1 Unconstrained Optimization Problems

UniCalc can solve both global and local optimization problems. To find global extrema of a smooth function, one can use an algorithm from [2] that finds the exact bounds of interval expressions (this algorithm is implemented in UniCalc). After finding the exact bounds of the objective function, we can find $x$ by applying UniCalc to the condition that the objective function $f(x)$ belongs to a small interval containing the corresponding bound. If, as a result of this computation, we get several different roots, then we can apply the root locating tool to separate different extremum points.

Another way to find internal extrema (global and local) of a smooth function is to use the known necessary condition for a point $x$ to be a extremum: the first order partial derivatives of the objective function must equal to zero, and the diagonal elements of the Hessian matrix (that contains derivatives of second order) must have appropriate signs. So, to find local extrema, we can formulate these conditions as equations (for the first derivatives) and inequalities (for the second derivatives), apply UniCalc to the resulting system, and thus find the intervals for which the desired conditions hold. If necessary, we can then separate extremum points by using the root locating process.

If we are interested in the global extrema, then we must add the condition that we have already used in the first global optimization algorithm to the conditions described above: that $f(x)$ belongs to the small interval around the corresponding bound.

For example, let the following function [6] be given:

$$f(x_1, x_2) = (x_1^2 + x_2 - 11)^2 + (x_1 + x_2^2 - 7)^2 + 3.$$

We wish to find its global minimum.

In order to find this, we supplement the function equation with the following equations and inequalities for the first and second order derivatives:

$$\begin{aligned} 4x_1 \cdot (x_1^2 + x_2 - 11) + 2x_1 + 2x_2^2 - 14 &= 0 \\ 2x_1^2 + 2x_2 + 4x_2 \cdot (x_1 + x_2^2 - 7) - 22 &= 0 \\ 12x_1^2 + 4 \cdot x_2 - 42 &> 0 \\ 4x_1 + 12x_2^2 - 26 &> 0 \end{aligned}$$

As a result of the calculations we find that the function values are in the interval $[3, M]$, where $M = Max\mathbb{R}$ is the largest real number that can be represented on the given computer. Therefore, we may suggest that the global minimum is in the interval $[3, 4]$. If we put this limitation on the function values, we will find the corresponding intervals for the variables:

$$x_1 = [-3.9312, 3.9312], \; x_2 = [-3.4541, 3.4542].$$

If the global minimum were not in this interval, the algorithm would terminate with a message about the system's incompatibility. Applying the root locating process yields four minimizing points:

$$\begin{array}{llllll} 1) & x_1 = & -3.7793, & x_2 & = & -3.2832, \\ 2) & x_1 = & -2.8051, & x_2 & = & -3.1313, \\ 3) & x_1 = & 3, & x_2 & = & 2, \\ 4) & x_1 = & 3.5844, & x_2 & = & -1.8481. \end{array}$$

## 5.2 Constrained Optimization Problems

The approach described above is not directly applicable to constrained optimization problems. However, UniCalc can still be applied. Namely, if all the constraints are inequalities, then we can apply UniCalc to solve the given system of constraints, and thus compute boxes for variables for which these constraints are satisfied. Then, we can use these boxes to estimate the intervals of possible values of the objective function. Taking the union of the resulting intervals, we get the bounds of the objective function on the set of all $x$ that satisfy the given constraints. Then, we can find the desired conditional extremum points $x$ by applying UniCalc to the condition that $f(x)$ belongs to a small interval centered in the resulting bound (if there are several such $x$, we have to apply the root locating process). If the problem also has equality constraints, then we can use symbolic variable substitutions to reduce the dimension of the problem (in particular, if the problem only has equality constraints, we thus reduce it to an unconstrained optimization problem).

Our experiments showed that this approach does solve constrained optimization problems; in some cases, however, it took substantial computation time.

# 6 FUTURE DEVELOPMENTS

Optimization problems are very important in practice, so it is necessary to have a solver for these problems. UniCalc is able to solve different optimization problems, but for some of them, the required computation time is too long. To speed up computations, we plan to combine our approach with the existing interval-based optimization techniques [5], and especially with parallel interval techniques [9]. Our goal is to design a solver that combines UniCalc's algorithm and user-friendly interface with other optimization techniques.

Another important directions of our future research is parallelizing the root locating process. Sometimes, this process takes much computation time, because we must consider many sub-boxes. To increase its efficiency, we can run computations with different sub-boxes on different processors in parallel (removing sub-boxes that contain no roots). This parallelization can drastically speed up the process. We are planning to perform experiments with different parallelization strategies, and choose the best strategy.

# REFERENCES

[1] G. Alefeld and Ju. Herzberger, *Introduction in Interval Computations*, Academic Press, N. Y., 1983.

[2] N. S. Asaithambi, Shen Zuhe, and R. E. Moore, "On computing the range of values", *Computing*, 1982, Vol. 28, pp. 225–237.

[3] A. B. Babichev, O. B. Kadyrova, T. P. Kashevarova, A. S. Leshchenko, and A. L. Semenov, "UniCalc, A Novel Approach to Solving Systems of Algebraic Equations", *Proceedings of the International Conference on Numerical Analysis with Automatic Result Verifications. Lafayette, Louisiana, USA, February - March, 1993. Interval Computations*, 1993, No. 2, pp. 29–47.

[4] E. Davis, "Constraint propagation with interval labels", *Artificial Intelligence*, 1987, Vol. 32, pp. 99–118.

[5] E. Hansen, *Global Optimization Using Interval Analysis*, Marcel Dekker Inc., N. Y., 1992.

[6] H. K. Jacobsen and H.C. Pedersen, *An optimization system for teaching purposes*, Eksamenproject, IMSOR, DTH, Lyngby, 1976.

[7] T. P. Kashevarova and A. L. Semenov, "Solving Subdefinite Problems for Systems of Ordinary Differential Equations of the First Order in the UniCalc System", *All-Union Scientific-Technical Conference 'Intelligent Systems in Machine Building', Abstracts,* Samara, 1991, pp. 21–24.

[8] R. B. Kearfott, "Some Tests of Generalized Bisection", *ACM Transactions on Mathematical Software*, 1987, No. 3, pp. 197–220.

[9] K. Madsen, *Parallel Algorithms for Global Optimization* Report 91–07, Institute for Numerical Analysis, The Technical University of Denmark, Lyngby, Denmark, 1991.

[10] K. Meintjes and A. P. Morgan, "Chemical Equilibrium Systems as Numerical Test Problems", *ACM Transactions on Mathematical Software*, 1990, No. 2, pp. 143–151.

[11] J. J. More, B. S. Garbow, and K. E. Hillstrom, "Testing Unconstrained Optimization Software", *ACM Transactions on Mathematical Software*, 1981, No. 1, pp. 17–41.

[12] A. S. Narin'yani, "Active data types for representing and processing of subdefinite information", In: *Actual Problems of the Computer Architecture Development and Computer System Software*, Novosibirsk, USSR Acad. of Sciences, Siberian Division, Computer Center, 1983, pp. 128–141.

[13] A. S. Narin'yani, "Subdefiniteness in knowledge representation and processing systems", *Transactions of USSR Acad. of Sciences, Technical Cybernetics*, 1986, No. 5, pp. 3–28.

[14] G. L. Nemhauser, A. H. G. Rinnooy Kan, and M. J. Todd (eds.), *Handbooks in Operational Research and Management Science, Vol. 1. Optimization*, North-Holland, 1989.

[15] C. C. Petersen, "Computational experience with variants of the Balas algorithm applied to the selection of R&D Projects", *Management Science*, 1967, Vol. 13, No. 9, pp. 736–750.

[16] A. L. Semenov, *Solving Integer/Real Nonlinear Equations by Constraint Propagation*, Technical Report No. 12, Institute of Mathematical Modelling, The Technical University of Denmark, Lyngby, Denmark, 1994.

[17] A. L. Semenov and A. S. Leshchenko, "Interval and Symbolic Computations in the UniCalc Solver", *International Conference on Interval and Computer-Algebraic Methods in Science and Engineering INTERVAL-94, Abstracts*, St-Petersburg, Russia, March, 1994, pp. 206–208.

[18] A. L. Semenov and D. V. Petunin, "The Use of Multiintervals in the UniCalc Solver", *The International Symposium on Scientific Computing, Computer Arithmetic and Validated Numerics SCAN-95. Wuppertal, Germany, September, 1995* (to appear).

[19] V. V. Telerman, *Active data types*, Preprint 792, USSR Acad. of Sciences, Siberian Division, Computer Center, Novosibirsk, 1988.

# 10

# AUTOMATICALLY VERIFIED ARITHMETIC ON PROBABILITY DISTRIBUTIONS AND INTERVALS

**Daniel Berleant**

*Dept. of Computer Systems Engineering, University of Arkansas, Fayetteville, AR 72701, email: djb@engr.uark.edu*

## ABSTRACT

In this chapter we address two related problems:

1. Representing and operating on operands which are probability distribution functions; and
2. Representing and operating on operands when one is a distribution function and the other is an interval.

We discuss how to do these operations and get automatically verified results, using a method based on histograms. Histograms are composed of bars, each with an associated interval and probability mass. A histogram discretizes a probability distribution function in that each bar represents a portion of it, with the area of the bar equal to the area of the distribution function over that part of its domain corresponding to the interval-valued portion of the $x$-axis under the bar.

## 1 INTRODUCTION

The earliest use of histograms for representing and operating on distribution functions we have found appeared in 1968 (Ingram et al. [8]). That method was further developed by Colombo and Jaarsma in 1980 [4] and subsequently generated attention mostly in reliability analysis [9, 1, 5, 16] although the technique

*For completeness in presentation, this chapter contains some passages similar in content to passages appearing in [2].



*R. B. Kearfott and V. Kreinovich (eds.), Applications of Interval Computations, 227–244.*

itself is a general one. The method used was not automatically verifying. The method also served as a foundation for Kaplan's (1981 [10]) popular variation, which has generated over 50 citations in Science Citation Index over the years. However it is unclear how to further develop Kaplan's method into one which is automatically verifying. Moore (1984 [14]) independently developed another variation in which results are expressed as cumulative distribution functions (CDFs), which (because they are the integrals of probability density functions) begin at zero and rise to one over a typically S-shaped curve. R. E. Moore (1984 [14]) and A. S. Moore (1985 [13]) address non-trivial applications, but do not address automatic verification. A more recent variation on histogram arithmetic is described by Gerasimov et al. (1991 [6]).

The approximating nature of the method of Ingram et al. and Colombo and Jaarsma is due to assumptions about how the mass of any given bar of the histogram is distributed over the interval-valued portion of the $x$-axis under the bar. For example, because a bar is depicted graphically with a flat top one might assume that the associated mass is distributed evenly over the interval for that bar.

We have developed an automatically verified way of operating on histograms by modifying the original histogram based approach of Ingram et al. and Colombo and Jaarsma. In the next section, we describe how the discretization imposed by a histogram may itself be automatically verified in that the discretization weakens the statement made by the probability distribution function (PDF), retaining correctness, rather than simply replacing the PDF with strong but approximate statement. Then, we describe how arithmetic operations on histograms can be automatically verifying. As a significant side benefit, automatically verified operations can be made on operands when one is a PDF and the other an interval, because both PDFs and intervals can be represented as histograms and arithmetic operations are done on histograms rather than on PDFs or intervals *per se.* While others have developed various methods for operating on PDFs such as using moments and defining special classes of PDFs, we are not aware of methods that work well when one of the operands happens to be an interval instead of a PDF.

# 2 CORRECTLY REPRESENTING PDFS AND INTERVALS WITH HISTOGRAMS

Histograms can be used to correctly represent both probability distribution functions (PDFs) and intervals. We take an interval $I$ to be an incompletely specified statement about value. There is a probability of one that the value falls within the interval. Hence, an interval $I$ defines a class of PDFs consisting of all PDFs $f(x)$ for which $f(x) = 0$ for $x \notin I$.

Let us examine first how to represent a PDF correctly using a histogram, and then how to represent an interval correctly using a histogram.

## 2.1 Correctly Representing a PDF with a Histogram

A histogram discretizes the underlying PDF. The discretization may be viewed as introducing information loss rather than error. To see this, consider a histogram discretization of a PDF to divide the PDF's domain on the $x$-axis into a set of intervals, each associated with a probability mass equal to the integral (area) of the PDF over that interval on the $x$-axis. The graphical depiction of a histogram is then simply a convention for describing this set of intervals and their associated masses, a set which we will call an *intermediate distribution.*

A histogram graphic (Figure 1, top) is user friendly but has the disadvantage that a histogram bar is conventionally shown with a flat top, which might give the impression that the probability mass associated with the interval for that bar is assumed to be distributed evenly over its range. In fact, in our technique no assumption is made about how the probability mass associated with an interval is distributed over that interval, hence a given histogram bar is consistent with any such distribution, not only the PDF that the histogram might have been intended to discretize but also an infinite number of other PDFs.

We can interpret a histogram as representing a PDF correctly, not approximately, by sacrificing some information about the PDF in representing it with a histogram. To do this, simply view the flat top of each bar as being an artifact. If one takes the perspective that a bar does *not* define how its area is distributed within its interval, then information loss by the discretization is implied since the original PDF *does* define exactly how its area is distributed.

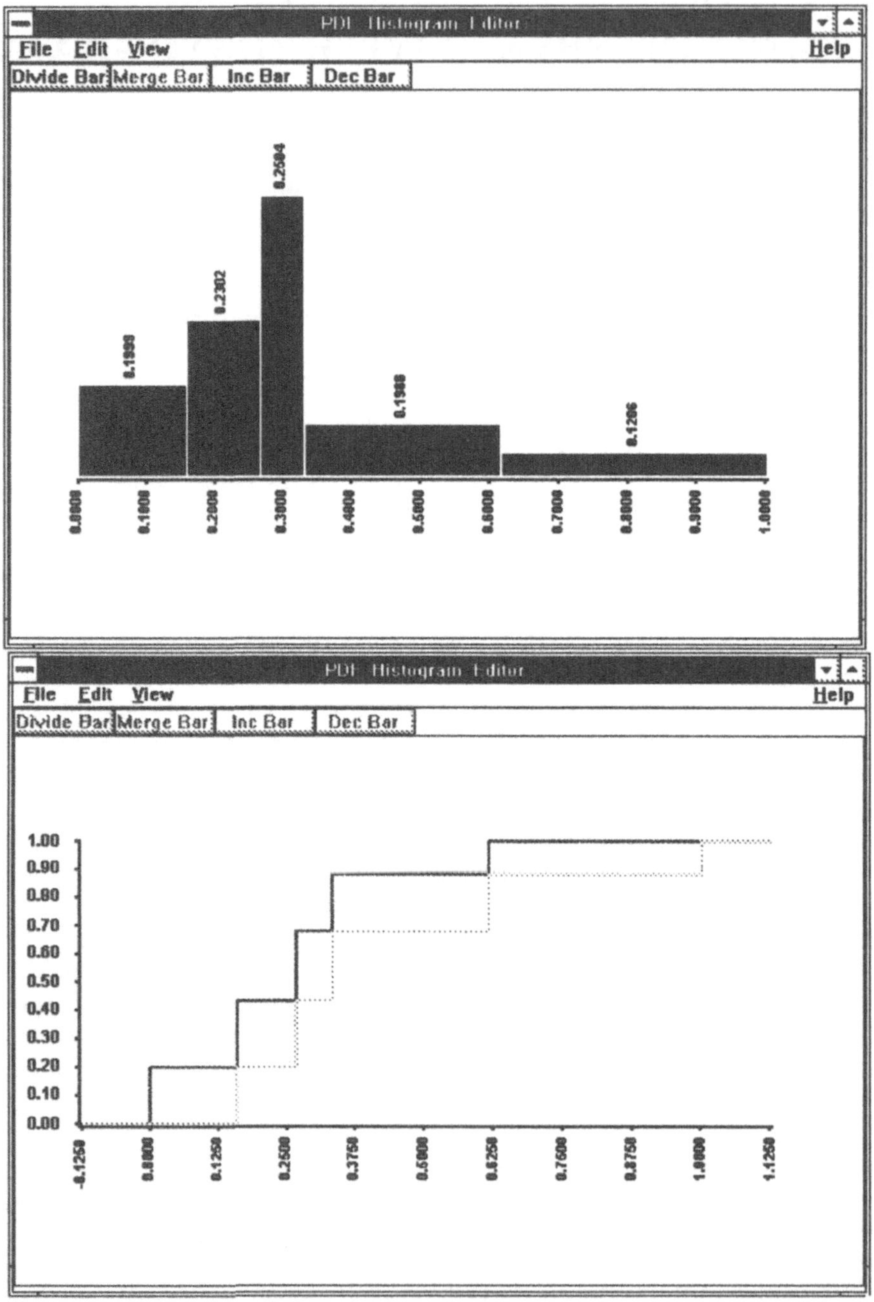


**Figure 1** A histogram and its two corresponding CDF bounds (one solid and one dotted).

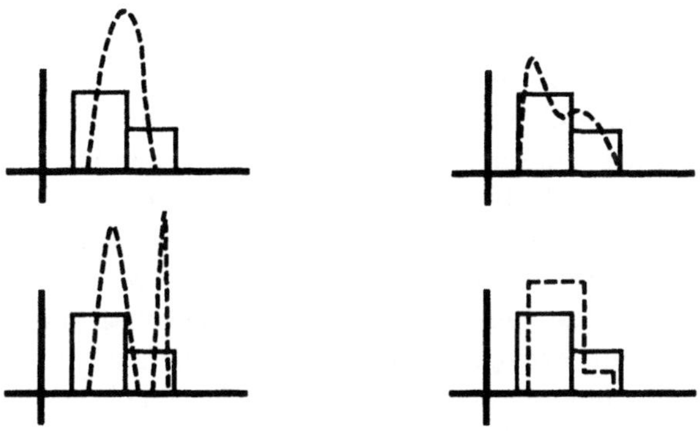

**Figure 2** Four out of an infinite set of probability density functions representable with the same histogram. A histogram bar spans an interval $I$, and a height. Consequently it has an area which encodes a probability mass. This area equals the area (and the probability mass) of the PDF over interval $I$. (Figure from [2].)

However, the limited statement the bar does make, that a certain area occurs under the PDF over a certain interval, is now correct. This is shown by Figure 2, which contains dissimilar PDFs which are all correctly describable using the same 2-bar histogram.

## *Integrations of Histograms can Also be Shown Graphically.*

A histogram (representing a family of PDFs) can be integrated to produce two cumulative distribution functions (CDFs) that bound the family of CDFs corresponding to the family of PDFs. The higher of the two assumes the extreme case where the probability mass associated with each bar of the histogram is distributed in a very skewed way — all at the lower bound of its associated interval. The lower of the two bounding CDFs assumes the other extreme, that each probability mass is concentrated at the upper bound of its corresponding interval. Any other of the infinite number of distributions consistent with the histogram, when integrated, results in a curve that stays within the two stair-step shaped bounding CDFs exemplified in Figure 1. The space between these bounding CDFs reflects the information loss associated with the discretization.

## 2.2 Correctly Representing an Interval with a Histogram

We take an interval to bound a value such that it has probability one of being within the interval, and probability zero of being outside it. An interval is represented in our technique as a histogram with one bar. To represent the interval using bounding CDFs, we must determine the space of CDFs consistent with the interval. The two most extreme CDFs that bound this space are the one that rises faster than any other CDF consistent with the interval, and the one that rises slower than any other CDF consistent with the interval. The one that rises fastest will correspond to the PDF which is an impulse of mass 1 at the interval's lower bound (that is, the variable's value is certainly equal to the interval's lower bound). Likewise, the one that rises slowest will correspond to the PDF which is an impulse of mass 1 at the interval's upper bound. The interval is therefore representable as a family of CDFs whose two bounding CDFs each have one "step." Figure 3 shows an interval and its representation as a bounded family of CDFs; the CDF that bounds this family from above jumps from zero to one at the interval's lower bound, and CDF that bounds it from below jumps from zero to one at the interval's high bound.

# 3 ARITHMETIC OPERATIONS

We introduce arithmetic on PDFs in the context of an example (Figure 4a–d) (a more abstract presentation of the method appears in [2]). The interval arithmetic operation, whichever one it is, is applied to the interval for a bar $i$ of operand 1 (Figure 4a) and the interval for a bar $j$ of operand 2 (Figure 4b), producing a result interval $I_{ij}$, for every possible $i$ and $j$ (Figure 4c). In other words, the operation is applied to each member of the Cartesian product of the bars in operand 1 and the bars in operand 2. The result is a set of intervals $I_{ij}$ collectively describing the result, called an *intermediate distribution* because it is neither a PDF nor a CDF, but is instead a list of intervals and their associated probability masses. The probability mass associated with each $I_{ij}$ in this result is the probability that the value of operand 1 is in bar $i$, which we will call $p_i$, and that the value of operand 2 is in bar $j$, which we will call $p_j$. Assuming independent operands, the probability associated with resulting interval $I_{ij}$ is then $p_i p_j$, as shown in the table of Figure 4c. The independence assumption is a limitation which could be eliminated by extending the method to non-independent operands whose joint distribution is known.

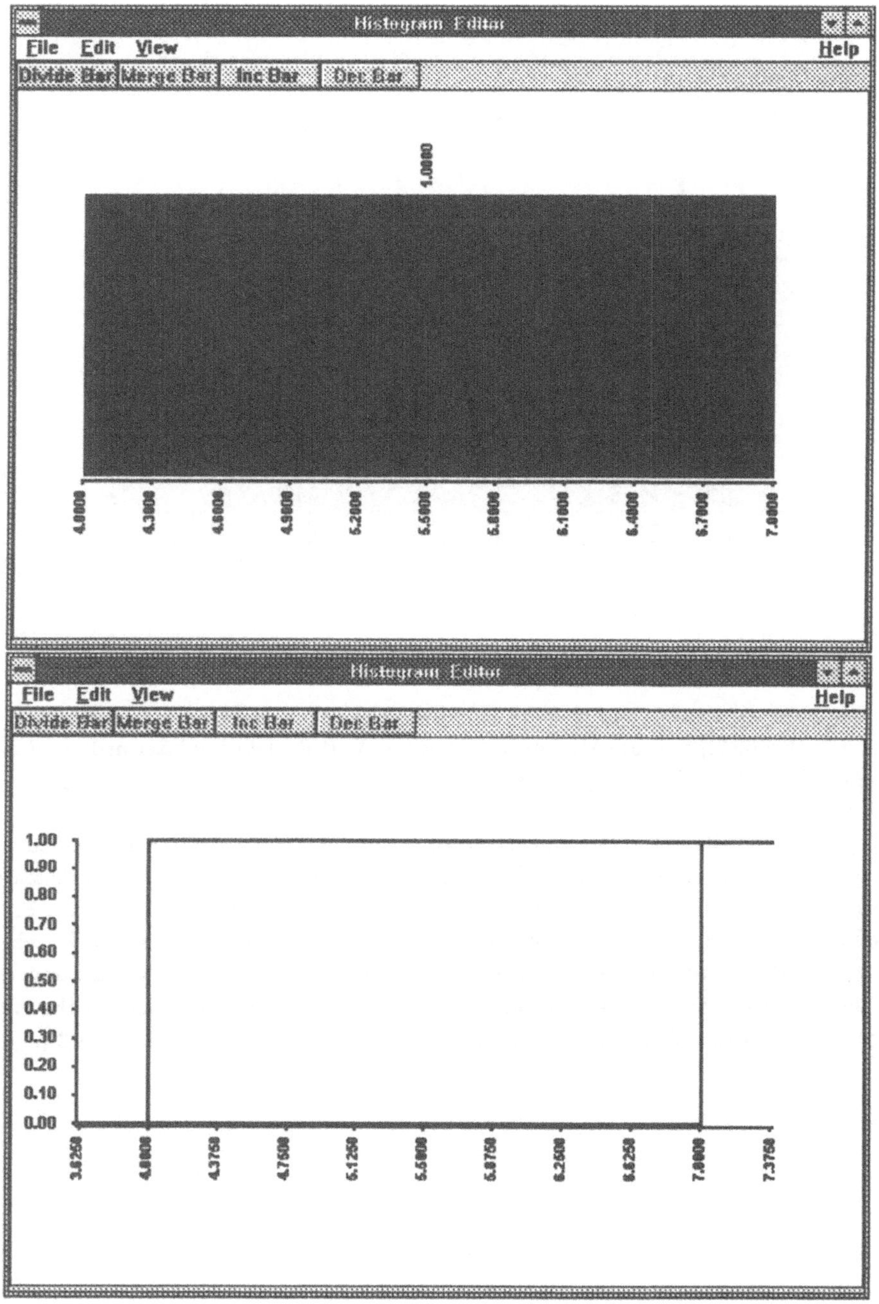


**Figure 3** The interval [4, 7] and its representation as a 1-bar histogram (top) and as two CDFs bounding the family of all CDFs consistent with the interval (bottom).

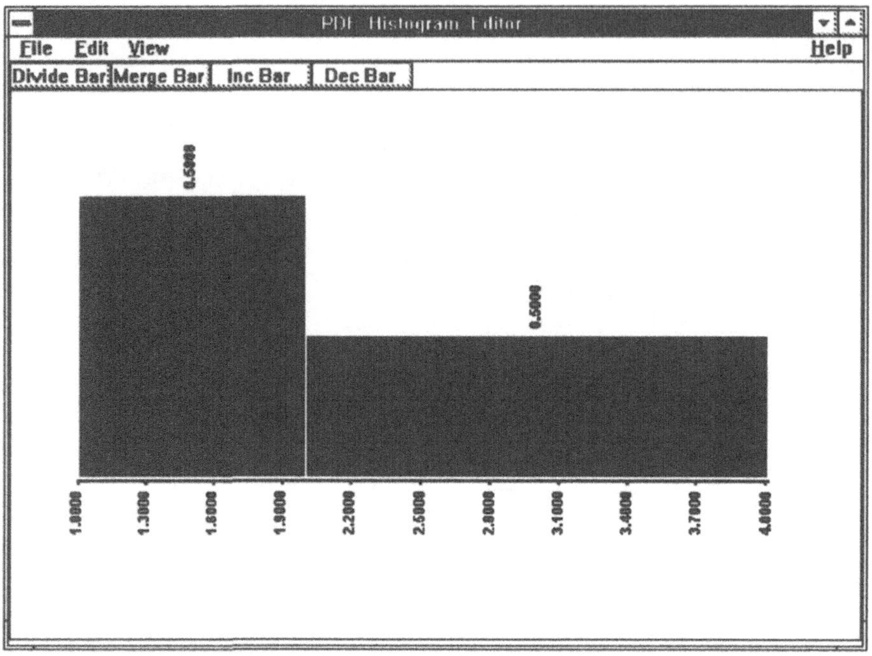


4a: Let this histogram describe uncertain value $X$. It depicts the intermediate distribution $\{p([1,2]) = \frac{1}{2}, p([2,4]) = \frac{1}{2}\}$.

**Figures** 4a–d illustrate the multiplication of two operands. The user created two histograms $X$ (Figure 4a) and $Y$ (Figure 4b). These histograms are represented internally as intermediate distributions, that is, lists of intervals and associated probability masses. The intermediate distributions are used to calculate $X \times Y$, whose intermediate distribution is shown in the last column of Figure 4c. Integrating this result produces the CDF bounds shown in Figure 4d. (The dotted lines in the lower left portion of that figure exemplify the effect of excess width, see section 3.2.)

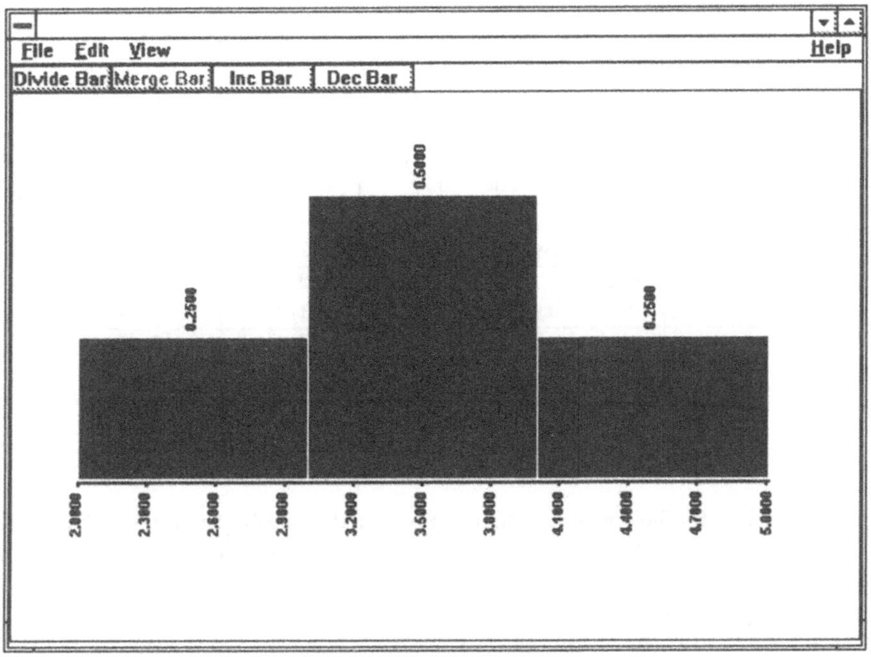


**4b: Let this histogram describe uncertain value *Y*. It depicts the intermediate distribution $\{p([2,3]) = \frac{1}{4}, p([3,4]) = \frac{1}{2}, p([4,5]) = \frac{1}{4}\}$.**

| Cartesian product term # | | Operand$_1$ ($X_i$) | Operand$_2$ ($Y_j$) | Result intermediate distribution |
|---|---|---|---|---|
| #1 | interval<br>probability | $[1, 2]$<br>$1/2$ | $[2, 3]$<br>$1/4$ | $[2, 6]$<br>$1/8$ |
| #2 | interval<br>probability | $[1, 2]$<br>$1/2$ | $[3, 4]$<br>$1/2$ | $[3, 8]$<br>$1/4$ |
| #3 | interval<br>probability | $[1, 2]$<br>$1/2$ | $[4, 5]$<br>$1/4$ | $[4, 10]$<br>$1/8$ |
| #4 | interval<br>probability | $[2, 4]$<br>$1/2$ | $[2, 3]$<br>$1/4$ | $[4, 12]$<br>$1/8$ |
| #5 | interval<br>probability | $[2, 4]$<br>$1/2$ | $[3, 4]$<br>$1/2$ | $[6, 16]$<br>$1/4$ |
| #6 | interval<br>probability | $[2, 4]$<br>$1/2$ | $[4, 5]$<br>$1/4$ | $[8, 20]$<br>$1/8$ |

**4c: Multiplying the intermediate distributions that describe the intermediate distributions for $X$ and $Y$ leads to another intermediate distribution, shown in the last column. (This intermediate distribution contains overlapping intervals, as is typical of intermediate distributions resulting from arithmetic operations.)**

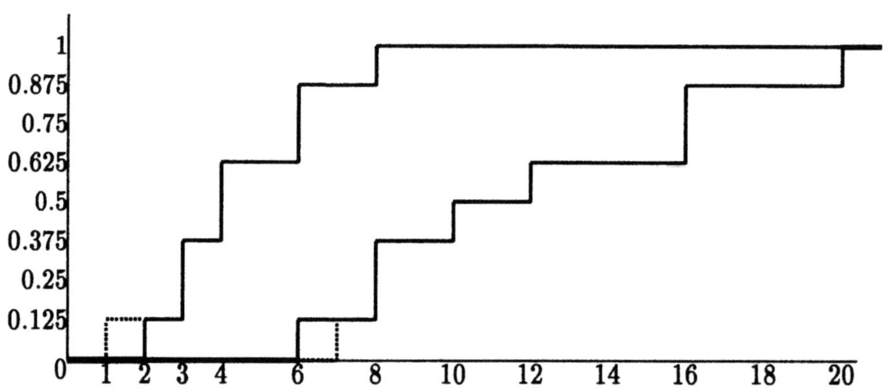


**4d: Integrating the resulting intermediate distribution in the last column of Figure 4c produces these stair-step shaped, bounding CDFs. (Figure from [2].)**

**Since no assumption may be made about the distribution of mass within any interval in either operand, no assumption may be made about the distribution of mass within any interval of the result. Therefore the integral of the intermediate result collection cannot be fully defined. However we can bound that integral**

with upper and lower CDFs. These CDFs bound a space of CDFs containing all CDFs corresponding to some distribution of the intermediate resulting interval probability masses over their respective intermediate resulting intervals.

This is how the bounding CDFs are derived. For each member of the resulting intermediate distribution, its integral can rise faster or slower, depending on its distribution of probability mass. By assuming each member's integral rises as fast as possible we get one bounding CDF. By assuming each member's integral rises as slow as possible we get the other bounding CDF.

Illustrating the integration of the resulting intermediate distribution of Figure 4c to produce Figure 4d, first consider row #1. The part of the resulting intermediate distribution derived in that row specifies probability $\frac{1}{8}$ distributed somehow over interval $[2,6]$. If that $\frac{1}{8}$ mass was distributed by having it all over the lower bound of the interval, 2, then the integral of the intermediate distribution comprising the result would rise immediately to $\frac{1}{8}$ as the integral curve passed the value 2 on the $x$-axis. This constitutes the fastest possible rise that interval $[2,6]$ could lead to. Looking next at the $\frac{1}{4}$ probability for row #2, the fastest possible rise it could cause would be if that probability mass was concentrated over the lower bound of its interval, 3, and so the CDF would discontinuously rise above its previous level by $\frac{1}{4}$ at a value of 3 on the $x$-axis. If all the other rows in Figure 4c are treated similarly, the integral will rise at the maximum possible rate and form the upper bounding CDF depicted in Figure 4d.

The lower bounding CDF is obtained the same way, except that the probability masses associated with the rows of the table are assumed to be concentrated at the upper bounds of their associated intervals, instead of the lower bounds as before. The resulting bounding CDFs are automatically verified, because any CDF corresponding to a PDF consistent with the histogram will fall within those bounds.

An approach to using CDF bounds in computer systems administration appears in Post and Diltz (1986 [15]) under the rubric of the *stochastic dominance* field [12], which recognizes the usefulness of bounded families of CDFs in fields such as finance. Another area of related work is robust statistics.

## 3.1 Combining a PDF with an Interval

Since an interval can be represented as a histogram with one bar, exactly the same process as is used to combine two PDF operands can be used to combine a PDF operand and an interval operand. As an example, consider describing $Z$, where $Z = X + Y$, $X$ is any PDF discretizable as the histogram shown earlier in Figure 4b, and $Y$ is the interval $[4, 7]$ (shown earlier as a histogram in Figure 3, top). Figure 5a shows the Cartesian product table with the intermediate distribution of the result shown in the last column. Integrated, this may be correctly depicted graphically, as shown in Figure 5b.

| Cartesian product term # | | Operand₁ ($X_i$) | Operand₂ ($Y_j$) | Result intermediate distribution |
|---|---|---|---|---|
| #1 | interval | $[2, 3]$ | $[4, 7]$ | $[6, 10]$ |
| | probability | 1/4 | 1 | 1/4 |
| #2 | interval | $[3, 4]$ | $[4, 7]$ | $[7, 11]$ |
| | probability | 1/2 | 1 | 1/2 |
| #3 | interval | $[4, 5]$ | $[4, 7]$ | $[8, 12]$ |
| | probability | 1/4 | 1 | 1/4 |

5a: Adding $X$ and $Y$ where $X$ is a histogram representation of a PDF, and $Y$ is a histogram representation of an interval.

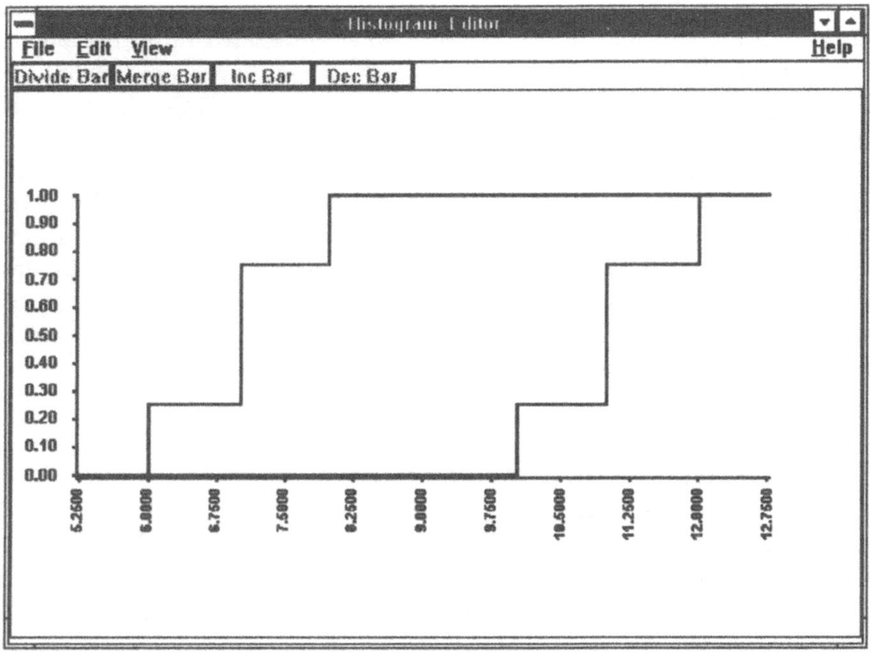


5b: CDF bounds on the result of adding a PDF and an interval.

**Figure 5.** Adding a PDF and an interval. Since the histogram representation can express both PDFs and intervals, the method described can do arithmetic on operands regardless of whether an operand is a PDF or an interval.

## 3.2 Excess Width

Interval calcuations are an important component of the method, and excess width is a possibility under some circumstances. For example, the desired operation on two histograms may be an expression that, when evaluated on the interval for a bar in one histogram and the interval for a bar on the other, results in an interval containing excess width. The resulting set of intervals would then contain intervals with excess width.

When intervals in the result of an operation have excess width, the intervals that constrain how probability masses are distributed are too wide, so that the process of finding the two most extreme distributions of mass within an interval to help form the upper and lower bounding CDFs produces extreme distributions that are too extreme. For example, concentrating the mass at an interval lower bound that is too low, when integrated to form a bounding CDF, makes the CDF rise too soon. Likewise, an upper bound that is too high leads to a bounding CDF that rises too slowly. As an example, consider row #1 of Figure 4c. The interval is $[2, 6]$, but suppose the calculation that produced it added excess width and instead estimated the interval as $[1, 7]$, wider than it really is. Then one extreme distribution of its mass of $\frac{1}{8}$ would have it concentrated at 1, and the other extreme distribution would have it concentrated at 7. This would widen the bounding CDFs as shown by dotted lines in Figure 4d. In general, excess width in calculations of intervals in the result of an operation creates weaker bounding CDFs. This means results are weaker, but still automatically verified.

## 3.3 Other Related Methods

One approach to operating on PDFs is to address only a limited class of PDFs, such as the class of lognormal curves. This method has the disadvantage over the method described here in that it does not apply to arbitrary PDFs.

A second approach is to use moments of PDFs to calculate moments of the result of combining them. This has usually meant approximating, although Rushdi and Kafrawy [16] present a variation providing "readily quantifiable accuracy that can be arbitrarily improved." However they do not address the issue of operations where one operand is a PDF and the other an interval.

A recent constraint satisfaction approach to the problem of operating on PDFs is given by Hyvönen (1995 [7]), who use a histogram-like representation to express envelopes around the PDF, intead of around the CDF as in the present work.

The most widely known and used method is the Monte Carlo method. A comparison with Monte Carlo is next (Table 1 summarizes).

### *Disadvantages of the Automatically Verified Histogram Method Compared to the Monte Carlo Method*

The automatically verified histogram method does not presently handle dependent operands, whereas the Monte Carlo approach can, if the dependency is known. However, the automatically verified histogram method could be ex-

| | Automatically verified histogram method | Monte Carlo methods |
|---|---|---|
| Handles dependent inputs? | (not currently) | √ |
| Automatically verifying? | √ (excess width possible) | × (insufficient width almost certain) |
| Handles interval inputs? | √ | √ |
| Handles PDF inputs? | √ | √ |
| Handles PDFs and intervals? | √ | × |

**Table 1** A comparison of the Monte Carlo and the automatically verified histogram methods.

tended to handle dependent inputs in that a joint distribution could be used to get the probabilities of the intervals in the intermediate distriabution resulting from an operation, instead of simply assuming independence and getting the probabilities of the intervals in the result by multiplying the probabilities of the operand intervals as in Figure 4c.

## *Advantages of the Automatically Verified Histogram Method Compared to the Monte Carlo Method*

Monte Carlo methods work by randomly choosing input values. Many sets of input values are chosen this way and used to get many corresponding output values. For inputs known to fall within intervals, the range of outputs generated by a sufficiently large random sample of possible inputs is used to estimate the range of possible outputs. However there is a possibility that those inputs that truly result in the most extreme outputs were never tried, so that the observed range of outputs is a subset of the actual range of outputs possible. Thus the observed range of outputs is likely to be too narrow, and hence the results of the Monte Carlo process are unguaranteed.

For inputs that are known sufficiently so that they can be characterized by PDFs, the inputs would be randomly sampled according to the PDFs so that for sufficiently large numbers of samples the distribution of samples will approximate the PDF used to generate them. The observed distribution of outputs corresponding to such a set of inputs then approximates the true PDF for the output. The likely deviation from that true PDF can be statistically characterized as an envelope around the CDF of the observed PDF. That envelope has a confidence level associated with it (e.g. Kolmogorov 1941) [11], not a guarantee. However, the automatically verified histogram method does provide a guarantee for the enveloping CDF bounds it produces.

Another advantage of the automatically verified histogram method applies when one input variable is a PDF and the other an interval. Monte Carlo methods have difficulty handling such a problem because of the difficulty of sampling an input space consisting of both intervals and PDFs. However the automatically verified histogram method works just as well when one operand is a PDF and the other an interval as it does when both operands are PDFs.

## Acknowledgements

The author thanks Hang Cheng for implementing the method in a software tool [3], and Bill Walster for pointing out that the method could be extended to dependent operands if their joint distribution is known.

## REFERENCES

[1] S. Ahmed, R. E. Clark, and D. R. Metcalf, "A Method for Propagating Uncertainty in Probabilistic Risk Assessment," *Nuclear Technology*, 1982, Vol. 59, pp. 238–245.

[2] D. Berleant, "Automatically Verified Reasoning with Both Intervals and Probability Density Functions," *Interval Computations*, 1993, No. 2, pp. 48–70.

[3] H. Cheng and D. Berleant, "A Software Tool for Automatically Verified Operations on Intervals and Probability Distributions," submitted to the International Journal *Reliable Computing*.

[4] A. G. Colombo and R. J. Jaarsma, "A Powerful Numerical Method to Combine Random Variables", *IEEE Transactions on Reliability*, 1980, Vol. R-29, No. 2, pp. 126–129.

[5] F. Corsi, "Mathematical Models for Marginal Reliability Analysis," *Microelectronics and Reliability*, 1983, Vol. 23, No. 6, pp. 1087–1102.

[6] V. A. Gerasimov, B. S. Dobronets, and M. Yu. Shustrov, "Numerical Calculations for Histogram Operations and their Applications," *Avtomatika i Telemekhanika*, 1991, Vol. 52, No. 2.

[7] E. Hyvönen, *Constraint Reasoning on Probability Distributions*, manuscript (1995), author is at VTT Information Technology, P.O. Box 1201, 02044 VTT, Finland.

[8] G. E. Ingram, E. L. Welker, and C. R. Herrmann, "Designing for Reliability Based on Probabilistic Modeling Using Remote Access Computer Systems", *Proceedings 7th Reliability and Maintainability Conference*, American Society of Mechanical Engineers, 1968, pp. 492–500.

[9] P. S. Jackson, R. W. Hockenbury, and M. L. Yeater, "Uncertainty Analysis of System Reliability and Availability Assessment", *Nuclear Engineering and Design*, 1981, Vol. 68, 1981, pp. 5–29.

[10] S. Kaplan, "On the Method of Discrete Probability Distributions in Risk and Reliability Calculations — Application to Seismic Risk Assessment," *Risk Analysis*, 1981, Vol. 1, No. 3, pp. 189–196.

[11] A. Kolmogoroff (*a.k.a.* Kolmogorov), "Confidence Limits for an Unknown Distribution Function," *Annals of Mathematical Statistics*, 1941, Vol. 12, No. 4, pp. 461–463.

[12] H. Levy, "Stochastic Dominance and Expected Utility: Survey and Analysis," *Management Science*, 1992, Vol. 38, No. 4, pp. 555–593.

[13] A. S. Moore, "Interval Risk Analysis of Real Estate Investment: A Non-Monte Carlo Approach," *Freiburger Intervall-Berichte*, 1985, Vol. 85/3, pp. 23–49.

[14] R. E. Moore, "Risk Analysis Without Monte Carlo Methods," *Freiburger Intervall-Berichte*, 1984, Vol. 84/1, pp. 1–48.

[15] G. V. Post and J. D. Diltz, "A Stochastic Dominance Approach to Risk Analysis of Computer Systems," *Management Information Systems Quarterly*, 1986, Vol. 10, pp. 363–375.

[16] A. M. Rushdi and K. F. Kafrawy, "Uncertainty Propagation in Fault-Tree Analyses Using an Exact Method of Moments," *Microelectronics and Reliability*, 1988, Vol. 28, No. 6, pp. 945–965.

# 11

# NESTED INTERVALS AND SETS: CONCEPTS, RELATIONS TO FUZZY SETS, AND APPLICATIONS

**Hung T. Nguyen* and Vladik Kreinovich****

** Department of Mathematical Sciences, New Mexico State University, Las Cruces, NM 88003-8001, email hungueyn@nmsu.edu*

*** Department of Computer Science, University of Texas at El Paso, El Paso, TX 79968, email vladik@cs.utep.edu*

## ABSTRACT

In data processing, we often encounter the following problem: Suppose that we have processed the measurement results $\tilde{x}_1, ..., \tilde{x}_n$, and, from this processing, have obtained an estimate $\tilde{y} = f(\tilde{x}_1, ..., \tilde{x}_n)$ for a quantity $y = f(x_1, ..., x_n)$; we know the intervals $\mathbf{x}_i$ of possible values of $x_i$, and we want to find the interval $\mathbf{y}$ of possible values of $y$. Interval computations are one of the main techniques for solving this problem.

In some cases, for each $i$, in addition to the *guaranteed* interval $\mathbf{x}_i$ of possible values, we have a smaller interval that an expert believes to contain $x_i$. There may be several such *nested* intervals. In these cases, in addition to the guaranteed interval $\mathbf{y}$, it is desirable to know the possible intervals of $y$ that correspond to the opinions of different experts.

Techniques of such *nested interval computations* and real-life applications of these techniques are described in this paper.

*This work was partially carried out while Hung T. Nguyen was on sabbatical leave at the University of Southern California, Los Angeles, Spring 1995



*R. B. Kearfott and V. Kreinovich (eds.), Applications of Interval Computations*, 245–290.

# 1 INTRODUCTION

## 1.1 Estimating Errors of Indirect Measurements: One of the Main Problems in Data Processing

One of the important problems in data processing is to estimate the error of an indirect measurement. Namely, we want to know a value of the physical quantity $y$ that is difficult or impossible to measure directly. So, to estimate the values of $y$, we measure other quantities $x_1, ..., x_n$ that are related to $y$, and reconstruct $y$ from the results $\tilde{x}_i$ of these measurements. This procedure is called an *indirect measurement.*

It is usually known how to reconstruct $y$ from $x_i$. In other words, we usually know an algorithm $f$ that, if applied to the actual values of $x_i$, would result in the actual value of $y = f(x_1, ..., x_n)$. Instead of the actual values $x_i$, we only know the measurement results $\tilde{x}_i$ that (due to measurement errors) may be slightly different from the actual values. As a result, when we apply the algorithm $f$ to these measurement result, the resulting estimate $\tilde{y} = f(\tilde{x}_1, ..., \tilde{x}_n)$ for $y$ is, generally speaking, different from the actual values of $y$.

If we want to know what the actual value of $y$ can be, we must know how different $y$ can be from this estimate $\tilde{y}$. In other words, we need to estimate the error $\Delta y = \tilde{y} - y$ of this indirect measurement.

This error $\Delta y$ comes from the errors $\Delta x_i = \tilde{x}_i - x_i$ of measuring $x_i$. To measure $x_i$, we use regular measuring instruments. For a measuring instrument to make sense, the manufacturer must provide it with some guaranteed error bounds $\Delta_i$. (Otherwise, whatever value $\tilde{x}_i$ we measure, the actual value $x_i$ can be arbitrarily far from $\tilde{x}_i$, i.e., no information about $x_i$ would come from the measurement itself.) If we know the bound $\Delta_i$, and the result of the measurement is $\tilde{x}_i$, then the actual value must belong to the interval $[\tilde{x}_i - \Delta_i, \tilde{x}_i + \Delta_i]$.

In some cases, the manufacturer also provides the user with the *probabilities* of different values of error $\Delta x_i$, but in many real-life cases, the bounds $\Delta_i$ are the only information about the errors that we have. In these cases, we have the following problem:

*We know:*

- *the intervals* $\mathbf{x}_i = [\tilde{x}_i - \Delta_i, \tilde{x}_i + \Delta_i]$ *that contain the actual values of* $x_i$*;*
- *an algorithm* $f$ *that transforms* $n$ *values* $x_1, ..., x_n$ *into a real number.*

*We must compute: The set* $\mathbf{y}$ *of possible values of* $y$*:* $\mathbf{y} = f(\mathbf{x}_1, ..., \mathbf{x}_n) = \{f(x_1, ..., x_n) | x_1 \in \mathbf{x}_1, ..., x_n \in \mathbf{x}_n\}$.

*Comments.*

1. This is the *main problem* to be considered in the present paper. In addition to the problem, we will also consider other problems, such as identification, optimization, design, and control. These problems will be briefly outlined in Section 3.

2. In some cases, the manufacturer can guarantee that some values of error are impossible. In this case, the set $\mathbf{x}_i$ of possible actual values can be a subset of the interval $[\tilde{x}_i - \Delta_i, \tilde{x}_i + \Delta_i]$.

3. A measuring instrument can measure several different quantities $y_1, ..., y_k$ at a time. In this case, in addition to the information about the possible errors of each measurement, the manufacturer can guarantee that certain combinations of errors are impossible: e.g., it can happen that $\Delta y_1$ attains its largest possible value $\Delta$, and it can happen that $\Delta y_2$ attains its largest possible value $\Delta$, but they can never attain these extreme values at the same time, because of the restriction that $\Delta y_1^2 + \Delta y_2^2 \le \Delta^2$. (This situation happens, e.g., if we measure geographical coordinates of a point.) Such information can be described by a set $\mathbf{y} \subset R^k$ of all the tuples that the manufacturer believes can be possible values of errors $(\Delta y_1, ..., \Delta y_k)$.

## 1.2 Input Intervals are Often Overestimated

Input intervals come from the manufacturer's guarantee. The manufacturer does not want to risk his/her reputation, and to be liable to lawsuits. So, when in doubt, the manufacturer prefers to overestimate the error bounds $\Delta_i$. The actual designers and testers of the measuring instrument, when you ask their opinion, usually assign much better quality to their instruments. They can say something like "We can only *prove* that the error is $\le 0.1$, but all our experiments have shown that this error is $\le 0.05$."

Cautious overestimation is a reasonable strategy when the error may be life-threatening: e.g., in measuring the reactor characteristics of the nuclear power stations, or in computing the trajectory of the spaceship. But this caution makes no sense if we are, e.g., estimating the amount of oil in a given area. In such cases, instead of an absolutely accurate and absolutely guaranteed estimate like $100 \pm 50$ mln tons, it would be nice to know that with high reliability, we are able to extract from the same measurement results a much narrower interval (e.g., $90 \pm 10$).

## 1.3 In Addition to the Main (Guaranteed) Interval, it is Reasonable to Consider its Subintervals

In these situations (when nothing life threatening can come from an under-estimation of a measurement error), it is thus reasonable, in addition to the intervals that are *guaranteed* by the manufacturer, to consider *subintervals* that the manufacturer cannot, strictly speaking, guarantee, but which the experts (designers and/or producers of the manufacturing instrument) claim to be true.

In other words, inside each interval $\mathbf{x}_i$, there is another interval $\mathbf{x}_i(c) \subseteq \mathbf{x}_i$.

**Example.** Suppose that we have a measuring instrument with guaranteed accuracy $\Delta = 0.1$ and actual accuracy 0.06. (We give this value to illustrate the idea; in real life, this value is not exactly known neither to the manufacturer of the measuring instrument, nor to the user.) Suppose also that, in addition to the guaranteed error bound 0.1, we have an expert's estimate that the error bound cannot exceed 0.08. Suppose that we have performed the measurement, and the measurement result is $\tilde{x} = 1.1$. In this case, the guaranteed estimate $\mathbf{x} = [\tilde{x} - \Delta, \tilde{x} + \Delta] = [1.0, 1.2]$. The actual (unknown) set of possible values is $X_{\text{act}} = [1.1 - 0.06, 1.1 + 0.06] = [1.04, 1.16]$. The set of possible values according to the expert is equal to $\mathbf{x}(c) = [1.1 - 0.08, 1.1 + 0.08] = [1.02, 1.18]$.

Given these additional $n$ intervals, we would like, in addition to the *guaranteed* interval $\mathbf{y} = f(\mathbf{x}_1, .., \mathbf{x}_n)$, to compute the smaller (*claimed*) interval for $y$: $\mathbf{y}(c) = f(\mathbf{x}_1(c), ..., \mathbf{x}_n(c))$.

A first time reader may ask himself/herself: "Why consider subintervals that are not guaranteed by the manufacturer? Are there other motivations for using nested sets other than experts suspecting the manufacturer's error bounds are overestimates?" To explain the practical usefulness of nested sets, let us give a couple of real-life examples where such subintervals are useful (in Section 4, we will show how nested interval computations help to solve the corresponding real-life problems).

**Case study: seismic engineering** (Dong et al. [13]). Earthquake are extremely difficult to predict. Because of that, if we only use the equations and the measured data $x_i$, we get very wide intervals of possible magnitudes and frequencies $y$, intervals that are known to be much wider than the interval of actual values. So, if we use these wide intervals $[y^-, y^+]$ to design a building that is guaranteed to withstand a typical earthquake, these buildings will be unnecessarily (and often unrealistically) expensive. To make the requirements more realistic, in addition to guaranteed measurement results, *expert estimates* are normally used that lead to narrower intervals.

In the second example, measurements are so bad that we have to rely mainly on expert estimates:

**Case study: AIDS epidemics** (Berleant et al. [6]). Equations that describe how AIDS is transferred are known. To solve these equations, we must know the initial conditions and the the numerical values of the parameters (such as the death rate among AIDS patients, etc). The majority of these parameters can be determined more or less accurately from the known statistical data. However, there is one important parameter about which it is very difficult to get a reliable statistical estimate: the number of intravenous drug abusers. So, in the absence of a reliable statistical estimate, we may use the expert estimates. Different experts can give different intervals of possible values, so, we have nested intervals for describing the expert knowledge.

## 1.4 There can be Several Subintervals

In the previous subsections, we assumed that there are two intervals that correspond to two degrees of confidence:

- an interval guaranteed by the manufacturer; this interval corresponds to an *absolute* confidence (100% guarantee); and
- its subinterval, that contains only the values that the expert (e.g., designer) believes to be possible.

This is, however, an idealized situation. Different experts can have different ideas of what values are possible and what values are not. More optimistic experts will choose smaller intervals, while more cautious experts will prefer larger ones. As a result, for each variable $x$, we may have several different intervals of possible values $\mathbf{x}(\alpha)$ that depend on an expert $\alpha$ (i.e., we have a *nested* interval). For each $\alpha$, we end up with a different interval $\mathbf{y}(\alpha)$. So, we get a nested interval for the result. In this paper, we describe the methods of computing nested intervals, and applications of these computations.

At first, we consider the case when experts can be ordered in a sequence form the most cautious one to the most risky one. Then, we will consider the general case when some experts may be incompatible.

# 2 NESTED INTERVALS AND NESTED SETS

## 2.1 Nested Sets: Motivation of the Definition

For simplicity, let us first consider the case when the experts can be ordered in a sequence from the most cautious to the most risky. (In Section 2.5, we will mention what happens if we do not make this assumption.)

Let us assume that we have the results $\tilde{x}$ of one or several measurements. We want to know what information about the actual values $x$ we can deduce from $\tilde{x}$. In other words, we want to know what values of $x$ could possibly lead to these measurement results. In more mathematical terms, we want to know the set $X_{\rm act}$ of all values of $x$ that could lead to these measurement results.

We do not know this set precisely. Instead, we know the following: first, from the manufacturer's guarantee, we know some set that definitely contains $X_{\rm act}$ (i.e., a set with the property that every possible value $x$ definitely belongs to it). For a single measurement with an error bound $\Delta$, such a set is an interval $[\tilde{x}-\Delta, \tilde{x}+\Delta]$. In general, we will describe this set by $X(\alpha_0)$. (The reason for choosing this notation will be given later.)

We believe that $X(\alpha_0)$ is possibly an *overestimate* for $X_{\rm act}$, so that the actual set of possible value is probably smaller than $X(\alpha_0)$. This belief is shared by several experts, who propose their opinions on the set of possible values. Let us denote the number of experts by $d$. We have assumed that the experts can be ordered in the order of their cautiousness, from the most cautious one (# 1 in this ordering), to the second cautious (# 2 in this ordering), ..., to the riskiest expert (who in this ordering gets # $d$). Let us denote the set of all values $x$ that the $j$-th expert believes to be possible by $X(\alpha_j)$. Then, the larger $j$, the riskier the expert's statement becomes, and therefore, the fewer elements are called possible. In mathematical terms, this means that we have the following relationship between these sets: $X(\alpha_0) \supseteq X(\alpha_1) \supseteq X(\alpha_2) \supseteq ... \supseteq X(\alpha_d)$.

If an element $x$ belongs to the set $X(\alpha_j)$, this means that not only the $j$-th expert believes this element to be possible, but actually $j$ experts (1, 2, ..., $j$) all agree that $x$ is a possible value. So, elements from the set $X(\alpha_j)$ can be described as elements about which at least $j$ out of $d$ experts agree that $x$ is possible. So, if we would estimate the degree of possibility of an element $x$ by polling experts, elements from $X(\alpha_j)$ would be those for which at least $j$ experts voted. The more experts vote for $x$ to be possible, the larger is the expert's community *degree of belief* that $x$ is possible. In these terms, we can say that the sets $X(\alpha_j)$ describe different *degrees of belief* in the possibility of $x$.

In addition to an *overestimate* $X(\alpha_0)$, we may also know that some values $x$ are *definitely possible*. For example, we may estimate $\Delta$ by adding up the upper bounds for component errors, but we may also be sure that a certain error component is definitely there (e.g., thermal noise), so we may know a lower bound for the largest possible error. If we know a set $D$ whose values are definitely possible, then, of course, experts' estimates $X(\alpha_j)$ must contain this set $D$. It is therefore smaller than all $X(\alpha_j)$, i.e., we have $X(\alpha_0) \supseteq X(\alpha_1) \supseteq ... \supseteq X(\alpha_d) \supseteq D$. This ordering prompts a natural notation $X(\alpha_{d+1})$ for this set $D$.

**Example.** Let us illustrate these notions on the same example as we used before to explain subintervals. Suppose that we have a measuring instrument with guaranteed accuracy $\Delta = 0.1$ and (unknown) actual accuracy 0.06. The only lower bound for the error bound is 0. Suppose also that we have two experts: a cautious one that estimates the error bound as 0.08, and a risky one that estimates it as 0.05. Here, $d = 2$. Suppose that we have performed the measurement, and the measurement result is $\tilde{x} = 1.1$. In this case, the guaranteed estimate $X(\alpha_0) = [\tilde{x} - \Delta, \tilde{x} + \Delta] = [1.0, 1.2]$. The actual (unknown) set of possible values is $X_{\text{act}} = [1.1 - 0.06, 1.1 + 0.06] = [1.04, 1.16]$. The sets of possible values according to the two experts are: $X(\alpha_1) = [1.1 - 0.08, 1.1 + 0.08] = [1.02, 1.18]$ and $X(\alpha_2) = [1.1 - 0.05, 1.1 + 0.05] = [1.05, 1.15]$. Finally, the set $D = X(\alpha_3)$ that contains elements that are guaranteed to be possible consists of a single element $X(\alpha_3) = \{1.1\}$.

Suppose now that we make a second measurement of the same quantity, and get a new result 0.99. This means that we can guarantee that

$$x \in [0.99 - 0.1, 0.99 + 0.1] = [0.89, 1.09].$$

Combining this information with the guaranteed information $x \in [1.0, 1.2]$, we can conclude that $x \in [1.0, 1.2] \cap [0.89, 1.09] = [1.0, 1.09]$, so, $[1.0, 1.09]$ is our new set $X(\alpha_0)$. For the first expert, the same procedure leads to $X(\alpha_1) = [1.02, 1.18] \cap [0.99 - 0.08, 0.99 + 0.08] = [1.02, 1.07]$. For the new set $D$, we have the following: if we only had the result of the first measurement, then we could guarantee that the value 1.1 is possible. However, now that we have the additional information, the set of possible values shrinks, and there is no guarantee that 1.1 is possible anymore. (Actually, it is not, because the error is bounded by 0.06, so if the actual value was 1.1, we could not have 0.99 as a measurement result.) Mathematically speaking, from the first measurement, we get $D = \{1.1\}$, from the second, we get $D = \{0.99\}$. In this case, the intersection $\{1.1\} \cap \{0.99\}$ is empty, and therefore, there is no value that we can guarantee to be possible. So, in this case, $X(\alpha_3)$ is empty. Similarly, $X(\alpha_2) = [1.05, 1.14] \cap [0.99 - 0.05, 0.99 + 0.05] = \emptyset$ (which means that the opinion of the second expert turned out to be incorrect).

*Comment.* Different values of $\alpha$ may sometimes describe the opinion not of different experts, but of one and the same expert: if we ask for an expert's most reliable opinion, the expert may give a much broader interval, but if we ask the expert for what she really thinks (thus allowing her to err), the same expert may give us a narrower interval.

Let us describe nested sets and nested intervals formally.

## 2.2 Definition and the Formulation of the Main Problem

**Definition 1.**

- *Let an integer $d$ be fixed. This number will be called the **number of experts**.*
- *Let a finite linearly ordered set $A = \{\alpha_0 < \alpha_1 < \alpha_2 < \ldots < \alpha_d < \alpha_{d+1}\}$ be fixed. The elements of this set will be called **degrees of belief**.*
- *Let a set $U$ be given. This set will be called the **Universal set**, or, the **Universe of discourse**.*

- *By a piece of information (or, a nested set), we mean a non-increasing mapping $X$ from $A$ to the set $2^U$ of all subsets of $U$, i.e., a tuple $(X(\alpha_0), X(\alpha_1), ..., X(\alpha_{d+1}))$, for which $X(\alpha_{d+1}) \subseteq X(\alpha_d) \subseteq ... \subseteq X(\alpha_1) \subseteq X(\alpha_0) \subseteq U$.*
- *Let $u$ be an element of $U$, and $\alpha \in A$ be a degree of belief. We say that $u$ belongs to the nested set $X$ with a degree of belief $\geq \alpha$ if $u \in X(\alpha)$.*

*Historical comment.* The notion of a nested set was first introduced in Gentilhomme [20] under the French name "ensemble flou". It was further analyzed by Negoita et al [51, 52].

## *Two Problems: Finding the Input Nested Sets and Computing the Nested Set for the Result*

Our ultimate goal is to describe the nested set $Y$ that characterizes $y$. To do that, we must find the nested sets $X_i$ for all input variables $x_i$, and then transform these input nested sets into a nested set $Y$ for $y$. Before we describe the algorithms, we need to describe these problems in mathematical terms.

## *Finding the Input Nested Set*

Finding the nested set $X$ that describes the input variable $x$ may not be a problem if we have a single measurement result. However, in some cases, we measure the same quantity several times, and thus obtain several measurement results that lead to several pieces of information (nested sets) $X$, $X'$, ... So, we must combine them to get a nested set that describes all the information about the input variable $x$.

If we will be able to combine two nested sets, then, we will be able to combine any number of them $X$, $X'$, $X''$, ...:

- we start by combining the first two nested sets ($X$ and $X'$);
- then, we combine their combination with the third nested set ($X''$), etc.

Therefore, in the following text, we will only describe how to combine two nested sets $X$ and $X'$. This combination can be easily done: Indeed, let us consider a degree of belief $\alpha_j$. If $1 \leq j \leq d$, then this degree of belief describes the opinion of the $j$–th expert. For this expert, according to the first piece of information, the actual value $x$ belongs to the set $X(\alpha_j)$, and according to the second piece of information, $x$ belongs to $X'(\alpha_j)$. As a result, for this expert, the set of possible values of $x$ coincides with $X(\alpha_j) \cap X'(\alpha_j)$. A similar formula can be derived for the case when $\alpha$ describes the guaranteed set of possible values. So, we arrive at the following definitions:

**Definition 2.** *By an intersection $X \cap X'$ of nested sets $X$ and $X'$, we mean a nested set with $(X \cap X')(\alpha) = X(\alpha) \cap X'(\alpha)$.*

In some cases, the reason we perform several measurements is, e.g., that one of them may err, so that only one of them is reliable. For two measurements, this means that after we get pieces of information $X$ and $X'$, we can only be sure that the actual value $x$ belongs to one of the corresponding sets. For an expert $\alpha$, this means that set of possible values of $x$ is a *union* $X(\alpha) \cup X'(\alpha)$. So, we can define a union of nested sets:

**Definition 3.** *By a union $X \cup X'$ of nested sets $X$ and $X'$, we mean a nested set with $(X \cup X')(\alpha) = X(\alpha) \cup X'(\alpha)$.*

There are also situations when another set operation naturally arises: complementation of the two sets. These situations are slightly more complicated. As an example, let us assume that the measurement error is caused by noise. We know that this noise is always present, but we do not know whether this noise is positive or negative. In this case, the set of possible values of the noise is a union of two intervals: one that corresponds to the positive values, and one that corresponds to the negative values. For simplicity, let us assume that this set of symmetric, i.e., it is of the form $[-\Delta, -\delta] \cup [\delta, \Delta]$ for some $0 < \delta < \Delta$. This set can be represented as a difference between the two intervals: $[-\Delta, \Delta] \setminus [-\delta, \delta]$. For each of these intervals, we can have more cautious and more risky estimates in the above sense. A more cautious estimate for $\Delta$ means a larger $\Delta$; similarly, a more cautious (in the sense of Definition 1) estimate for $\delta$ also means a larger $\delta$. However, the most cautious estimate for the difference corresponds to the case when we take the *largest* possible $\Delta$ and the *smallest* possible $\delta$. In other words, to get the most cautious estimate of the difference $[-\Delta, \Delta] \setminus [-\delta, \delta]$, we subtract the *riskiest* estimate for $[-\delta, \delta]$ from the *most cautious* estimate for $[-\Delta, \Delta]$. So, to describe negative information (for set complement), the expert uses not his/her own degree of belief $\alpha$, but some other degree of belief $\alpha^{\perp}$: the most cautious $\alpha$ uses the riskiest $\alpha^{\perp}$, the more cautious $\alpha$ uses the riskier $\alpha^{\perp}$, and the riskiest $\alpha$ uses the most cautions $\alpha^{\perp}$.

This description works for the values $\alpha_j$ that correspond to different experts. A similar idea works for the values $j = 0$ and $j = d+1$ that describe the upper and lower bounds for the set of possible values: Indeed, we can view the sets $X^\perp(\alpha_j)$ as a description of the complement $X_{act}^\perp$ to the (unknown) set $X_{act}$. The only information that is guaranteed for $X_{act}$ is that $X(\alpha_{d+1}) \subseteq X_{act} \subseteq X(\alpha_0)$. From these inclusions, we can conclude that $X^\perp(\alpha_0) \subseteq X_{act}^\perp \subseteq X^\perp(\alpha_{d+1})$. So, $X^\perp(\alpha_0)$ is a guaranteed lower bound for the desired (unknown set) $X_{act}^\perp$, and $X^\perp(\alpha_{d+1})$ is a guaranteed upper bound. So, we can take $X^\perp(\alpha_0)$ as $(X^\perp)(\alpha_{d+1})$, and $X^\perp(\alpha_{d+1})$ as $(X^\perp)(\alpha_0)$. In other words, for $\alpha = \alpha_0$ and $\alpha = \alpha_{d+1}$, we define $(X^\perp)(\alpha) = X^\perp(\alpha^\perp)$, where $\alpha_0^\perp = \alpha_{d+1}$ and $\alpha_{d+1}^\perp = \alpha_0$. Let us summarize this description in mathematical terms.

**Definition 4.** *Let $\alpha \to \alpha^\perp$ denote a strictly decreasing function $A \to A$ for which $\alpha_0^\perp = \alpha_{d+1}$ and $\alpha_{d+1}^\perp = \alpha_0$. By a* complement *$X^\perp$ to a nested set $X$, we mean a nested set $(X^\perp)(\alpha) = X^\perp(\alpha^\perp)$ (where $A^\perp$ means a complement $U \setminus A$). By a* difference *$X \setminus X'$ between nested sets $X$ and $X'$, we mean a nested set $X \cup (X')^\perp$.*

The desired functions $\alpha \to \alpha^\perp$ can be easily described:

**PROPOSITION 1.** *If $\alpha \to \alpha^\perp$ is a strictly decreasing function $A \to A$ for which $\alpha_0^\perp = \alpha_{d+1}$ and $\alpha_{d+1}^\perp = \alpha_0$, then $\alpha_j^\perp = \alpha_{d+1-j}$.*

(For the reader's convenience, all proofs are placed in a special Appendix).

Thus defined complement operations have some properties of the set-theoretic complement: e.g., De Morgan laws $(X \cup X')^\perp = X^\perp \cap (X')^\perp$ and $(X \cap X')^\perp = X^\perp \cup (X')^\perp$ are easy to prove. However, some properties of this complement operation may sound counterintuitive: for example, it is possible that $X \setminus X \neq \emptyset$ and $X \cup X^\perp \neq U$. This contradiction with traditional set theoretic intuition can be easily explained if we use Definition 4: e.g., $X \setminus X$ does not mean that we subtract a set $X$ from itself. To get $(X \setminus X)(\alpha)$, we subtract from $X(\alpha)$ the set $X(\alpha^\perp)$ that corresponds to a *different* degree of belief, and can thus be *smaller* than $X(\alpha)$ (if $\alpha^\perp > \alpha$).

## *Computing the Nested Set for y*

The result of processing nested sets can be defined expert-by-expert:

**Definition 5.** *Let $X_1, ..., X_n$ be nested sets with a common set $A$, and let $f : U \times ... \times U \to U$ be a function. Then, we define $f(X_1, ...X_n)$ as a nested set $Y$ with $Y(\alpha) = f(X_1(\alpha), ..., X_n(\alpha))$.*

The ***main problem*** of nested set processing can be thus formulated as follows: ***given*** the pieces of information $X_1, ..., X_n$ that describe inputs, ***compute*** the nested set $Y$ that describes the output.

## 2.3 Preliminary Step: Checking whether Expert Knowledge is Consistent

An expert may be wrong. In some cases, it is possible to check that: namely, if we know that the measured quantities $x_i$ must satisfy some constraints (equalities or inequalities). If for $x_i \in X_i(\alpha)$, these constraints cannot be satisfied, this means that an expert $\alpha$ is wrong. For example, assume that we are measuring three currents, and we know that due to Kirkhoff's law, $x_3 = x_1 + x_2$. Suppose that as a result of measuring these currents, we got the values $\tilde{x}_1 = 1.0$, $\tilde{x}_2 = 2.0$, and $\tilde{x}_3 = 2.9$. If the measurement accuracy is 0.1, then the constraint can still be satisfied: e.g., for $x_1 = 1$, $x_2 = 2$, and $x_3 = 3$. Assume, however that an expert $\alpha$ believes that the actual accuracy of the measuring instrument is 0.01. Then, for this expert, $X_1(\alpha) = [0.99, 1.01]$, $X_2(\alpha) = [1.99, 2.01]$, and $X_3(\alpha) = [2.89, 2.91]$. In this case, $X_1(\alpha) + X_2(\alpha) = [2.97, 3.03]$, so $X_3(\alpha) \cap X_1(\alpha) + X_2(\alpha) = \phi$, and these conditions are inconsistent. So, this expert is wrong. In Mitra et al. [44], an algorithm is proposed that checks consistency for the case when $x_i$ corresponds to measuring time.

## 2.4 Methods of Solving the Main Problem

### *A Natural Approach: Computing $Y(\alpha)$ Set-by-Set (Expert-by-Expert). Advantages of the Natural Approach*

A natural way of computing $Y$ is to compute $Y(\alpha_j)$ for $i = 0, 1, ..., d+1$ with traditional interval techniques (if the sets $X_i(\alpha)$ are intervals), or generalized interval techniques (if the sets $X_i(\alpha)$ are more complicated than intervals). This method is described, e.g., in Gerasimov et al. [21]. This approach has two advantages:

- it uses well-known techniques for which good software tools have been developed; and
- it is easily parallelizable: if we have $d+2$ processors $P_0, ..., P_{d+1}$, then we can let each of them compute $Y(\alpha_j)$ for the corresponding $j$; if the number of processors $p$ is smaller than $d+2$, then we can divide $A$ into $p$ groups, and ask $m$–th processor to compute the values $Y(\alpha_j)$ for all $j$ from the $m$–th group.

## *An Alternative Approach*

To describe how nested sets are processed, we must first describe how they are stored in the computer. The natural approach assumes that we store each nested set as a sequence of sets $X(\alpha_0), ..., X(\alpha_{d+1})$. When all the sets $X(\alpha_j)$ are intervals, this is a reasonable idea: each interval is stored as two numbers (left- and right endpoints), so, to store $d+2$ intervals, we store $2(d+2)$ numbers. To describe a nested interval, we need exactly that many numbers, so, there is no redundancy in this storage scheme.

However, if the sets $X(\alpha_j)$ are more complicated than intervals, then this expert-by-expert storage scheme is not the best one. Indeed, how do we store a set $S$ in the computer? If the Universe of discourse $U$ is finite, then we can represent a set $S$ by storing, for each $u \in U$, the value 1 or 0 ("true" and "false") depending on whether $u$ belongs to the set $S$ or not. (This is, e.g., how set type in Pascal is represented.) In other words, a set $S$ is stored as the values of its *characteristic function* $\chi_S(u)$.

So, to store $d+2$ sets $X(\alpha_j)$, we must, for every $u$, store $d+2$ truth values $\chi_{X(\alpha_j)}(u)$. So, for each $u \in U$, we need $d$ bits to store all the information about this $u$. But this is wasting storage space. Indeed, we know that the sequence $X(\alpha_j)$ is monotonic: $X(\alpha_j) \subseteq X(\alpha_k)$ if $j > k$. Therefore, if for some element $j$, we know that $u \in X(\alpha_j)$, then we automatically know that $u \in X(\alpha_{j-1})$, ..., $u \in X(\alpha_0)$; we do not need to waste storage space to store these values. Similarly, if we know that $u \notin X(\alpha_j)$, then we can automatically deduce that $u \notin X(\alpha_{j+1})$, ..., $u \notin X(\alpha_{d+1})$.

In view of this monotonicity, for every $u \in U$, there are only $d+3$ possible situations:

- $u \notin X(\alpha_j)$ for all $j$;
- $u \in X(\alpha_j)$ and $u \notin X(\alpha_{j+1})$ for some $j = 0, ..., d$;
- $u \in X(\alpha_j)$ for all $j$.

Since there are $d+3$ cases, we can enumerate them by numbers from 0 to $d+2$, and describe the case in the computer by its number. To store a number from 0 to $d+2$ in its binary form, we need $\approx \log_2(d+2)$ bits. For large $d$, we have $\log_2(d+2) \ll d$ bits. So, this is a much better storage scheme.

Hence, a better way to *describe* (*store*) a nested set is for each $u \in U$, to store the number of the case that describes this particular $u$.

*Comment.* If a set $U$ is infinite, then we cannot store all the values $\chi_S(u)$ of the characteristic function. Instead, we have a *program* that, given $u$, returns the value of $\chi_S(u)$. For the new storage scheme, we will similarly need a function that would input $u$ and return the corresponding case number.

This alternative representation of nested sets not only saves computer *memory*, but it can also lead to *faster algorithms*. Indeed, in modern computers, the actual time to perform arithmetic operations with numbers that are already in the processor is negligible in comparison with the time that is necessary to fetch these numbers from the memory to the processor and move the results back to the memory. Data processing algorithms are often reasonably simple, so that few arithmetic operations are necessary; therefore, the major part of the computation time for these algorithms is taken by this fetching. For such data processing algorithms, the fewer memory cells we need to store the data, the faster the resulting algorithms. Therefore, for the alternative representation, the algorithms can be faster.

We should warn the reader that although we practically *proved* that the alternative representation leads (asymptotically) to better storage, the reduction in computation time is justified only by, so to say, *physical* (less formal) arguments. These arguments seem to infer that in some cases, the alternative representation leads to faster computations, but, on the other hand, in other cases (e.g., if the data processing algorithm is complicated enough) expert-by-expert computations may turn out to be faster.

There is also a practical argument that the alternative representation is often faster: In the following sections, we will describe how to define and process nested sets stored in this manner. It turns out that this alternative way of storing nested sets corresponds to *fuzzy sets* and operations with these sets. So, numerous successful applications of fuzzy techniques are a justification that this alternative representation is indeed often faster.

*Mathematical comment.* From the mathematical viewpoint, it would be nice to have a clear realistic example in which the computation time is estimated for both approaches, and in which the alternative leads to a faster processing algorithm; unfortunately, there is no such example in the literature. We feel that, no matter how successful the applications are, from the mathematical viewpoint, the absence of such an example is a major drawback of this approach. We hope therefore that such an example will eventually appear.

*Comment.* According to the alternative representation, to *describe* (*store*) a nested set, for each $u \in U$, we store the number of the case that describes this particular $u$. A natural way to enumerate these cases is to use the value $j$ (or $\alpha_j$) which is the largest $\alpha$ for which $u \in X(\alpha_j)$. This definition also covers the last case (when $u \in X(\alpha_j)$ for all $j$). However, it does not cover the first case, when no such $\alpha$ exists. For sets of real numbers, the supremum of the empty set is defined as $-\infty$, i.e., as a special value that is added to a real line, and that has the property that it is smaller than everything on the real line. We can make a similar addition to our ordered set $A$, and arrive at the following definitions:

## Mathematical Formulation of the New Approach

**Definition 6.** *For every ordered set $A$, let us define $A^*$ as $A \cup \{\alpha_{-1}\}$, where $\alpha_{-1} < \alpha$ for every $\alpha \in A$. For every set $B \subseteq A$, we define $\sup B$ as the largest element of $B$, and $\sup \emptyset$ as $\alpha_{-1}$. For every nested set $X$, we define $\mu_X(u)$ as $\sup\{\alpha | u \in X(\alpha)\}$. This function $\mu_X$ is called a* **membership function** *of the nested set $X$.*

In other words, $\mu_X(u)$ is the largest value $\alpha$ for which the degree of belief that $u \in X$ is $\geq \alpha$. The largest value with this property is clearly the degree of belief itself. So, the values $\mu_X(u)$ is called the *degree of belief* that $u \in X$.

In other words, we represent a nested set by a function $\mu : U \rightarrow A^*$. That this description is not redundant can be seen from the fact that such functions are in 1-1-correspondence with nested sets:

**PROPOSITION 2.** *Every function $\mu : U \rightarrow A^*$ is a membership function of some nested set.*

To use this type of storage, we must be able to do the following:

- First, we must reformulate all operations with nested sets in terms of membership functions.
- Second, we must be able to reconstruct the nested set from its membership function.

**PROPOSITION 3.**
$\mu_{X\cap X'}(u) = \min(\mu_X(u), \mu_{X'}(u))$;
$\mu_{X\cup X'}(u) = \max(\mu_X(u), \mu_{X'}(u))$,
$\mu_{X^\perp}(u) = f_\neg(\mu_X(u))$, *where* $f_\neg(\alpha_j) = \alpha_{d-j}$.

**PROPOSITION 4.** (Nguyen [53])

$$\mu_{f(X_1,\ldots,X_n)}(u) = \max_{u_1,\ldots,u_n : f(u_1,\ldots,u_n)=u} \min(\mu_{X_1}(u_1), \ldots, \mu_{X_n}(u_n)).$$

**PROPOSITION 5.** *For every nested set* $X$, $X(\alpha) = \{u|\mu_X(u) \geq \alpha\}$.

*Historical Comment.* In view of these propositions, a nested set can be defined as a function $\mu : U \rightarrow A^*$. For several nested sets, the union and intersection can be defined as in Proposition 3, and the result of applying a function to nested sets – as in Proposition 4. These definitions have been originally proposed by L. Zadeh (on different grounds) in [36], with the only difference that what we called *nested set* Zadeh called a *fuzzy set*. The set $\{u|\mu_X(u) \geq \alpha\}$ is called an *$\alpha$–level set*, or *$\alpha$–cut* of the fuzzy set $X$. Formula from Proposition 4 is called *extension principle*. It was proposed by Zadeh in his pioneer paper [36] (see also [79, 37, 16, 68, 33]).

Thus, we get an *explanation* of Zadeh's formulas. Different parts of this explanation appeared in Gentilhomme [20], Tanaka et al. [63, 64], Negoita et al. [51, 52], Orlovsky [56, 57], Nguyen [53], Dubois et al. [16, 17], Kaufmann et al. [29], Artbauer [1], Gu et al. [27], Goodman et al. [24, 25], Klement [31], Lai [36], Tamura et al. [62], Uehara et al. [66], Bojadziev et al. [11], Kruse et al. [35], Grabisch et al. [26], Klir et al. [33], Mantaras [42].

*Comments.*

- Thus, nested sets provide a new justification of the formulas of fuzzy set theory, formulas that have been originally proposed on a more intuitive basis, without a convincing mathematical justification. We hope that this new justification will be more convincing, at least for specialists in interval computations.

- Instead of describing a function with finitely many values, it is often more convenient to approximate it by a continuous function. In other words, we can generalize it to the case of *infinite* set $A$. Generalization of Proposition 4 to infinite $A$ and related problems are described in Nguyen [53] and Fuller et al. [18].

## How to Store Degrees of Belief in the Computer

Different degrees of belief can be interpreted as ranging from "absolutely false" ($\alpha_{-1}$) to "absolutely true ($\alpha_{d+1}$). In the computer, "false" is usually represented by 0, and "true" by 1. So, it is reasonable to represent intermediate degrees of belief as real numbers from 0 to 1 in such a way that $\alpha_{-1}$ corresponds to 0, and $\alpha_{d+1}$ to 1. The most natural arrangement is when the resulting values are equally spaced, i.e., if $\alpha_{-1} = 0$, $\alpha_0 = 1/(d+2)$, ..., $\alpha_i = (i+1)/(d+2)$, ..., $\alpha_{d+1} = 1$. These values can be interpreted as follows: Indeed, before we consider our $d$ experts with their additional knowledge, we have only two options: either we can be pessimistic, and assume that every value that can potentially be possible is actually possible (in this case, our set of possible values is $X(\alpha_0)$), or we can be extremely optimistic and assume that only those values about which we are 100% sure that they are possible are actually possible, and none else (in this case, our set of possible values is $X(\alpha_{d+1})$). After we add $d$ experts, we now have $d+2$ possible viewpoints: a pessimistic one, an optimistic one, and $d$ viewpoints of different experts.

It is easy to see that for every value $u \in U$ and for every nested set $X$, $\mu_X(u)$ is equal to the ratio of the two numbers:

- the numerator $i+1$ is the number of viewpoints according to which $u$ is a possible value of $x$, and
- the denominator is the total number $d+2$ of possible viewpoints.

Indeed, if $\mu_X(u) = \alpha_i = (i+1)/(d+2)$, $1 \le i \le d$, this means that the $i$–th is the most optimistic expert who believes that $u$ is possible. Therefore, since we have assumed that the experts are ordered according to their cautiousness, $i$ experts $1, ..., i$ believe that $u$ is possible, and also, in this case, a pessimist would believe that $u$ is possible. So, $u$ is possible according to $i+1$ out of $d+2$ viewpoints. Similarly, this equality can be checked for $i=-1$, $i=0$, and $i=d+1$.

This equality leads to the possibility to determine $\mu_X(u)$ by ***polling***. To do the polling, we need to get experts who represent different points of view: both experts whose expertise leads to some additional knowledge and those who simply know the field, but do not add extra knowledge to it. There are only $d+2$ possible viewpoints, so, we have $N = d+2$ people who represent $N$ different viewpoints. If $m$ out of these $N$ agree with the statement that "$u$ is a possible value of $x$", then we take $\mu_X(u) = m/N$ as the degree of belief in this statement. This idea is indeed one of the most widely used ways of estimating the values of a membership function; see, e.g., Blin et al. [9, 8], Dubois et al. [16], Section IV.1.d, Klir et al. [10].

## Average Number of Elements

When the universal set $U$ is finite, we can ask the following natural question: how many elements are possible? For an expert $\alpha$, the number of possible elements is equal to $|X(\alpha)|$, where by $|S|$, we denote the number of elements in a set $S$. As a natural measure of size, we can take the arithmetic ***average*** over all experts. It turns out that this average can be described in terms of the membership function $\mu_X$:

**PROPOSITION 6.** *Let $\alpha_i = (i+1)/(d+2)$. Then, for every nested set $X$,*

$$\frac{1}{d+2}\sum_{\alpha\in A}|X(\alpha)| = \sum_{u\in U}\mu_X(u).$$

***Historical comment.*** For fuzzy sets, this number was introduced by Zadeh and called ***sigma-count*** [81, 82].

## The Simplest Case of $d=0$: Tetrvals and Twins

**Intervals with intervally uncertain bounds.** If experts that were initially unsure make their decision, then the resulting degree of belief in $S$ can take any value from the ***interval*** $[d(S), pl(S)]$: the value $d(S)$ corresponds to the case when they all decide against $S$, and the right-hand side $pl(S)$ corresponds to the case when all these experts choose to believe in $S$. We have already mentioned that the ratio $N(S)/N$ depends on $N$ and is therefore, not a very good description of the experts' belief. To get a better description, we can use an *interval* $\mathbf{d}(S) = [d^-(S), d^+(S)]$ of possible values (e.g., $d^{\pm}(S) = (N(S) \pm 0.5)/N$. Similarly, we get an interval $\mathbf{pl}(S)$ of possible values of plausibility. As a result, we get a representation of the expert's belief in $S$ by a *pair* of intervals $[\mathbf{d}(S), \mathbf{pl}(S)]$. This representation was proposed in Atanassov et al. [2, 3, 2], and called *interval-valued intuitionistic fuzzy logic*. This description is equivalent to the following one: instead of a single interval $\mathbf{d}(S) = [d(S), pl(S)]$ to represent the expert's degree of belief, we now have *two* nested intervals: a more cautious one $\mathbf{D}(S) = [d^-(S), pl^+(S)]$, and a more risky one $\mathbf{d}(S) = [d^+(S), pl^-(S)] \subseteq \mathbf{D}(S)$.

For intervals with intervally uncertain bounds, we have a modified problem:

- For *expert systems*, we have a knowledge base that consists of statements $S_1, ..., S_N$; for a given query $Q$, we also have an algorithm (obtained by using the inference engine) $f(d_1, ..., d_N)$ that transforms the degrees of belief $d(S_1), ..., d(S_N)$ in the statements from the knowledge base into a degree of belief $d(Q) = f(d(S_1), ..., d(S_N))$ in $Q$.
- For *intelligent control*, we have an algorithm that transforms these same degrees of belief $d(S_1), ..., d(S_N)$ into the recommended control value $\bar{u} = f(d(S_1), ..., d(S_N))$.

Suppose now that we know only the *nested intervals* $\mathbf{d}(S_1) \subseteq \mathbf{D}(S_1)$, ..., $\mathbf{d}(S_N) \subseteq \mathbf{D}(S_N)$ that describe the expert's degrees of belief. From this information, we want to compute the nested interval $\mathbf{d}(Q) \subseteq \mathbf{D}(Q)$ (or $\mathbf{u} \subseteq \mathbf{U}$) for the result. Here,

$$\mathbf{d}(Q) = f(\mathbf{d}(S_1), ..., \mathbf{d}(S_N)) = \{f(d_1, ..., d_N) | d_i \in \mathbf{d}(S_i)\},$$

$$\mathbf{D}(Q) = f(\mathbf{D}(S_1), ..., \mathbf{D}(S_N)) = \{f(d_1, ..., d_N) | d_i \in \mathbf{D}(S_i)\}$$

(and similarly for $\mathbf{u}$ and $\mathbf{U}$). So, one way to compute the desired nested interval is to apply interval computations techniques twice and compute both intervals. In some case, the function $f$ is monotonic in each of the variables, so, for each $\alpha$, we can simply compute the values of $f$ at the corresponding endpoints. (For nested intervals in intelligent control, this idea is described in Dubois et al. [15].)

Another possibility is to apply algorithms that handle the nested interval as a single object. Such methods were proposed in Gardeñes, Zyuzin, Musaev [19, 89, 90, 45, 46], where the nested interval is called a *twin* or a *tetrval*.

*Comment.* A similar notion of "uncertainty of systematic uncertainty" has also been proposed in Loo [38] in slightly different terms, but, as shown in Artbauer [1], it is essentially a tetrval. A similar idea of describing uncertainty by two intervals was proposed by Narin'yani [47, 48, 49].

## 2.5 Nested Sets and Nested Intervals: General Case

In the previous text, we considered the case when the experts could be linearly ordered according to their cautiousness. In real life, an expert can be more cautious in one aspect, but more risky in another, so comparing the experts' riskiness does not necessarily lead to a linear order. We can still define an ordering by saying that an expert $\beta$ is *riskier* than expert $\alpha$ (and denoting it by $\alpha \leq \beta$) if for every quantity $x$, the set $X(\alpha)$ of possible values of $x$ according to $\alpha$ contains all the values $X(\beta)$ that experts $\beta$ consider possible (i.e., if $X(\alpha) \supseteq X(\beta)$), but this ordering is no longer linear.

When all experts were linearly ordered, our choice was limited: we could choose one of them and consider whatever this expert says to be true. In the general situation, we have more choices. For example, we can choose two experts $\alpha$ and $\beta$ whom we consider the most credible, and believe everything that each of these two experts say. Therefore, if for some quantity $x$, these two experts describe the sets $X(\alpha)$ and $X(\beta)$ of possible values, then according to this description of beliefs, we believe that the actual value of $x$ must belong to the intersection $X(\alpha) \cap X(\beta)$. It is reasonable to describe the resulting belief situation as $\alpha \cap \beta$. On the other hand, we may not completely trust each of these experts individually, but we can believe that one of them is right. For this degree of belief, possible values of $x$ form the set $X(\alpha) \cup X(\beta)$. This situation can be described as $\alpha \cup \beta$.

In general, we can take more complicated descriptions of beliefs: e.g., we can say that we believe something if either expert 1 says it, or experts 2 and 3 both agree on it. The resulting "degrees of beliefs" satisfy the following property: whatever degrees of beliefs $\alpha$ and $\beta$ we have, we can always come up with the degree of belief $\alpha \cap \beta$ (that corresponds to believing in everything in both $\alpha$ and $\beta$), and with the degree of belief $\alpha \cup \beta$ (that corresponds to believing that one of the beliefs $\alpha$ and $\beta$ was correct). In mathematical terms, the set of possible degrees of belief thus becomes a *lattice*, in which the *union* $\vee$ (the least upper bound) corresponds to $\cap$, and the *meet* $\wedge$ (the greatest lower bound) corresponds to $\cup$.

For the basic quantities, we have $X(\alpha \cap \beta) = X(\alpha) \cap X(\beta)$ and $X(\alpha \cup \beta) = X(\alpha) \cup X(\beta)$. However, these properties are not necessarily true for derivative quantities. For example, assume that the first expert $\alpha_1$ believes that $x \in X(\alpha_1) = \{-1, 0\}$, and the second expert $\alpha_2$ believes that $x \in X(\alpha_2) = \{0, 1\}$. Then, if we believe both experts, we can conclude that $x \in X(\alpha_1 \cap \alpha_2) = X(\alpha_1) \cap X(\alpha_2) = \{0\}$. However, for $y = x^2$, this property is no longer true: for the first expert, we have $Y(\alpha_1) = \{0, 1\}$. Similarly, for the second expert, we have $Y(\alpha_2) = \{0, 1\}$. However, if we believe both experts, then $x = 0$, $y = 0$, and hence, $Y(\alpha_1 \cap \alpha_2) = \{0\} \neq Y(\alpha_1) \cup Y(\alpha_2)$.

Similarly to the linearly ordered case, in addition to the degrees of belief that describe combinations of expert opinions, we have an element $\alpha_0$ for which $X(\alpha_0)$ describes the set of elements that can be described as possible without the necessity to believe any of the experts, and an element $\alpha_{d+1}$ for which $X(\alpha_{d+1})$ describes the elements ***proven*** to be possible. For every other degree of belief $\alpha$, $X(\alpha_0) \supseteq X(\alpha) \supseteq X(\alpha_{d+1})$. Therefore, if we define $\alpha_0 \cup \alpha$ and $\alpha_{d+1} \cup \alpha$ as $X(\alpha_0 \cup \alpha) = X(\alpha_0) \cup X(\alpha)$ and $X(\alpha_{d+1} \cup \alpha) = X(\alpha_{d+1}) \cup X(\alpha)$, we will conclude that $\alpha_0 \cup \alpha = \alpha_0$ and $\alpha_{d+1} \cup \alpha = \alpha$. Similarly, $\alpha_0 \cap \alpha = \alpha$ and $\alpha_{d+1} \cap \alpha = \alpha_{d+1}$. In algebraic terms, this means that in the lattice of degrees of belief, elements $\alpha_0$ and $\alpha_{d+1}$ play the role of 0 and 1. As a result, we arrive at the following definition.

**Definition 1′.**

- *Let an integer $d$ be fixed. This number will be called the **number of experts.***
- *Let a finite lattice $A$ be fixed, with elements $\alpha_0$ and $\alpha_{d+1}$ that are 0 and 1 (i.e., for which $\alpha_0 \vee \alpha = \alpha_0$, $\alpha_{d+1} \vee \alpha = \alpha$, $\alpha_0 \wedge \alpha = \alpha$, and $\alpha_{d+1} \wedge \alpha = \alpha_{d+1}$). Elements of $A$ will be called **degrees of belief.***
- *Let a set $U$ be fixed. This set $U$ will be called the **Universal set**, or the **Universe of discourse**.*
- *By a **piece of information** (or, a **nested set**), we mean a non-decreasing mapping $X$ from $A$ to the set $2^U$ of all subsets of $U$, i.e., a mapping for which if $\alpha \leq \beta$, then $X(\alpha) \supseteq X(\beta)$.*
- *Let $u$ be an element of $U$, and $\alpha \in A$ be a degree of belief. We say that $u$ **belongs to the nested set $X$ with a degree of belief** $\geq \alpha$ if $u \in X(\alpha)$.*

We can define intersection and union of nested sets by Definitions 2 and 3. It is easy to check that if $X$ and $X'$ are nested sets, then their union and intersection are nested sets as well. The result of processing nested sets can be described by Definition 5. Since $A$ is a finite lattice, every set has a supremum, and we can use Definition 6 to define a membership function $\mu_X(u)$. However, in this case, Proposition 5 will no longer be true. For example, in the above example, $Y(\alpha_1) = Y(\alpha_2) = \{0,1\}$, so, $\mu_Y(1) = \sup\{\alpha_1, \alpha_2\} = \alpha_1 \vee \alpha_2$. So, for $u = 0$ and $\alpha = \alpha_1 \vee \alpha_2 = \alpha_1 \cap \alpha_2$, we have $\mu_Y(u) \geq \alpha$. However, as we have already mentioned, $1 = u \notin Y(\alpha) = Y(\alpha_1 \vee \alpha_2) = \{0\}$.

*Historical comment.* Membership functions with values in a lattice were originally proposed in Goguen [22, 23].

In such cases, an ideal way to store the experts' opinion is to store the conclusions of all the experts. However, in many cases, this takes too much storage space. To reduce the storage space, we may want to follow the idea applied above: forget about who said what, and only store the number of experts who have voted for a given statement. This approach was used by Kohout et al. (see, e.g., [34] and references therein).

The problem with this approach is as follows: When we store the complete information about the experts' opinion, then for every two statements $A$ and $B$, from the information about $A$ and the information about $B$, we can determine the experts' opinion about $B$. However, if we only know the number of experts who believe in $A$, and the number of experts who believe in $B$, then we cannot determine uniquely the number of experts who believe in $A\&B$. For example, if half of the the experts believe in $A$, and half believe in $B$, then it could be that the same experts who believe in $A$ also believe in $B$; in this case, exactly half of the experts believe in $A\&B$. It can also happen that none of the experts who believe in $A$ believe in $B$, in which case, none of the experts believe in $A$ and $B$ at the same time. So, instead of a single value of $d(A\&B)$, we get an *interval* of possible values of $d(A\&B)$. This situation is described in the surveys Nguyen et al. [54, 55] (see also references therein).

# 3 OTHER PROBLEMS WHERE NESTED SETS AND NESTED INTERVALS CAN BE USED: IDENTIFICATION, OPTIMIZATION, CONTROL, AND DECISION MAKING

In the above text, we mainly concentrated on the problem of estimating accuracy of the results of indirect measurements. In this section, we will show how nested sets and nested intervals can be (and are) used in other application problems: identification, optimization, control, and decision making.

## 3.1 Identification Problem for Nested Sets and Nested Intervals

In general, the identification problem can be formulated as follows: we know that a physical quantity $y$ depends on the quantities $x_1, ..., x_n$; we want to find the dependency $y = f(x_1, ..., x_n)$. In order to find the dependency, we measure $x_i$ and $y$ in several situations. As a result, we get finitely many tuples $(x_1^{(k)}, ..., x_n^{(k)}, y^{(k)})$, $1 \le k \le N$ (they are usually called *patterns*), and we must find a function $f$ that satisfies $N$ conditions $y^{(k)} = f(x_1^{(k)}, ..., x_n^{(k)})$. There are infinitely many functions that satisfy these conditions. In order to choose one of them, we must have some additional knowledge about the function $f$ (i.e., we must restrict $f$ to some finite-dimensional class). Often, this class consists of linear functions or, in a slightly more general setting, functions that are linear combinations of known ones, i.e., functions of the type $f(x_1, ..., x_n) = c_1\phi_1(\vec{x}) + ... + c_m\phi_m(\vec{x})$, where the base functions $\phi_j$ are given, and the coefficients $c_j$ have to be determined from the experiments.

If for every measurement, we only know the interval of possible values of error, then we get an interval identification problem. If, in addition to the guaranteed estimates, we have experts' estimates of these same errors, then the possible values of $x_i$ and $y$ in each measurement are characterized by *nested sets* (in particular, by nested intervals). In this case, we have a nested set (interval) identification problem (see, e.g., Tamura et al. [62]): we have nested intervals $X_i^{(k)}$ and $Y^{(k)}$, and we want to find a function $f$ for which $f(X_1^{(k)}, ..., X_n^{(k)}) = Y^{(k)}$ for some mapping $f$ from tuples of nested intervals into nested intervals. A nested interval $X$ can be described by the bounds $x \pm (\alpha)$ of its intervals $X(\alpha)$,

$\alpha \in A$. So, in general, each value $y^{\pm}(\alpha)$ is a function of $2n(d+2)$ variables $x_i^{\pm}(\beta)$ for all $\beta \in A$.

This complicated regression problem can be simplified, if we take into consideration that the nested interval identification problem makes sense only if we use a single measuring instrument for each of the variables $x_i$ and $y$: for example, if we use two different measuring instruments (with different accuracies) to measure $y$, then, even if we know the results of measuring $x_i$, we cannot predict the interval $Y(\alpha)$, because we do not know which of the instruments was used.

In case a single instrument is used, for every $\alpha$ and $\beta$, the larger $x^-(\alpha)$, the larger the measurement result $\tilde{x}$ and therefore, the larger $x^{\pm}(\beta)$. So, for every quantity $x$, there is a 1-1 correspondence between the variables $x^-(\alpha)$ and $x^{\pm}(\beta)$. In view of this correspondence, there is no need to consider all $2(d+2)$ values $x_1^{\pm}(\alpha_i)$ as inputs for the function $f$: one of these variables will be sufficient.

It is convenient, when predicting $y^{\pm}(\alpha)$, to choose the variables $x_i^{\pm}(\alpha)$ that correspond to this same degree of belief $\alpha$ and to this same bound (left if $-$, right if $+$). As a result, we have the numerical identification problem $y^{\pm}(\alpha) = f^{\pm}(\alpha, x_1^{\pm}(\alpha), ..., x_n^{\pm}(\alpha))$. In particular, if we know that the dependency between $y$ and $x_i$ is linear, it is natural to choose each function $f^{\pm}(\alpha)$ to be also linear in its variables, i.e., $y^{\pm}(\alpha) = c_0^{\pm}(\alpha) + c_1^{\pm}(\alpha) \cdot x_1^{\pm}(\alpha) + ... + c_n^{\pm}(\alpha) \cdot x_n^{\pm}(\alpha)$ for some coefficients $c_i^{\pm}(\alpha)$. So, we have $2(d+2)$ numerical regression problems: two (corresponding to $+$ and $-$) for each $\alpha \in A$. Similarly to the main problem, we have two possibilities here: One possibility is to solve the resulting regression problems one by one. Another is to try to combine them together. The values $c_i^{\pm}(\alpha)$ resemble the bounds $x_i^{\pm}(\alpha)$ that form a nested interval. However, for the coefficients, it is not necessarily true that $c_i^-(\alpha) \le c_i^+(\alpha)$ or that the resulting intervals $[c_i^-(\alpha), c_i^+(\alpha)]$ are decreasing in $\alpha$. So, to describe the coefficients $c_i^{\pm}(\alpha)$ as a single object, we must describe some generalization of the notion of the nested set. Such a generalization was proposed in [62].

## 3.2 Optimization for Nested Sets and Nested Intervals

Suppose that we must optimize the function $f(\vec{x}, \vec{c})$ under the conditions that $g_i(\vec{x}, \vec{c}) \le 0$. Here, $\vec{x} = (x_1, ..., x_n)$ are the desired variables, and $\vec{c} = (c_1, ..., c_m)$ are parameters that are known with uncertainty (i.e., only nested sets $C_i$ are known). Then, for each degree of belief $\alpha$, we believe that $c_i \in C_i(\alpha)$.

If we fix $\vec{c}$, we get a constrained optimization problem; let us define the set of its solutions $(x_1, ..., x_n)$ by $X(\vec{c})$. (This set can consists of a single solution, or it can contain several, if the optimum is not attained at a single point $\vec{x}$.)

For each degree of belief $\alpha$, we can define the set $S(\alpha)$ of possible solutions to the optimization problem as the set of all solutions $\vec{x}$ that are possible under this degree of belief; in mathematical terms, $S(\alpha)$ is the union of the solution sets $X(\vec{c})$ for all $\vec{c}$ for which $c_i \in C_i(\alpha)$. These sets $S(\alpha)$ form a *nested set* that can thus be viewed as a nested solution of the optimization problem.

Optimization problems for fuzzy (=nested) sets have been first formulated in Bellman et al. [5].

Optimization has been done using intervals for a long time (see, e.g., a survey by Kearfott [30]). The majority of the methods of nested set optimization are, in effect, reasonably straightforward generalizations of these interval optimization techniques.

For a detailed exposition of the modern state of research in this field, the see, e.g., Zimmermann, Orlovsky et al. [83, 56, 84, 57, 85, 86, 87, 88].

## 3.3 Control and Decision Making

### *Decision Making with Finitely Many Alternatives*

Suppose that we must choose between two or finitely many alternatives based on a single criterion (i.e., cost or reliability). To make a choice, we must compute the value of this criterion for each parameter, and compare the resulting values. The desired value of the criterion depends on some parameters that we must measure. Due to uncertainty of measurements, for each alternative, we get not a single value of the criterion, but an *interval* of possible values. If for alternative $a$ the resulting interval $[a^-, a^+]$ is guaranteed to be larger than the interval $[b^-, b^+]$ that corresponds to the parameter $b$ (i.e., if $a^- \geq b^+$), then we definitely prefer $a$ to $b$. But what should be done if these two intervals intersect, and therefore, $a$ can be better or worse than $b$? In some cases, in addition to the guaranteed interval, we have expert estimates that lead to narrower intervals. In these cases, Loui [39] proposes the following: if guaranteed estimates do not enable us to make a unique decision, let us try to make the decision based on the intervals provided by the first (more cautious) expert $\alpha_1$; if this still does not work, let us use the interval provided by the next expert $\alpha_2$, etc, until finally, we find an expert whose interval is narrow enough for the decision to be made.

## *Decision Making: General Case*

In control and decision making, we may look for the best decision. For example, we may look for a decision $u$ for which $J(x, u) \to \max$ for some objective function $J$, where $x$ is a parameter (or parameters) that we measure. Since measurements are not absolutely precise, we will end up with the interval $X(\alpha_0)$ of possible values of $x$. For each $x \in X(\alpha_0)$, we can find the value $u$ that is optimal for this particular $x$. The values corresponding to different $x \in X(\alpha_0)$ form a *set* of possible controls $U(\alpha_0)$. In many cases, this set is an interval, but sometimes, its structure is more complicated.

An expert $\alpha \in A$ can believe that a smaller set $X(\alpha)$ of values of $x$ is possible. This expert will thus come up with a set $U(\alpha)$ that is a subset of $U(\alpha_0)$. So, we have a ***nested set*** of possible values of control. We need to choose a single value of $u$. For decision making, we can simply present the nested set to the decision maker and let the decision maker choose. For automatic control, we must automatically select a number from the set. In fuzzy set language, the transition from a nested (fuzzy set) to a number is called ***defuzzification***.

There are two basic approaches to defuzzification:

- First, since we need to select a single value, it makes sense to go with the bravest of the experts (the one whose choice is the narrowest); if this expert has already selected a single control value, we will choose it, otherwise, we will choose from the set $X(\alpha)$ provided by this expert.
- The second approach is to take into the consideration the opinion of all the experts.

Let us analyze these two approaches. We will start with the first approach. Let us assume (for simplicity) that the set from which we must choose is an interval. So, the question becomes: how to choose from an interval? We must describe a ***choice function*** $s$ that transforms an interval $\mathbf{a} = [a^-, a^+]$ into a number $a = s(\mathbf{a}) \in \mathbf{a}$. This function must satisfy the following natural properties. First, if we change the starting point for measuring $a$ (time and temperature are good examples where such a change is possible), then all the measured values are shifted ($x \to x + c$ for a fixed $c$). This is a formal change that does not affect our knowledge, like a change from Kelvin to centigrade in measuring temperature. Therefore, it is natural to assume that the choice provided by the function $s$ should not change after a change; in other words, that if we first make a choice (i.e., go from $\mathbf{a}$ to $s(\mathbf{a})$) and then change the starting point (i.e., go from $s(\mathbf{a})$ to $s(\mathbf{a}) + c$), the result must be the same as if we first change the scale (from $[a^-, a^+]$ to $[a^- + c, a^+ + c]$) and then make a choice: i.e., $s([a^- + c, a^+ + c]) = s([a^-, a^+]) + c$.

Similarly, If we change the unit in which we measure $a$, the choice must not change. If we change the unit (e.g, go from inches to centimeters), then, the numerical values of this quantity change as $x \to \lambda x$ for some $\lambda > 0$ (= the ratio of the old and the new units). So, we arrive at the formula $s([\lambda \cdot a^-, \lambda \cdot a^+]) = \lambda \cdot s([a^-, a^+])$.

It is also natural to assume that the choice function should be invariant w.r.t. changing the sign of $a$: e.g., if $a$ is the angle on which to steer the wheel, then the sign is just a convention, and the result should not depend on whether we decided to call clockwise angles positive or negative. So, we arrive at the following additional invariance: $s([-a^+, -a^-]) = -s([a^-, a^+])$. Combining these three conditions together, we obtain the following result:

**Definition 7.** *By a choice function, we mean a function $s$ that maps every interval $[a^-, a^+]$ into a point from that interval, and that has the following properties for every interval and for every $c$ and $\lambda > 0$:*

- $s([a^- + c, a^+ + c]) = s([a^-, a^+]) + c$ *(shift-invariance);*
- $s([\lambda \cdot a^-, \lambda \cdot a^+]) = \lambda \cdot s([a^-, a^+])$ *(unit-invariance);*
- $s([-a^+, -a^-]) = -s([a^-, a^+])$ *(symmetry).*

**PROPOSITION 7.** *Every choice function has the form*

$$s([a^-, a^+]) = \frac{a^- + a^+}{2}.$$

So, we take the *midpoint* of the interval of possibly best decisions. In fuzzy terms, this defuzzification method (midpoint of the smallest interval that corresponds to the largest $\alpha$) is called *mean of maxima*. This approach can be interpreted as follows: we consider all possibly best decisions, and we take the *average* of them (with equal weight). In mathematical terms, in the discrete case, this average would be equal to the ratio

$$\left(\sum_{x\in X} x\right) \Big/ \left(\sum_{x\in X} 1\right),$$

and in the continuous case, it is equal to the ratio

$$\left(\int_X x\,dx\right) \Big/ \left(\int_X dx\right).$$

If we have *several* experts who make different predictions, then in this averaging process, we can ask every expert to write down all the values that this expert considers possible, and average them. In this new averaging, every value is counted as many times as many experts consider it possible. In other words, we propose the following ratio:

$$\left(\sum_{\alpha\in A} \int_{X(\alpha)} x\,dx\right) \Big/ \left(\sum_{\alpha\in A} \int_{X(\alpha)} dx\right).$$

This ratio can be expressed in terms of the membership function $\mu_X(x)$:

**PROPOSITION 8.**

$$\left(\sum_{\alpha\in A} \int_{X(\alpha)} x\,dx\right) \Big/ \left(\sum_{\alpha\in A} \int_{X(\alpha)} dx\right) = \left(\int x\cdot\mu_X(x)dx\right) \Big/ \left(\int \mu_X(x)dx\right).$$

This formula is called *center of gravity defuzzification.*

These two formulas work well for *nested intervals*, but they do not always help for nested sets that are not intervals. For example, if the car is approaching an obstacle, then we can either turn to the left, or turn to the right. If the situation is absolutely symmetric, then the possible set of controls is, e.g., a union of two intervals $[-\Delta, -\delta] \cup [\delta, \Delta]$. Because of symmetry, the average of this set is 0, so, our choice will lead us straight ahead to the obstacle.

When the set from which we choose is a union of several intervals, Yen et al [75, 76, 77] propose taking the largest interval, and then taking the midpoint of this largest interval. For nested sets for which $X(\alpha)$ is a union of intervals, we can choose an interval that captures the most votes (i.e., for which $\int \mu_X(x)dx$ is the largest), and apply the formula described above (center-of-gravity defuzzification) to that interval only.

# 4 APPLICATIONS OF NESTED SETS AND NESTED INTERVALS

In this section, we survey the existing applications of nested sets and nested intervals.

## 4.1 Computing the Nested Interval for the Result of Data Processing

*Mathematical comment.* In this subsection, we survey applications of nested sets and nested intervals to the problem of obtaining the *true range* of a function, the problem that we described in the Introduction as one of the main problems of data processing. In mathematical terms, for each degree of belief $\alpha$, we have a function $f(x_1, ..., x_n)$ and $n$ intervals $X_i(\alpha)$, and we want to find the interval $Y(\alpha) = [y^-(\alpha), y^+(\alpha)]$ of the possible values of $f(x_1, ..., x_n)$ for $x_i \in X_i(\alpha)$. From the mathematical viewpoint, this problem is effectively an *optimization problem*: indeed:

- $y^-(\alpha)$ is the solution of the constrained optimization problem: $f(x_1, ..., x_n) \to \min$ under the constraints $x_i \in X_i(\alpha)$.
- $y^+(\alpha)$ is the solution of a similar constrained optimization problem: $f(x_1, ..., x_n) \to \max$ under the constraints $x_i \in X_i(\alpha)$.

Therefore, these applications can be viewed as particular cases of *applied optimization.*

**Case study: seismic engineering** (Dong et al [13]). As we have already mentioned in the introduction, earthquakes are extremely difficult to predict. Because of that, if we only use the equations and the measured data $x_i$, we get very wide intervals of possible magnitudes and frequencies $y$, intervals that are known to be much wider than the interval of actual values. So, if we use these wide intervals $[y^-, y^+]$ to design a building that is guaranteed to withstand a typical earthquake, these buildings will be unnecessarily (and often unrealistically) expensive. To make the requirements more realistic, in addition to guaranteed measurement results, expert estimates are normally used. For every parameter $x_i$ (e.g., for the frequency of earthquakes), estimates by different experts form a nested interval (sometimes, a nested set). In [13], interval methods are used for an expert-by-expert processing of these nested intervals.

Namely, to estimate the range of $f$ on a box $X_1(\alpha) \times ... \times X_n(\alpha)$, the box is subdivided into several sub-boxes small enough that each sub-box contains at most one local extremum point; for each sub-box, the range is estimated. Then, the union of the estimates for sub-box ranges is taken as an estimate for the total range. On each sub-box, the maximum of $f$ is attained either at an interior point (in which case this point is a local extremum of $f$), or on one of the sides. For each side, we can repeat a similar procedure, and end up with a conclusion that the maximum is attained either at one of the corners, or at a local extremum point of one of the sides of smaller dimension. There are at most $\approx 2^n$ vertices and local extremum points in each sub-box, so, for small $n$, we can simply enumerate them all. The resulting nested intervals $Y$ are presented to the decision makers.

**Case study: AIDS epidemics** (Berleant et al [6]). This is another example of usefulness of nested sets presented in the Introduction. Equations that describe how AIDS is transferred are known. To solve these equations, we must know the initial conditions and the the numerical values of the parameters (like the death rate among AIDS patients, etc). The majority of these parameters can be determined more or less accurately from known statistical data. However, there is one important parameter about which it is very difficult to get a reliable statistical estimate: the number of intravenous drug abusers. So, in the absence of a reliable statistical estimate, we may use estimates of experts. Different experts can give different intervals of possible values, so we have nested intervals for describing the expert knowledge (that can be described by a membership function). Applying interval computations to intervals corresponding to several levels, we get a nested interval for the predicted number of AIDS patients in the year 2000 [6]. Due to the serious input uncertainty, all level intervals are so wide that we cannot simply neglect quadratic terms and apply linearization techniques; therefore, we have to use interval computation techniques.

To get a better estimate, it is reasonable to *combine* these expert estimates with the statistical estimates. A natural way to combine expert knowledge with statistical knowledge is to represent the statistical information (that is given in terms of a probability density function) in terms of nested intervals; then, we will simply take an intersection of the two nested intervals that represent expert and statistical knowledge. This combination can be done as follows: crudely speaking, $X(\alpha)$ describes the interval in which $\alpha-$th part of the experts believe. So, to get a similar interval from the probability density function $\rho$, we can take the set $S = \{u|\rho(u) \geq p_0\}$ for some $p_0$ for which the probability to belong to $S$ is exactly $\alpha$. This approach has lead to a reasonable prediction [6].

**Case study: engineering design** (Yang et al [69]). In design, we would like to predict the properties of a designed system without actually designing it in hardware. So, we need to estimate the values $y = f(x_1, ..., x_n)$ for different functions $f$ and for different nested intervals $x_i$. In manufacturing measurements, the accuracy is limited by cost considerations, so the intervals $X_i(\alpha)$ are not very small, and therefore, we cannot neglect quadratic terms and use linearization techniques. Instead, methods of interval computations are used.

**Case study: decision analysis and risk analysis** (Dong et al [14]). Usually, we have several criteria $c_1, ..., c_m$ to choose between several alternatives $a_1, ..., a_n$ (e.g., when we are choosing a car, we may opt for speed, or for fuel efficiency, or for cost, etc). We may choose the alternative $a_i$ for which the first criterion $c_1$ takes the largest possible value, i.e., for which $c_1(a_i) \to \max_i$, or we can take the alternative for which $c_2(a_i) \to \max_i$, etc. A natural way to handle this multi-criteria situation is to assign to every alternative a *weighted average* of the values $c_1(a_i), ..., c_i(a_i)$, i.e., the value $r_i = (w_1 c_1(a_i) + ... + w_n c_1(a_n))/(w_1 + ... + w_n)$. The values $c_j(a_i)$ are obtained from measurements, so they are represented by nested intervals. The weights are also obtained from analysis of experiments (or simply by asking experts), so the weights are also nested intervals. So here, the $x_i$ are values $x_1 = c_1(a_i), ..., x_n = c_1(a_n), x_{n+1} = w_1, ..., x_{2n} = w_n$, and $f(x_1, ..., x_{2n}) = (x_1 x_{n+1} + x_2 x_{n+2} + ... + x_n x_{2n})/(x_{n+1} + ... + x_{2n})$. (Sometimes, a more complicated combination function $f$ is used.) The nested intervals for $c_{ji}$ and $w_i$ are reasonably large (especially in the design stage), so we cannot apply linearization techniques. In [14], a degree-by-degree approach is used. Under many reasonable assumptions, the function $f$ is monotonic. For a monotonic function $f$, for every $\alpha$, the maximum and minimum for $x_i \in X(\alpha) = [x^-(\alpha), x^+(\alpha)]$ are attained for $x_i = x^\pm(\alpha)$. Therefore, for every $\alpha$, it is sufficient to consider the values of $f$ for $2^{2n}$ "corners" $(x_1^\pm, ... x_{2n}^\pm)$. For small $n$, this a reasonable procedure, much faster than other known numerical techniques [14].

If we know that a function $f$ is monotonically increasing in each variable, then we can only consider two points instead of $2^n$: $f(x_1^-(\alpha), ..., x_n^-(\alpha))$ is the lower bound of the desired interval $Y(\alpha)$, and $f(x_1^+(\alpha), ..., x_n^+(\alpha))$ is the upper bound for this interval (Dubois et al [16]). Another case when a simpler algorithm is possible is when $f$ is a convex function (see Schmucker [59]).

*Comment.* Several methods and examples of processing fuzzy inputs are discussed in Mareŝ [43].

## 4.2 Optimization

*Mathematical comment.* We have already mentioned that, in the above examples, although we start with a practical problem that is not formulated in terms of optimization, the problem is equivalent to an optimization problem. In other cases, the applied problem itself is already formulated as an optimization problem. Examples of such applications are given in this subsection.

**Case study: agriculture** (Shaocheng [60]). When we plan to run a chicken farm, we must know how much money will be required to feed the chickens. Experts can tell us how many calories, proteins, vitamins, etc, a chicken needs per day; how many of these are contained in each type of available chicken food (e.g., soya, millet, etc), and what is the cost per unit for each type of food. We must find the cost of the cheapest combination that provides for the chickens. In mathematical terms, if we denote the total number of available types of food by $n$, and the total number of necessary food ingredients that the menu must contain by $m$, then the problem is: find the amounts $x_i$, $1 \le i \le n$, of different components that minimize the total cost $c = c_1 x_1 + ... + c_n x_n$ under the conditions that the total amount of every ingredient is sufficient, i.e., that the total amount $a_{j1}x_1 + ... + a_{jn}x_n$ of this ingredient in the resulting mixture must exceed the chicken's demand $d_j$: $a_{j1}x_1 + ... + a_{jn}x_n \ge d_j$ for all $j$. At any given moment of time, we will know the exact current values of the parameters $c_i$, $a_{ij}$, and $d_j$, and we will be able to find the optimal values of $x_i$ and the corresponding cost $c$ by solving a linear programming problem. However, in the planning stage, we can only have *interval* predictions of the costs and of, say, protein contents of the future crop. In other words, from experts, we can get *intervals* $\mathbf{c}_i = [c_i^-, c_i^+]$, $\mathbf{a}_{ij} = [a_{ij}^-, a_{ij}^+]$, $\mathbf{d}_j = [d_j^-, d_j^+]$ of possible values of the corresponding parameters. In this case, we can only get an *interval* estimate for the cost. From the mathematical viewpoint, we must solve the *interval linear programming* problem: describe the set of all possible costs for all linear programming problems that can be obtained by selecting numbers from the given intervals. For this particular problem, the desired interval $\mathbf{c} = [c^-, c^+]$ is easy to describe: the lowest overall cost corresponds to when the cost of the components is lowest, the contents is highest, and the demand is the lowest possible. In other words, $c^-$ is the solution of the linear programming problem $c^- = c_1^- + ... + c_n^- x_n \to \min$ subject to the constraint $a_{j1}^+ x_1 + ... + a_{jn}^+ x_n \ge d_j^-$. Similarly, the largest possible optimal cost is a solution of the linear programming problem $c^+ = c_1^+ + ... + c_n^+ x_n \to \min$ subject to the constraint $a_{j1}^- x_1 + ... + a_{jn}^- x_n \ge d_j^+$. The problem with this approach is that some experts may give more cautious interval estimates while some give riskier (narrower) interval estimates. In other words, the knowledge of these experts can be described by *nested* intervals. In this case, the solution is also a nested interval, with levels that correspond to different experts. In [60], both expert-by-expert and membership function approaches are used to describe the nested interval that corresponds to the solution.

Similar mathematical problems (leading to nested interval linear programming) arise in several different applications that are described in the remaining part of this subsection. (We highlighted the chicken feed application, because in the corresponding paper, the relationship between the applied methods and interval computations are explicitly explained; in the other papers, this relationship may be obscured. For a general mathematical formalism, see, e.g., Yazenin [70, 71, 72, 73, 74] and Lodwick [37].)

**Case study: air pollution** (Sommer et al. [61]): Nested interval linear programming is used to determine how to regulate different manufacturing processes with different pollutants so that the cost is minimized under the condition that the overall concentration of different pollutants does not exceed the given limits.

**Case study: media selection** (Wiedy et al. [67]): Each medium (newspaper ads, TV, radio, etc) targets a special audience. Nested interval linear programming describes how to target the largest number of people at a minimal cost.

**Case study: scheduling** (Deporter et al. [12]): The parameters to select are starting times of different stages of the large project (and/or the percentage of resources that are allocated to different stages of the project in different moments in time). The restrictions describe the necessity to finish some stages so we can start others, and limited resources that prevent us from doing too many stages at the same time. The goal is to minimize the total time of the project. Uncertainty is caused by the impossibility to predict exactly the duration of each stage, and the future amount of resources.

**Case study: reliability problems** (Sasaki et al. [58]): This involves choosing the cheapest design that guarantees the desired reliability of different regimes. Here, the parameters $x_i$ describe the reliability of different components (namely, the average number of failures per time unit). The overall reliability of different regimes can be described as a linear function of component reliabilities. The cost of the corresponding components can also be described as a linear function of $x_i$ (if we can neglect the terms that are quadratic in $x_i$).

**Case study: design of a manufacturing plant** (Tsai [65]). Here, the parameters $x_i$ describe the number of manufacturing units of different types. We need sufficiently many to be able to successfully perform several different types of operations (i.e., the total ability $a_{j1}x_1 + ... + a_{jn}x_n$ to perform $j$–th operation must exceed the desired number $d_j$), and the goal is to minimize the total cost.

**Case study: controlling a highway network** (Bit et al. [7]). Here, the control parameters $\vec{x} = (x_1, ..., x_n)$ are car flows sent to different routes. The goal is to increase the total car flow. Restrictions describe the finite capacity of the streets and roads. The interval and nested interval uncertainty is caused by such difficult-to-predict parameters such as sudden weather changes, unexpected soccer results, or an amazing TV chase that glues everyone to their homes.

## 4.3 Control and Decision Making

**Case study: intelligent control.** Application of expert knowledge to control problems is called *intelligent control*, or *fuzzy control* (The latter term is due to the fact that fuzzy representation is most often used in control situations).

The expert system approach leads us a nested set for describing possible values of control. To use this nested set, we must choose a single point. The methods of choosing a point that we have described above (mean-of-maximum and center-of-gravity) are indeed the ones that are most frequently used in intelligent control. These approaches were initiated in Mamdani [40, 41], and have since been applied to cement kilns, appliances, subways, helicopters, robots, etc. For a latest list of applications of intelligent control, see, e.g., Kandel et al. [28] and Kruse [35].

**Case study: robotics** (Bojadziev et al. [10, 11]). Some robots are designed for the manufacturing floor, where all tasks and conditions are well-defined. But other robots (e.g., robots for space exploration or for handling emergency situations in nuclear power stations) are designed for environments whose properties are not completely known. In this case in addition to the guaranteed intervals that contain possible values, we have narrower intervals that correspond to expert estimates for the unknown parameters. Thus, we have a nested interval. To compare different control strategies, we can compare their quality on intervals $X(\alpha)$ that correspond to different degrees of belief. For example, if we are interested in the worst-case behavior, i.e., in the smallest possible value of the objective function $J(x)$, we can report to the decision maker the values

$$J(\alpha) = \min_{x \in X(\alpha)} J(x)$$

that correspond to different $\alpha \in A$.

## 4.4 Expert Systems

In Uehara et al. [66], nested intervals describe experts' knowledge about different values.

In Zadeh, Nazaraki et al. [81, 82, 50], nested sets are used to interpret expert informal statements of the type $Q(X \to Y)$ for some informal "quantifier" $Q$ and properties $X$ and $Y$. For example, if $Q =$"many", $X =$"small", and $Y =$"strong", we get a statement "many small people are strong". We would like to interpret this statement as meaning that among people who have the property $X$, many also have a property $Y$, i.e., that the ratio $|X \cap Y|/|X|$, where $|X|$ denotes the number of people that satisfy the property $X$, is large enough to correspond to "many". To formalize this idea, the authors of the cited papers interpret $|X|$ and $|X \cap Y|$ as sigma-counts and choose an interval $\mathbf{q} = [q^-, q^+]$ of values of the ratio that corresponds to "many" (or to any other "quantifier" $Q$). Then, $Q(X \to Y)$ is interpreted as $|X \cap Y|/|X| \in \mathbf{q}$. This idea leads to reasonable description of our common-sense reasoning.

# APPENDIX A

# PROOFS

**Proof of Proposition 1.** Since the function $\alpha \to \alpha^{\perp}$ is strictly decreasing, it is 1-1, i.e., different $\alpha$ lead to different $\alpha^{\perp}$. Hence, this function has exactly $d+2$ different values for $d+2$ inputs $\alpha_j \in A$. But the set $A$ has exactly $d+2$ elements. Therefore, each element from $A$ is an image of this map. Due to monotonicity, the values $\alpha_0^{\perp}, ..., \alpha_{d+1}^{\perp}$ must go from $\alpha_{d+1}$ to $\alpha_0$. If we skip one of the values from the sequence $\alpha_{d+1}, \alpha_d, ..., \alpha_0$, then we have fewer than

$d+2$ elements in the image. So, when $\alpha$ passes through the values $\alpha_0, \ldots, \alpha_{d+1}$, $\alpha^\perp$ must pass through the values $\alpha_{d+1}, \ldots, \alpha_0$. In other words, $\alpha_j^\perp = \alpha_{d+1-j}$. Q.E.D.

**Proof of Propositions 2 and 5** is straightforward.

**Proof of Proposition 3.** Let us first prove that $\mu_{X\cap X'}(u) \geq \min(\mu_X(u), \mu_{X'}(u))$. Indeed, let us denote $\mu_X(u)$ by $\alpha$, $\mu_{X'}(u)$ by $\alpha'$, and $\mu_{X\cap X'}(u)$ by $\alpha_\cap$. Without loss of generality, we can assume that $\alpha \leq \alpha'$. By definition of $\mu_X(u)$ as the largest value $\alpha$ from $A$ for which $u \in X(\alpha)$, we conclude that $u \in X(\alpha)$ and, similarly, that $u \in X'(\alpha')$. Since $\alpha \leq \alpha'$, and the sequence of sets $X(\alpha)$ is decreasing in $\alpha$, we can conclude that $X'(\alpha') \subseteq X'(\alpha)$. Since $u \in X'(\alpha')$, we can conclude that $u \in X'(\alpha)$.

So, $u \in X(\alpha)$ and $u \in X'(\alpha)$, hence, $u \in X(\alpha) \cap X'(\alpha)$. By definition of an intersection of nested sets, this means that $u \in (X \cap X')(\alpha)$. By definition of $\mu$, this means that $\alpha_\cap = \sup\{\alpha | u \in X(\alpha)\} \geq \alpha = \min(\alpha, \alpha')$. So, $\mu_{X\cap X'}(u) \geq \min(\mu_X(u), \mu_{X'}(u))$.

Let us now prove the inverse inequality $\mu_{X\cap X'}(u) \leq \min(\mu_X(u), \mu_{X'}(u))$. Indeed, by definition of $\mu$, we have $u \in (X \cap X')(\alpha_\cap) = X(\alpha_\cap) \cap X'(\alpha_\cap)$. Therefore, $u \in X(\alpha_\cap)$ and $u \in X'(\alpha_\cap)$. From $u \in X(\alpha_\cap)$ and the definition of $\mu$, we conclude that $\alpha_\cap \leq \alpha = \sup\{\alpha | u \in X(\alpha)\}$. Similarly, $\alpha_\cap \leq \alpha'$. Therefore, $\alpha_\cap \leq \min(\alpha, \alpha')$, i.e., $\mu_{X\cap X'}(u) \leq \min(\mu_X(u), \mu_{X'}(u))$.

From $\mu_{X\cap X'}(u) \leq \min(\mu_X(u), \mu_{X'}(u))$ and $\mu_{X\cap X'}(u) \geq \min(\mu_X(u), \mu_{X'}(u))$, we conclude that $\mu_{X\cap X'}(u) = \min(\mu_X(u), \mu_{X'}(u))$.

The proof for $\cup$ is similar. Indeed, let us denote $\alpha = \mu_X(u)$, $\alpha' = \mu_{X'}(u)$, and $\alpha_\cup = \mu_{X\cup X'}(u)$, and let us assume (without losing generality) that $\alpha \leq \alpha'$. Then, from the definition of $\alpha'$, it follows that $u \in X'(\alpha')$, hence, $u \in X(\alpha') \cup X'(\alpha') = (X \cup X')(\alpha')$, and $\alpha_\cup = \sup\{u | u \in (X \cup X')(\alpha)\} \geq \alpha' = \max(\alpha, \alpha')$.

The inverse inequality can be proven in a similar manner: from the definition of $\alpha_\cup$, it follows that $u \in (X \cup X')(\alpha_\cup) = X(\alpha_\cup) \cup X'(\alpha_\cup)$. Since $u$ belongs to the union of two sets, it must belong to one of them. If $u \in X(\alpha_\cup)$, then $\alpha_\cup \leq \alpha$. If $u \in X'(\alpha_\cup)$, then $\alpha_\cup \leq \alpha'$. In both cases, $\alpha_\cup \leq \max(\alpha, \alpha')$.

So, $\alpha_\cup \leq \max(\alpha, \alpha')$ and $\alpha_\cup \geq \max(\alpha, \alpha')$, hence, $\alpha_\cup \leq \max(\alpha, \alpha')$, i.e., $\mu_{X\cup X'}(u) = \max(\mu_X(u), \mu_{X'}(u))$.

Finally, let us prove that $\mu_{X^\perp}(u) = f_\neg(\mu_X(u))$, where $f_\neg(\alpha_j) = \alpha_{d-j}$. Indeed, by definition of $X^\perp$, $u \in X^\perp(\alpha) \leftrightarrow u \notin X(\alpha^\perp)$, where $\alpha_i^\perp = \alpha_{d+1-i}$. So, $u \in X^\perp(\alpha_i) \leftrightarrow u \notin X(\alpha_{d+1-i})$. Hence, $\mu_{X^\perp}(u) = \sup\{\alpha_i | u \in X^\perp(\alpha_i)\} = \sup\{\alpha_i | u \notin X(\alpha_{d+1-i})\}$. Since $\alpha_i$ is a increasing sequence, we can write that $\mu_{X^\perp}(\alpha_i) = \alpha_m$, where $m = \max\{i | u \notin X(\alpha_{d+1-i})\}$. This formula can be simplified if we introduce a new variable $j = d - i$. Then, the largest $i$ corresponds to the smallest value of $j$, i.e.,

$$m = \max\{i | u \notin X(\alpha_{d+1-i})\} = d - \min\{j | u \notin X(\alpha_{j+1})\}.$$

By definition, $\mu_X(u)$ is $\alpha_j$ for the largest $j$ for which $u \in X(\alpha_j)$. For $\alpha = \alpha_{j+1}, \alpha_{j+2}, \ldots$, we have $u \notin X(\alpha)$. Therefore, the smallest $\alpha$ for which $u \notin X(\alpha)$, is $\alpha_{j+1}$, where $\alpha_j = \mu_X(u)$. Hence, $\min\{j | u \notin X(\alpha_{j+1})\}$ is exactly the $j$ for which $\alpha_j = \mu_X(u)$. Hence, $\mu_{X^\perp}(u) = \alpha_{d-j} = f_\neg(\alpha_j) = f_\neg(\mu_X(u))$. Q.E.D.

**Proof of Proposition 4.** Let us first prove that the left-hand side (LHS) of the desired equality is smaller than or equal to its right-hand side (RHS). Indeed, from $\alpha = \mu_{f(X_1,\ldots,X_n)}(u)$, it follows that $u \in f(X_1, \ldots, X_n)(\alpha)$. By Definition 5, this means that $u \in f(X_1(\alpha), \ldots, X_n(\alpha))$, i.e., that $u = f(u_1, \ldots, u_n)$ for some $u_i \in X_i(\alpha)$. From $u_i \in X_i(\alpha)$, it follows that $\mu_{X_i}(u_i) \geq \alpha$. Therefore, $\min(\mu_{X_1}(u_1), \ldots, \mu_{X_n}(u_n)) \geq \alpha$, and hence,

$$\max_{u_1,\ldots,u_n: f(u_1,\ldots,u_n)=u} \min(\mu_{X_1}(u_1), \ldots, \mu_{X_n}(u_n)) \geq \alpha = \mu_{f(X_1,\ldots,X_n)}(u),$$

i.e., RHS$\geq$LHS.

For the converse, let $\alpha$ be equal to the RHS. Since $A$ is a finite set, the supremum is attained for some $u_i$, so, there exist $u_1, \ldots, u_n$ for which $u = f(u_1, \ldots, u_n)$ and $\min(\mu_{X_1}(u_1), \ldots, \mu_{X_n}(u_n)) = \alpha$. Hence, $\alpha_i = \mu_{X_i}(u_i) \geq \alpha$ for all $i$. By definition of $\mu$, we have $u_i \in X_i(\alpha_i)$. Since $X(\alpha)$ is a decreasing sequence of sets, we have $u_i \in X_i(\alpha)$. Hence, $u = f(u_1, \ldots, u_n) \in f(X_1(\alpha), \ldots, X_n(\alpha)) = (f(X_1, \ldots, X_n))(\alpha)$. Therefore, this $\alpha$ is smaller than or equal to the largest of all $\alpha$ for which $u \in (f(X_1, \ldots, X_n))(\alpha)$, i.e., $\alpha \leq \mu_{f(X_1,\ldots,X_n)}(u)$. So, RHS$\leq$LHS. We have already proved that RHS$\leq$LHS; hence, RHS=LHS. Q.E.D.

**Proof of Proposition 6.** For every $\alpha$,

$$|X(\alpha)| = \sum_{u \in X(\alpha)} 1.$$

Therefore,

$$\frac{1}{d+2} \sum_{\alpha \in A} |X(\alpha)| = \sum_{u \in U} (\frac{1}{d+2} \cdot \sum_{\alpha \in A \& u \in X(\alpha)} 1).$$

For every $i$, if $\mu_X(u) = \alpha_i = (i+1)/(d+2)$, then we have exactly $i+1$ values $(\alpha_0, \alpha_1, ..., \alpha_i)$ for which $u \in X(\alpha)$. Therefore,

$$\sum_{\alpha \in A \& u \in X(\alpha)} 1 = i+1,$$

$$\frac{1}{d+2} \cdot (\sum_{\alpha \in A \& u \in X(\alpha)} 1) = \frac{i+1}{d+2} = \mu_X(u),$$

and so,

$$\frac{1}{d+2} \sum_{\alpha \in A} |X(\alpha)| = \sum_{u \in U} (\frac{1}{d+2} \cdot \sum_{\alpha \in A \& u \in X(\alpha)} 1) = \sum_{u \in U} \mu_X(u).$$

Q.E.D.

**Proof of Proposition 7.** Since $s$ is a choice function, we have $s([0,1]) \in [0,1]$. Let us denote this number $s([0,1])$ by $\beta$. Let us prove that for every interval $[a^-, a^+]$, $s([a^-, a^+]) = \beta \cdot a^+ + (1-\beta)a^-$. We will consider two cases: when the interval is degenerate, and when it is not.

- For a degenerate interval $[a, a]$, from $s([a, a]) \in [a, a]$, we can conclude that $s([a, a]) = a$. But for these intervals, $\alpha a^+ + (1-\beta)a^- = \beta \cdot a + (1-\beta)a = a$, so the formula is correct.
- To get $s([a^-, a^+])$ for an arbitrary non-degenerate interval $[a^-, a^+]$, we can use shift- and unit-invariance:
  - First, we use shift-invariance to prove that
    $$s([a^-, a^+]) = s([(a^+ - a^-) + a^-, 0 + a^-]) = s([a^+ - a^-, 0]) + a^-.$$
  - Then, we use unit-invariance to compute
    $$s([a^+ - a^-, 0]) = s([(a^+ - a^-) \cdot 1, (a^+ - a^-) \cdot 0]) =$$
    $$(a^+ - a^-)s([0,1]) = \beta(a^+ - a^-).$$
  - Finally, we combine these two formulas:
    $$s([a^-, a^+]) = s([a^+ - a^-, 0]) + a^- = \beta(a^+ - a^-) + a^- =$$
    $$\beta a^+ + (1-\beta)a^-.$$

Finally, from symmetry, we conclude that $\beta = 1/2$. Q.E.D.

**Proof of Proposition 8** is similar to the proof of Proposition 6.

## Acknowledgements

This work was partially supported by NSF Grants No. CDA-9015006 and EEC-9322370, and by NASA Grant No. NAG 9-757. The authors are thankful to R. Baker Kearfott and to the anonymous referees for their help.

## REFERENCES

[1] O. Artbauer, "Application of interval, statistical, and fuzzy methods to the evaluation of measurements", *Metrologia*, 1988, Vol. 25, pp. 81–86.

[2] K. T. Atanassov, *Review and new results on intuitionistic fuzzy sets*, Preprint IM-MFAIS-1-88, Sofia, Bulgaria, 1988.

[3] K. T. Atanassov, "Operations over interval valued intuitionistic fuzzy sets", *Fuzzy Sets and Systems*, 1994, Vol. 64, pp. 159–174.

[4] K. T. Atanassov and G. Gargov, "Interval valued intuitionistic fuzzy sets", *Fuzzy Sets and Systems*, 1989, Vol. 31, No. 3, pp. 343–349.

[5] R. E. Bellman, L.A. Zadeh, "Decision making in a fuzzy environment", *Management Science*, 1970, Vol. 17, No. 4, pp. B 141–B 164.

[6] D. Berleant, R. R. Goforth, and A. Khanna, "Comparisons of simulation with pdf's vs. fuzzy values", *Proceedings of the Symposium on Computer Simulation in Industrial Engineering and in Problems of Urban Development*, Nov. 18–19, 1992, Mexico City.

[7] A. K. Bit, M. P. Biswal, S. S. Alam, "Fuzzy programming approach to multicriteria decision making transportation problem", *Fuzzy Sets and Systems*, 1992, Vol. 50, pp. 135–141.

[8] J. M. Blin, "Fuzzy relations in group decision theory", *J. of Cybernetics*, **4**, 17–22, 1974.

[9] J. M. Blin and A. B. Whinston. "Fuzzy sets and social choice", *J. of Cybernetics*, **3**, 28–36, 1973.

[10] G. Bojadziev and G. Bojadzieva, "Liapunov design avoidance control for an open-end two-link robot manipulator", In: M. Jamshidi and M. Saif (eds.), *Robotics and Manufacturing*, ASME Press, N.Y., 1990, Vol. 3, pp. 307–314.

[11] G. Bojadziev and G. Bojadzieva, "Avoidance control of an open-end robot manipulator with two links subject to uncertainties of fuzzy nature", In: Mohammad Jamshidi, Charles Nguyen, Ronald Lumia, and Junku Yuh (Editors), *Intelligent Automation and Soft Computing. Trends in Research, Development, and Applications. Proceedings of the First World Automation Congress (WAC'94), August 14–17, 1994, Maui, Hawaii*, TSI Press, Albuquerque, NM, 1994, Vol. 1, pp. 439–444.

[12] E. L. Deporter, K. P. Ellis, "Optimization of project networks with goal programming and fuzzy linear programming", *Computers in Industrial Engineering*, 1990, Vol. 19, pp. 500–504.

[13] W. M. Dong, W. L. Chiang, H. C. Shah, "Fuzzy information processing in seismic hazard analysis and decision making", *International Journal of Soil Dynamics and Earthquake Engineering*, 1987, Vol. 6, No. 4., pp. 220–226.

[14] W. Dong and F. Wong, "Fuzzy weighted averages and implementation of the extension principle", *Fuzzy Sets and Systems*, 1987, Vol. 21, pp. 183–199.

[15] D. Dubois, M. Grabisch, and H. Prade, "Gradual rules and the approximation of control laws", In: H. T. Nguyen, M. Sugeno, R. Tong, and R. Yager (eds.), *Theoretical aspects of fuzzy control*, J. Wiley, N.Y., 1995, pp. 117–146.

[16] D. Dubois and H. Prade. *Fuzzy sets and systems: theory and applications*, Academic Press, N.Y., London, 1980.

[17] D. Dubois, H. Prade, "Random sets and fuzzy interval analysis", *Fuzzy Sets and Systems*, 1991, Vol. 42, pp. 87–101.

[18] R. Fuller, T. Keresztfalvi, "On generalization of Nguyen's theorem", *Fuzzy Sets and Systems*, 1990, Vol. 4, pp. 371–374.

[19] E. Gardeñes, A. Trepat, J. M. Janer, "Approaches to simulation and to the linear problem in SIGLA system", *Freiburger Intervall-Berichte*, 1981, No. 81/8.

[20] Y. Gentilhomme, "Les ensembles flou en linsguistique", *Cahiers de Ling. Théor. et Appl.*, 1968, Vol. 5, pp. 47–65.

[21] V. A. Gerasimov, M. Yu. Shustrov, "Numerical operations with fuzzy objects", In: S. P. Shary and Yu. I. Shokin (editors), *Interval Analysis*, Krasnoyarsk, Academy of Sciences Computing Center, Technical Report No. 17, 1990, pp. 11–15 (in Russian).

[22] J. A. Goguen, "$L$–fuzzy sets", *Journal of Mathematical Analysis and Applications*, 1967, Vol. 18, pp. 145–174.

[23] J. A. Goguen, "The logic of inexact reasoning", *Synthese*, 1969, **19**, 325–373, 1969 (reprinted in D. Dubois, H. Prade, R. Yager (eds.), *Reading in Fuzzy Sets for Intelligent Systems*, Morgan Kaufmann, San Mateo, CA, 1994, pp. 417–441).

[24] I. R. Goodman, "Algebraic and probabilistic bases for fuzzy sets and the development of fuzzy conditioning", In: I. R. Goodman, M. M. Gupta, H. T. Nguyen, and G. S. Rogers (eds.), *Conditional logic in expert systems*, North Holland, Amsterdam, 1991.

[25] I. R. Goodman, M. M. Gupta, H. T. Nguyen, and G. S. Rogers (eds.), *Conditional logic in expert systems*, North Holland, Amsterdam, 1991.

[26] M. Grabisch, H. T. Nguyen, and E. A. Walker, *Fundamentals of uncertainty calculi with applications to fuzzy inference*, Kluwer, Dordrecht, Netherlands, 1995.

[27] Z. X. Gu and H. Q. Yang, *Basic theory of fuzzy information process*, Chengdu Institute of Radio Engineering Publishing House, P.R. China, 1989 (in Chinese).

[28] A. Kandel, G. Langholtz (editors), *Fuzzy Control Systems*, CRC Press, Boca Raton, FL, 1994.

[29] A. Kaufmann and M. M. Gupta, *Introduction to fuzzy arithmetic*, Van Nostrand Reinhold Co., N.Y., 1985.

[30] R. B. Kearfott, "A Review of Techniques in the Verified Solution of Constrained Global Optimization Problems," *This Volume.*

[31] E. P. Klement, B. Kovalerchuk, "Interval mode of fuzzy control", In: *Proceedings of the International Conference on Interval and Stochastic Methods in Science and Engineering INTERVAL'92*, Moscow, 1992, Vol. 2, pp. 45–46.

[32] G. J. Klir and T. A. Folger, *Fuzzy sets, uncertainty, and information*, Prentice-Hall, U.K., 1988.

[33] G. J. Klir and Bo Yuan, *Fuzzy sets and fuzzy logic. Theory and applications*, Prentice Hall, NJ, 1995.

[34] L. J. Kohout, "Fuzzy interval-valued inference system with para-consistent and grey set extensions", *Reliable Computing*, 1995, Supplement (Extended Abstracts of APIC'95: International Workshop on Applications of Interval Computations, El Paso, TX, Feb. 23–25, 1995), pp. 107–110.

[35] R. Kruse, J. Gebhardt, and F. Klawonn, *Foundations of fuzzy systems*, J. Wiley & Sons, Chichester, England, 1994.

[36] Y.-J. Lai, C.-L. Hwang, *Fuzzy mathematical programming. Method and application*, Springer-Verlag, Berlin, 1992.

[37] W. A. Lodwick, "Analysis of structure in fuzzy linear programs", *Fuzzy Sets and Systems*, 1990, Vol. 38.

[38] S. G. Loo, *Cybernetica*, 1977, Vol. 20, pp. 201–210.

[39] R. P. Loui, "Interval-based decisions for reasoning systems", In: L. N. Kanal and J. F. Lemmer, *Uncertainty in Artificial Intelligence*, Elsevier, North Holland, 1986, pp. 459–472.

[40] E. H. Mamdani, "Application of fuzzy algorithms for control of simple dynamic plant", *Proceedings of the IEE*, 1974, Vol. 121, No. 12, pp. 1585–1588.

[41] E. H. Mamdani, "Application of fuzzy logic to approximate reasoning using linguistic systems", *IEEE Transactions on Computing*, 1977, Vol. 26, pp. 1182–1191.

[42] R. Lopez De Mantaras, "From intervals to possibility distributions: adding flexibility to reasoning under uncertainty", *Reliable Computing*, 1995, Supplement (Extended Abstracts of APIC'95: International Workshop on Applications of Interval Computations, El Paso, TX, Feb. 23–25, 1995), pp. 151–152.

[43] M. Mareš, *Computation over fuzzy quantities*, CRC Pres, Boca Raton, FL, 1994.

[44] D. Mitra, M. L. Gerard, P. Srinivasan, and A. E. Hands, "A possibilistics interval constraint problem: fuzzy temporal reasoning", *FUZZ-IEEE'94*, IEEE Press, 1994, Vol. 2, pp. 1434–1439.

[45] E. A. Musaev, "The support of interval computations in high-level languages", In: *Proceedings of the 1st Soviet-Bulgarian Seminar on Numerical Processing, Pereslavl-Zalessky, October 19–24, 1987*, Pereslavl-Zalessky, VINITI, 1988, Publ. No. 2634-B89, pp. 110–121 (in Russian).

[46] E. A. Musaev, "Unexpected aspect of interval approach in a task of optimal currency exchange rates", In: *International Conference on Interval and Computer-Algebraic Methods in Science and Engineering (Interval'94), St. Petersburg, Russia, March 7-10, 1994, Abstracts*, pp. 179–180.

[47] A. S. Narin'yani, *Subdefinite sets - a new datatype for knowledge representation*, Academy of Sciences, Siberian Branch, Computing Center, Novosibirsk, Technical Report No. 232, 1980 (in Russian).

[48] A. S. Narin'yani, "Tools that simulate data incompleteness, and their usage in knowledge representation", In: *Knowledge representation and simulation of the understanding process*, Academy of Sciences, Siberian Branch, Computing Center, Novosibirsk, 1980 (in Russian).

[49] A. S. Narin'yani, "Subdefiniteness in knowledge representation and processing systems", *Transactions of USSR Acad. of Sciences, Technical Cybernetics*, 1986, No. 5, pp. 3–28 (in Russian).

[50] H. Nazaraki and I. B. Türksen, "An integrated approach for syllogistic reasoning and knowledge consistency level maintenance", *IEEE Transactions on Systems, Man, and Cybernetics*, 1994, Vol. 24, No. 4, pp. 548–563.

[51] C. V. Negoita and D. A. Ralescu, "Representation theorems for fuzzy concepts", Kybernetes, 1975, Vol. 4, pp. 169–174.

[52] C. V. Negoita and D. A. Ralescu, *Applications of fuzzy sets to systems analysis*, John Wiley and Sons, N.Y., Toronto, 1975.

[53] H. T. Nguyen, "A note on the extension principle for fuzzy sets", *J. Math. Anal. and Appl.*, 1978, Vol. 64, pp. 359–380.

[54] H. T. Nguyen and V. Kreinovich, "Interval Sessions at NAFIPS/IFIS/NASA'94," *Reliable Computing*, 1995, Vol. 1, No. 1, pp. 93–98.

[55] H. T. Nguyen, V. Kreinovich, and Qiang Zuo, "Interval-Valued Degrees of Belief: Applications of Interval Computations to Expert Systems and Intelligent Control" (submitted to *Reliable Computing*).

[56] S. A. Orlovsky, "On programming with fuzzy constraint sets", *Kybernetes*, 1977, Vol. 6, No. 3, pp. 197–201.

[57] S. A. Orlovsky, "Problems of decision-making in case of uncertainty of initial information", Moscow, Nauka Publ., 1981 (in Russian).

[58] M. Sasaki, M. Gen, K. Ida, "A method for solving reliability optimization problems by fuzzy multiobjective 0-1-linear programming", *Electronics and Communications*, 1991, Part 3, Vol. 74, pp. 106–116 (in Japanese).

[59] K. J. Schmucker, *Fuzzy sets, natural language computations, and risk analysis*, Computer Science Press, Rockville, MD, 1984.

[60] Tong Shaocheng, "Interval number and fuzzy number linear programmings", *Fuzzy Sets and Systems*, 1994, Vol. 66, pp. 301–306.

[61] G. Sommer, M. A. Pollatschek, "A fuzzy programming approach to an air pollution regulation problem", *European Journal of Operations Research*, 1978, Vol. 10, pp. 303–313.

[62] N. Tamura and K. Horiuchi, "VSOP fuzzy numbers and fuzzy comparison relations", *Proceedings of the Second IEEE International Conference on Fuzzy Systems*, San Francisco, CA, March 28–April 1, 1993, Vol. II, pp. 1287–1292.

[63] H. Tanaka, T. Okuda, K. Asai, "On fuzzy mathematical programming", *J. of Cybernetics*, 1973, Vol. 3, No. 4, pp. 37–46.

[64] H. Tanaka, K. Asai, "Fuzzy linear programming problems with fuzzy numbers", *Fuzzy Sets and Systems*, 1984, Vol. 13, pp. 1–10.

[65] C.-C. Tsai, C.-H. Chu, T. A. Barta, "Fuzzy linear programming approach to manufacturing cell formation", *FUZZ-IEEE'94*, IEEE Press, 1994, Vol. 2, pp. 1406–1411.

[66] K. Uehara, M. Fujise. "Fuzzy inference based on families of $\alpha$−level sets", *IEEE Transactions on Fuzzy Systems*, 1993, Vol. 1, No. 2, pp. 111–124.

[67] G. Wiedy, H. J. Zimmermann. "Media selection and fuzzy linear programming", *J. Oper. Res. Soc.*, 1978, Vol. 31, pp. 342–249.

[68] R. R. Yager, "A characterization of the extension principle", *Fuzzy Sets and Systems*, 1986, Vol. 18, No. 3, pp. 205–217.

[69] H. Q. Yang, H. Yao, and J. D. Jones, "Calculating functions of fuzzy numbers", *Fuzzy Sets and Systems*, 1993, Vol. 55, pp. 273-283.

[70] A. V. Yazenin, "Fuzzy variables and fuzzy mathematical programming", In: *Models of alternative selection in a fuzzy medium*, Riga Polytechnical Institute, Riga, Latvia, 1984 (in Russian).

[71] A. V. Yazenin, *Fuzzy mathematical programming*, Kalinin University, Kalinin, Russia, 1986 (in Russian).

[72] A. V. Yazenin, "Fuzzy and stochastic programming", *Fuzzy Sets and Systems*, 1987, Vol. 22.

[73] A. V. Yazenin, "Possibility programming models in fuzzy optimization", *Tekhnischeskaya Kibernetika*, 1991, Vol. 29, No. 5 (in Russian); English translation in *Journal of Computer and Systems Sciences International*, 1992, Vol. 31, No. 2.

[74] A. V. Yazenin, "Possibilistic and interval linear programming", *Tekhnischeskaya Kibernetika*, 1993, No. 5, pp. 149–155 (in Russian); English translation in *Journal of Computer and Systems Sciences International*, 1994, Vol. 32, No. 6, pp. 154–160.

[75] J. Yen and N. Pfluger, "Path planning and execution using fuzzy logic", In: *AIAA Guidance, Navigation and Control Conference*, New Orleans, LA, 1991, Vol. 3, pp. 1691–1698.

[76] J. Yen and N. Pfluger, "Designing an adaptive path execution system", In: *IEEE International Conference on Systems, Man and Cybernetics*, Charlottesville, VA, 1991.

[77] J. Yen, N. Pfluger, and R. Langari, "A defuzzification strategy for a fuzzy logic controller employing prohibitive information in command formulation", *Proceedings of IEEE International Conference on Fuzzy Systems, San Diego, CA, March 1992.*

[78] L. Zadeh, "Fuzzy sets", *Information and control*, 1965, Vol. 8, pp. 338–353.

[79] L. A. Zadeh, "Outline of a new approach to the analysis of complex systems and decision processes", *IEEE Transactions on Systems, Man and Cybernetics*, 1973, Vol. 3, pp. 28–44.

[80] L. A. Zadeh, "The concept of linguistic variable and its application to approximate reasoning", *Information Sciences*, 1975, Vol. 8, pp. 199–249.

[81] L. A. Zadeh, "A theory of common sense knowledge", In: H. J. Scala, S. Termini, and E. Trillas (eds.), *Issues of Vagueness*, Dordrecht, Reidel, 1984, pp. 257–296.

[82] L. A. Zadeh, "Syllogistic reasoning in fuzzy logic and its application to usuality and reasoning with dispositions", *IEEE Trans. SMC,*, 1985, Vol. SMC-15, pp. 754–763.

[83] H. J. Zimmermann, "Description and optimization of fuzzy systems", *Intl. J. of General Systems*, 1976, Vol. 2, No. 4, pp. 209–215.

[84] H. J. Zimmermann, "Fuzzy programming and linear programming with several objective functions", *Fuzzy Sets and Systems*, 1978, Vol. 3, pp. 45–55.

[85] H. J. Zimmermann, "Fuzzy mathematical programming", *Computers and Operations Research*, 1983, Vol. 10, pp. 291–298.

[86] H. J. Zimmermann, *Fuzzy set theory and its applications*, Kluwer, Dordrecht, 1985.

[87] H. J. Zimmermann, "Application of fuzzy set theory to mathematical programming", *Information Sciences*, 1985, Vol. 36, pp. 29–58.

[88] H. J. Zimmermann, *Fuzzy set theory and its applications*, Kluwer, 1991.

[89] V. S. Zyuzin, "An iterative method for solving a system of segment algebraic equations", In: *Differential equations and functions theory*, Saratov University Publ., Saratov, 1987, pp. 72–82 (in Russian).

[90] V. S. Zyuzin, "Twins and a method for solving systems of twin equations", In: *Interval Analysis*, Krasnoyarsk, Academy of Sciences Computing Center, Technical Report No. 6, 1988, pp. 19–21 (in Russian).

# 12

# FUZZY INTERVAL INFERENCE UTILIZING THE CHECKLIST PARADIGM AND BK-RELATIONAL PRODUCTS

**L. J. Kohout and W. Bandler**

*Department of Computer Science, Florida State University, Tallahassee, Florida 32306-4019, USA*

## 1 INTRODUCTION

Many-valued logic based interval reasoning plays an increasingly important role in fuzzy and other many-valued extensions of two-valued (crisp) logic. It is often overlooked that many-valued logic (MVL) reasoning has a richer inference rule base than the crisp (i.e. classical two-valued) logic. Indeed some rules that are not possible in the crisp logic (which is point-based) *do come* to existence when we accept intervals as the basic elements of the semantic space ($[0, 1]$ or a more general lattice) into which the logic expressions are valuated [11, 8, 37].

In the first part of this paper, we shall explain how an interval fuzzy membership function can be derived within the *checklist paradigm* [6, 13] as an approximation of the crisp (Boolean) logical vector of $n$ truth values simultaneously assigned to a proposition. This assignment can be interpreted, for example, as a vote of the community of agents concerning the truth status of a proposition. Given a crisp Boolean vector that represent this vote (a "fine structure"), a summarization function for each vector, and a measure $m(f)$ operating on a pair of such vectors, a "coarse structure" of many valued pair of interval connectives can be generated [6, 9, 13]. This coarse structure is captured algebraically as an *interval many-valued logic system*, consisting of pairs of logical connectives of the same logical type that is used to manipulate the interval valuations of logical sentences.

It will be seen that the checklist paradigm generates a variety of *pairs of* distinct connectives **con** of the same logical type (e.g. a pair of implication (**ply**) operators, **and** connectives, etc.). These pairs of many-valued logic connectives of the same type form the natural bounds **contop** and **conbot** within $[0, 1]$



*R. B. Kearfott and V. Kreinovich (eds.), Applications of Interval Computations*, 291–335.

fuzzy valuation space, thus determining the end points of intervals generated by fuzzy inference. The checklist paradigm provides the means for deriving systems of ***interval-valued*** fuzzy logics used for approximate inference. Thus a formal semantics for several kinds of interval logics has been derived using the exact mathematical method based on the checklist paradigm. The most relevant further references are [6, 9, 10, 13, 14, 44], each addressing a different aspect of the problem.

Interrelationships between different MVL-connectives in interval logic systems can be captured with advantage in a more abstract way by ***groups of logical transformations*** of the MVL connectives. The basics of the theory of logic transformations characterizing MVL systems are briefly outlined in section 3. It is also shown that some systems of the checklist paradigm generated interval logics can by described by $\mathcal{S}_{2\times2\times2}$ symmetry group.

Inference in Knowledge-Based Systems (KBS) is performed by means of fuzzy relational compositions of special kind, the so called triangle and square products of fuzzy relations. These are discussed in section 4. These fuzzy products use in their definitions many-valued logic systems as the base logics. Various pairs of many-valued connectives forming interval logics systems derived by means of the checklist paradigm can be used as the base logic systems for this purpose.

In addition to the theoretical analysis of interval logics presented here, an application of fuzzy interval inference in Knowledge Engineering is also discussed in this paper. How the fuzzy interval inference is practically utilized in the design of Fuzzy Relational KBS is outlined in Section 5.2. Important epistemological reasons for using fuzzy interval inference in the medical domain are also briefly outlined there. This is followed by the presentation of a small portion of the relational knowledge base and some formulae for fuzzy relational interval inference, in order to illustrate the basic mechanism as used by a medical knowledge based system called CLINAID. The section is concluded by an illustrative sample of CLINAID'S inference. The paper is concluded with a brief discussion of the link of interval inference to soft computing.

# 2 MANY-VALUED LOGICS FOR INTERVAL FUZZY INFERENCE BASED ON THE CHECKLIST PARADIGM

## 2.1 The Loss of Informational Variety, Intervals and the Checklist Paradigm

### *How the Loss of Logical Information Leads to Intervals*

Intervals come naturally where one would not expect them. For example, in the case when the fuzzy membership function $\mu(S)$ of a fuzzy proposition $S$ is interpreted as forming part of some logic of approximation or consensus, intervals may suddenly arise. This can best be elucidated by an example. This example illustrates how intervals arise in fuzzy logics, when the fuzzy membership function is an approximation of the crisp logical vector that represents the record of ***yes-no*** answers to $n$ questions concerning the truth status of a logic statement (for further details, formal definitions, and proofs see [6, 13]).

Let the fuzzy membership function be interpreted as the degree of consensus, the vote of a community of agents on the truth of a logic statement. Let us take a collection of $n$ questions concerning the truth of a given proposition $S$, where the permitted answer to each question by each voter is either ***yes*** or ***no***. Record the answers as a 2-valued Boolean (crisp) vector of logical values

$$\mathbf{S} = \langle s_1, s_2, ..., s_n \rangle.$$

Call this vector $\mathbf{S}$ the ***fine structure*** of the valuation of the proposition $S$. Count the number of ***yes*** answers and divide this by the total number of questions $n$. This gives the fuzzy membership value $\mu(S)$ of the proposition $S$, the degree of consensus on the truth of proposition $S$.

Using the same set of questions, do this for a different proposition, say $R$. This will yield the fine structure

$$\mathbf{R} = \langle r_1, r_2, ..., r_n \rangle.$$

Compute the corresponding fuzzy membership value of $R$.

To perform any logical inference, the propositions must be combined by logical connectives. Let us take the statement $S \rightarrow R$, where $\rightarrow$ is an ***implication*** (**ply**) operator. The meaning of this implication is "the degree to which consensus on the truth of $S$ implies the consensus on the truth of $R$". The fuzzy membership value of this statement can be computed from the fine structures of $S$ and $R$ component wise, using the Boolean implication. This yields the combined fine structure

$$\mathbf{S} \Rightarrow \mathbf{R} = \langle s_1 \rightarrow r_1, s_2 \rightarrow r_2, ..., s_n \rightarrow r_n \rangle.$$

Example 1, Part 1: *The fine structure of Vote 1:*

$$\mathbf{S} = (0,1,1,0); \quad \mu(S) = \frac{1}{n}\sum_{i=1}^{n} s_i = 0.5.$$

$$\mathbf{R} = (0,1,0,1); \quad \mu(R) = \frac{1}{n}\sum_{i=1}^{n} r_i = 0.5.$$

$$\mathbf{S} \rightarrow \mathbf{R} = (1,1,0,1); \quad \mu(S \rightarrow R) = \mu(\mathbf{S} \Rightarrow \mathbf{R}) = \frac{1}{n}\sum_{i=1}^{n}(s_i \rightarrow r_i) = 0.75.$$

If our computation is done with the summary fuzzy membership values $\mu(S)$ and $\mu(R)$, the question arises, what kind of many-valued implication operator should be used to combine the coarse (fuzzy) values of $S$ and $R$ to get the correct result, corresponding to the computation obtained via the fine structures. Let us continue with our example.

Example 1, Part 2: *Some many-valued logic (MVL) implications:*

- Lukasiewicz implication (**ply**) operator:

$$\mu(S \rightarrow_5 R) = \min(1 - \mu(S) + \mu(R), 1) = 1 - 0.5 + 0.5 = 1.$$

- Kleene-Dienes implication (**ply**) operator:

$$\mu(S \rightarrow_6 R) = \max(1 - \mu(S), \mu(R)) = \max(1 - 0.5, 0.5) = 0.5.$$

- Reichenbach implication (**ply**) operator (also called ):

$$\mu(S \rightarrow_{5.5} R) = 1 - \mu(S) + \mu(S) \cdot \mu(R) = 1 - 0.5 + (0.5 \cdot 0.5) = 0.75.$$

The exact value of the implication $S \rightarrow R$ computed from the fine structure was 0.75. The only operator in that gave us the same value when computed from the coarse, fuzzy values of $S$ and $R$ was Reichenbach or KDL **ply**, $\mu(S \rightarrow_{5.5} R)$. A new question now arises: ***Is this the "correct" MVL-ply operator that always gives the exact values, identical with those computed from the fine Boolean structure directly?*** The general answer to our question is *NO!*.

It turns out that when one uses the coarse fuzzy values, there is usually no precise correspondence between the fine and coarse structures. Some information is lost in the process of ***concentration*** of Boolean (crisp) vectors by $\sum$ applied to **S** and **R** separately. In general one can obtain only an interval bound on the precise value of $\mu(S \rightarrow R)$ as will be explained below. This is caused by the fact that the ***yes*** answer to ***any k*** questions out of $n$ questions will yield the membership value $k/n$. That is, any permutation over a fixed Boolean (crisp) vector, say $\mathbf{S} = \langle s_1, s_2, ..., s_n \rangle$, will give the same sum $\mu(S) = (1/n)\sum_i s_i$.

In this sense, the fuzzy membership value $\mu(S) \rightarrow \mu(R)$ of the combined proposition $S \rightarrow R$ is only an interval approximation of the exact value $\mu(S \rightarrow R)$ computed from the fine structures of the component propositions involved.

Example 2: *Two more examples of the fine and coarse structure of consensus votes.*

Vote 2: This is a fine structure different from that of Vote 1 above. Yet it gives the coarse values $\mu(P)$, $\mu(Q)$ and $\mu(\mathbf{P} \Rightarrow \mathbf{Q})$ identical to corresponding values of Vote 1. Hence, the coarse structures of propositions for Vote 1 are ***indistinguishable*** from those of the vote 2:

$$\mathbf{P} = (1,0,0,1); \quad \mu(P) = \frac{1}{n}\sum_{i=1}^{n} p_i = 0.5.$$

$$\mathbf{Q} = (1,0,1,0); \quad \mu(Q) = \frac{1}{n}\sum_{i=1}^{n} q_i = 0.5.$$

$$\mathbf{P} \Rightarrow \mathbf{Q} = (1,1,1,0); \quad \mu(P \rightarrow Q) = \mu(\mathbf{P} \Rightarrow \mathbf{Q}) = \frac{1}{n}\sum_{i=1}^{n}(p_i \rightarrow q_i) = 0.75.$$

Vote 3:

$$\mathbf{M} = (1,1,0,0); \quad \mu(M) = \frac{1}{n}\sum_{i=1}^{n} m_i = 0.5.$$

$$\mathbf{W} = (0,0,1,1); \quad \mu(W) = \frac{1}{n}\sum_{i=1}^{n} w_i = 0.5.$$

$$\mathbf{M} \Rightarrow \mathbf{W} = (1,1,1,0); \quad \mu(M \rightarrow W) = \mu(\mathbf{M} \Rightarrow \mathbf{W}) = \frac{1}{n}\sum_{i=1}^{n}(m_i \rightarrow w_i) = 0.5.$$

For Vote 3, we get the values for the Lukasiewicz, Kleene-Dienes and Reichenbach MVL-ply operators that are also identical to those obtained for Vote 1. The value of $\mu(\mathbf{M} \Rightarrow \mathbf{W})$ is 0.5. Hence, this value is identical with the value computed the Kleene-Dienes ply operator. It should be noted that the fine implication operator value for Vote 3 is different from the value 0.75 obtained for the corresponding statements of Vote 1 and Vote 2.

So we face another question, namely: *Can anything be said about the interrelationship of the fine and coarse structures of the above examples?* Fortunately, as we have proven elsewhere [6], the *coarse* structure imposes bounds upon the fine structure, without determining it completely, but ***yielding interval bounds.*** The bounds on the fine structures in the above examples will be

$$\mu(S) \rightarrow_{5.5} \mu(R) \geq \mu(S \rightarrow R) \geq \mu(S) \rightarrow_{6} \mu(R).$$

Thus the checklist paradigm produces the Lukasiewicz implication as the **plytop** operator. The other bound **plybot** is the Kleene-Dienes implication operator.

The interval bounds on the exact value of the fine structures for other connectives are given in [9, 10, 13]. These papers show that ***associated with the various Boolean logical connectives operating on the fine structure between propositions are their extreme values*** that represent the interval valuations of logical sentences.

For example,

$$\min(\mu(S), \mu(R) \geq \mu(S \textbf{ and } R) \geq \max(0, \mu(S) + \mu(R) - 1)).$$

## *Interval Systems of Other Connectives*

There are 16 such inequalities, as there are 16 *logical types* of different connectives **con**, 10 of which are nontrivially two-argument. Thus we obtain the inequality restricting the possible values of

$$\mathbf{S \text{ Con } R} = \langle s_1 \textbf{ con } r_1, s_2 \textbf{ con } r_2, ..., s_n \textbf{ con } r_n \rangle$$

for any of the 16 connectives:

$$\mu(S) \textbf{ contop } \mu(R) \geq \mu(S \textbf{ Con } R) \geq \mu(S) \textbf{ conbot } \mu(R)),$$

where **con** is a 2 valued (Boolean, crisp) connective, and **contop** and **conbot** is a pair of many-valued connectives.

Choosing for **con** the connective type **or** we get:

$$\min(1, \mu(S) + \mu(R)) \geq \mu(S \textbf{ or } R) \geq \max(\mu(S), \mu(R)).$$

We have shown [9, 13] that for a given contraction/approximation measure $m_1(F)$ (see below), there are 16 inequalities linking the **top** and **bot** types of connectives, as there are 16 logical types of connectives. Out of these 16 possible two-argument options, 10 are genuinely dependent on both arguments. The latter are of the following logical types:

- *conjunction* ($a$ **or** $b$),
- *disjunction* ($a$ **and** $b$),
- *implications* ($a \rightarrow b,\ b \rightarrow a$),
- *non-implications* ($\neg(a \rightarrow b),\ \neg(b \rightarrow a)$),
- *equivalence* ($a \equiv b$),
- *non-equivalence* ($a \bigoplus b$).

The minimal set of necessary conditions that the individual logical types of connectives have to satisfy are:

1. Duality of **and** and **or**.
2. A non-implication is obtained by negating an implication, and vice versa.

These conditions give general logical constraints on the systems of connectives. Taking a specific constraint measure, one obtains more specific results. For example, $m_1$ (defined, e.g., in [13]) will yield 16 pairs **conbot** $\leq$ **contop** specifying the end points of the interval for all the possible logical types of connective, some of which are listed below:

| Logical Type: | Valuation: |
|---|---|
| $\neg(b \rightarrow a)$ | $\max(b - a) \leq \min(1 - a, b)$; |
| $\neg(a \rightarrow b)$ | $\max(a - b) \leq \min(a, 1 - b)$; |
| $a \equiv b$ | $\max(a + b - 1, 1 - (a + b)) \leq \min(1 - a + b, 1 - b + a)$; |
| $a$ **eor** $b$ | $\max(a - b, b - a) \leq \min(a + b, 2 - (a + b))$; |
| $\neg(a$ **or** $b)$ | $\max(0, 1 - a - b) \leq \min(1 - a, 1 - b)$; |
| $\neg(a$ **and** $b)$ | $\max(1 - a, 1 - b) \leq \min(1, 2 - a - b)$ |

For the exhaustive listing of all 16 connectives see, e.g., [9, 13].

## *Interdefinability of Interval Connectives*

In classical two-valued logic, the interdefinability of the various operators is a well explored phenomenon. Either of **nor** (negated or) and **nand** (negated and) can generate all the others; and more commonly, negation together with any one of **and**, **or**, or **ply** can be used to determine all the others. This latter is true also of the multiple-valued operators considered in the present paper, but the relationships are complicated by certain factors.

To begin with, consider the classical interdefinitions:

1. $a$ **ply** $b = \neg a$ **or** $b = \neg(a$ **and** $\neg b)$.
2. $a$ **or** $b = \neg(\neg a$ **and** $\neg b) = \neg a$ **ply** $b$.
3. $a$ **and** $b = \neg(a$ **ply** $\neg b) = \neg(\neg a$ **or** $\neg b)$.

It is easily checked that the following matching systems of operators verify all these equations. Here we distinguish, to avoid confusion, Lukasiewicz ***algebra*** of connectives from the Lukasiewicz logical system, which we do not discuss:

**Table: Symmetric Matching within a Logic Algebra:**

- **plytop, ortop, andbot**; i.e.:
  - Lukasiewicz implication;
  - Bold union $a$ **ortop** $b = \min(1, a + b)$; and
  - Bold intersection: $a$ **andbot** $b = \max(0, a + b - 1)$.

  This set of operators is called the *Lukasiewicz logic algebra*. To define all three operators, we can start with the Lukasiewicz **ply** operator. Then:

  - we can define $a$ **and** $b$ as $a$ **and** $b = \neg(a$ **ply** $\neg b)$; as a result, we get Bold union;
  - then, we can use DeMorgan law to deduce Bold union from the Bold intersection.

- **plybot, orbot, andtop**; i.e.:
  - Kleene-Dienes implication;
  - max; and
  - min.

  This set of operators is called the *Kleene-Dienes (KDL) logic algebra*. To derive all three formulas, we can start with the intersection connective $a$ **and** $b = \min(a, b)$, and then:

  - define the KDL ply operator as $a$ **ply** $b = \neg(a$ **and** $\neg b)$;
  - use DeMorgan law to deduce the union $a$ **or** $b = \max(a, b)$ from the intersection.

In other cases [10] of interrelationship of the checklist paradigm generated interval connectives, the operators do not occur together in the matching ways. We do not see the symmetry that was present above.

**Table: Matching across Logic Algebras:**

- **plytop**, **orbot**, **andtop**; i.e.:
  - Lukasiewicz implication;
  - $a$ **orbot** $b = \max(a, b)$;
  - $a$ **andtop** $b = \min(a, b)$.

  Here, the Lukasiewicz implication operator interacts with two connectives of Kleene-Dienes algebra, namely:

  - the intersection connective $a$ **and** $b = \min(a, b)$ and
  - the union $a$ **or** $b = \max(a, b)$.

  So, the connectives are interdefined across two different logic algebras.

- **plybot**, **ortop**, **andbot**; i.e.:
  - Kleene-Dienes implication;
  - Bold union: $a$ **ortop** $b = \min(1, a + b)$;
  - Bold intersection: $a$ **andbot** $b = \max(0, a + b - 1)$.

  Here, the Kleene-Dienes (KDL) implication operator interacts with two connectives of the Lukasiewicz algebra:

  - Bold union $a$ **ortop** $b = \min(1, a + b)$ and
  - Bold intersection $a$ **andbot** $b = \max(0, a + b - 1)$.

  So, the connectives are again interdefined across two different logic algebras.

## *The IFF and EOR Operators and Their Deviation From Classical Logic*

In classical logic one usually defines the *equivalence* or **iff** operator either as mutual implication, or as a simultaneous presence or absence, that is, by one of the formulae

$$a \textbf{ iff } b = (a \textbf{ ply } b) \textbf{ and } (b \textbf{ ply } a);$$

$$a \textbf{ iff } b = (a \textbf{ and } b) \textbf{ or } (\neg b \textbf{ and } \neg a).$$

*Exclusive or*, or the **eor** operator, is also conveniently defined in two ways:

- either as *a without b, or b without a,*
- or else as *a or b but not both.*

Thus:

$$a \textbf{ eor } b = (a \textbf{ and } \neg b) \textbf{ or } (\neg a \textbf{ and } b);$$

$$a \textbf{ eor } b = (a \textbf{ or } b) \textbf{ and } \neg(a \textbf{ and } b).$$

Using these definitions together with the definitions of the previous section, easy calculations bring the results of the following table [10]:

**Table: Formulae for Equivalence (IFF) and Exclusive-Or (EOR):**

$$a \textbf{ ifftop } b = (a \textbf{ plytop } b) \textbf{ andtop } (b \textbf{ plytop } a);$$

$$a \textbf{ ifftop } b = (a \textbf{ andtop } b) \textbf{ ortop } (\neg b \textbf{ andtop } \neg a);$$

$$a \textbf{ iffbot } b = (a \textbf{ plybot } b) \textbf{ andbot } (b \textbf{ plybot } a);$$

$$a \textbf{ iffbot } b = (a \textbf{ andbot } b) \textbf{ orbot } (\neg b \textbf{ andbot } \neg a);$$

$$a \textbf{ eortop } b = (a \textbf{ andtop } \neg b) \textbf{ ortop } (\neg a \textbf{ andtop } b);$$

$$a \textbf{ eortop } b = (a \textbf{ ortop } b) \textbf{ andtop } \neg(a \textbf{ andbot } b);$$

$$a \textbf{ eorbot } b = (a \textbf{ andbot } \neg b) \textbf{ orbot } (\neg a \textbf{ andbot } b);$$

$$a \textbf{ eorbot } b = (a \textbf{ orbot } b) \textbf{ andbot } \neg(a \textbf{ andtop } b).$$

There are other equalities showing the mutual interaction of the checklist paradigm generated interval connectives. Lack of space does not permit us to treat this topic to its exhaustion in this paper. Another formula given below, however, indicates the intriguing nature of these interrelationships. We have:

$$(\neg a \textbf{ plytop } \neg b) \textbf{ andbot } (\neg b \textbf{ plytop } \neg a) =$$

$$(a \textbf{ andbot } b) \textbf{ ortop } (a \textbf{ norbot } b),$$

where $a \textbf{ nor } b = \neg(a \textbf{ or } b)$.

## 2.2 The Checklist Paradigm and its Generating Mechanisms

The interval inference is theoretically justified by the checklist paradigm. We have seen that the checklist paradigm generates ***pairs of*** distinct connectives **conbot**, **contop** of the same logical type **con**, that determine the end points of intervals, thus providing formally and epistemologically justified systems of ***interval-valued*** approximate inference. The most relevant references concerned with the checklist paradigm that also contain proofs and further epistemological justification are [6, 9, 13, 14, 44], each addressing a different aspect of the problem. Two recent surveys [36, 34] list further theoretical papers as well as applications and also link the checklist paradigm to the computational aspects and to Bandler and Kohout's papers on closures and interiors of relations and "fast fuzzy relational algorithms" [7, 16].

It is, however, useful to look at the mechanism of the checklist paradigm generating intervals in a greater depth. A brief explanation of the inner mechanism of the checklist paradigm and of its link to interval logic based inference is therefore given in this section.

**Notational convention.** Let us introduce the following notational conventions for the values of fuzzy relations and logic expressions. It is assumed that all these values are within the interval $[0, 1]$.

- Let $\mathcal{R}(X \rightsquigarrow Y)$ denote the lattice of all ***crisp relations*** from the set $X = \{x_1, x_2, ...\}$ to the set $Y = \{y_1, y_2, ...\}$, i.e., all subsets of the Cartesian product $X \times Y$.
- Let $\mathcal{R}_F(X \rightsquigarrow Y)$ denote the lattice of all ***fuzzy relations*** from the set $X = \{x_1, x_2, ...\}$ to the set $Y = \{y_1, y_2, ...\}$, i.e., the set of all fuzzy subset of $X \times Y$, or, all functions $\mu_R : X \times Y \to [0, 1]$. Given $R \in \mathcal{R}_F(X \rightsquigarrow Y)$, its ***fuzzy membership value*** is defined as $\mu_R(x, y) \in [0, 1]$.
- We shall use interchangeably for the value $\mu_R(x_i, y_j)$ either the expression $x_i R y_j$ or $R_{ij}$.
- If $A$ is a logic expression, its logical value will be denoted by $a$. The logical values of composite expressions will be denoted in a similar way, e.g., $\mu(A$ **and** $B)$ will be written as $a$ **and** $b$.

## *The Checklist and the Associated Fine and Coarse Structures*

**Definition.** *Let:*

$O$ *be a set; its elements are called* ***objects;***

$C$ *be a set; its elements are called* ***abstract constructs,*** *or* ***concepts;*** *and*

$P$ *be a set, whose elements will be called* ***items;***

$Q$ *be a (crisp) relation between* $O$ *and* $P$ *(*$Q \in \mathcal{R}(O \rightsquigarrow P)$*); if* $xQy$*, we say that* ***object*** $x$ ***is 0 with item*** $y$*;*

$R$ *be a crisp relation between* $C$ *and* $P$ *(*$R \in \mathcal{R}(C \rightsquigarrow P)$*; if* $xRy$*, we say that* ***construct*** $z$ ***is associated with item*** $y$*.*

*Then:*

- *For a given construct* $z$*, by its* ***afterset*** $zR$*, we mean the set of all items* $y$ *associated with* $z$*.*
- *For a given object* $x$*, by its* ***afterset*** $xQ$*, we mean the set of all items* $y$ *associated with* $x$*.*
- *By a* ***checklist,*** *we mean a pair* $(z, zR)$*, where* $z \in C$ *is a construct, and* $zR$ *is its afterset.*
- *An* ***object-description*** *is a pair* $(x, xQ)$*, where* $y$ *is an object, and* $yQ$ *is its afterset.*

**Example:**

$O$ : the set of observed patients.

$C$ : the set of concepts from the domain of clinical psychology. We will assume that $C = \{a, m\}$, where:

$a$ means "anxiety", and

$m$ means "motivation".

$P$ : the set of properties or attributes by which concepts can be characterized. In our example, $P = \{p_1, ..., p_6\}$, where:

$p_1$ means "afraid";

$p_2$ means "nervous";

$p_4$ means "panicky";

$p_6$ means "tense";

...

$R$ relates $a$ with $p_1$, $p_2$, $p_4$, and $p_6$;

In this example, the *anxiety checklist* is a pair

$$(a, aR) = (a, \{p_1, p_2, p_4, p_6\}) = (\text{anxiety}, \{\text{afraid, nervous, panicky, tense}\}).$$

**Definition.** *Let us assume that we are given a fixed object $x_k \in O$ and a fixed construct $z_i \in C$.*

- *By a **checklist valuation**, we mean a pair $\mathcal{V} = (z_i, x_k, D)$, where $D \subseteq P$ is the set of all items $y_j$ that confirm $z_i$, i.e., for which:*
  - *either both $x_k$ and $z_i$ are associated with $y_j$,*
  - *or both are not associated with $y_j$.*

  *Formally, $y_j \in D$ iff $D_j$ = "true", where $D_j$ is defined as $Q_{ij} \equiv R_{kj}$, and $\equiv$ is the (two-valued) equivalence connective.*

- *The **coarse** checklist valuation structure $\mathcal{W}$ **associated** with $\mathcal{V}$ is the triple $\mathcal{W} = (z_i, x_k, \underline{d})$, where $\underline{d}$ is the fraction of items that confirm $z_i$ for an object $x_k$, i.e.,*

$$\underline{d} := \frac{1}{n}\sum_{j=1}^{n} D_j.$$

*Comment.* We can simplify this formula, if we use the notation

$$\sigma(D_j) = \frac{1}{n}\sum_{j=1}^{n} D_j;$$

this notation is called *sigma count* in fuzzy set theory. In terms of sigma count, this formula takes the form $\underline{d} := \sigma(D)$.

Given two checklist valuation structures, one may ask about more complex fine structures that can be formed by logic operations over their values. For example given $\mathcal{V}^a = (a, x, D^a)$ and $\mathcal{V}^b = (b, x, D^b)$, can one construct $\mathcal{V}^{a \text{ and } b} = (a \text{ and } b, z, D^{a \text{ and } b})$ ? Obviously, we can take $D_j^{a \text{ and } b} := D_j^a \& D_j^b$. The value of the corresponding coarse structure can be computed the obvious way as the sigma count $\sigma(D_j^{a \text{ and } b})$.

If the fine structures $\mathcal{V}^a$ and $\mathcal{V}^b$ are not available, question arises, whether there exists a many-valued connective $\mathbf{and}_{MVL}$ by which the exact value of $\mathcal{W}^{a \text{ and } b}$ can be computed from $\mathcal{W}^a$ and $\mathcal{W}^b$.

To obtain the exact value for the composed coarse structure is unfortunately impossible in general, as the truth-functionality is lost in transition from the fine to the coarse structure. The rule of compositionality of Fregean logic does not hold in this case. We can, however find a pair of many-valued logic (MVL) connectives, **andtop** and **andbot** that give the upper and the lower bound, thus forming an interval in which all the *unknown* exact values of the coarse structure *must* be contained.

Let us introduce the following convenient notational conventions for future use:

- For 2-valued crisp fine structure components let us define: $\underline{a}_j := D_j^a$, etc.
- For the many-valued coarse structures introduce $\underline{a}$ by the definition

$$\underline{a} := \underline{d}^a = \sigma(D_j^a) = \frac{1}{n} \sum_{j=1}^{n} D_j^a.$$

Let us look now at other remaining cases. For any of 16 two-valued connectives [47], [25], say $F$, the exact value of the coarse structure $\underline{a \text{ CON } b}$ can be computed from the fine structure by the formula $\underline{a\ F\ b} := \sigma(\underline{a}_j\ F\ \underline{b}_j)$. If only $\underline{a}$ and $\underline{b}$ are available, it is not always possible to find a many-valued connective CON such that $\underline{a} \text{ CON } \underline{b} = \underline{a\ F\ b}$ would hold in general.

Fuzzy logics in general deal with uncertainty, imprecision and incompleteness of data. The transition from the fine to the coarse structure in usually accompanied by loss of information, or what can be called "loss of variety" in Ashby's sense. Hence, the coarse structure *approximates* the fine structure in such cases. One must to specify the in a more precise way, what kind of information loss is expected in order to derive a pair of interval many-valued logic connectives, **contop** and **conbot** that give the bounds within which the exact value must fall. This specification determines the character of approximation and the exact meaning of the phrase "$\mathcal{W}$ is associated with $\mathcal{V}$".

Above, the fine structure $\mathcal{V}$ and the coarse structure $\mathcal{W}$ associated with it were defined. The coarse structure was "associated" with the fine structure by the formula for computation of its exact values from the fine structure. In general, the form of this association of structures, and hence the formula yielding the values of $\mathcal{W}$, can be different. The way of describing this connection will be called a *measure*. In the case of association presented above, measure $m_1 = 1 - u_{10}$ was used [6], which yields **top**($\underline{d}$) $= \max(0, \underline{a} + \underline{b} - 1)$ and **bot**($\underline{d}$) $= \min(\underline{a}, \underline{b})$. To understand the meaning of this measure, the meaning of $u_{lm}$ has to be described first.

Pairs of binary values of two fine structure composed by $F$, yielding the composed binary values $\nu^j_{lm}$ can be combined in four distinct ways:

$$\nu^j_{00} = \neg \underline{a}_j \& \neg \underline{b}_j; \;\; \nu^j_{01} = \neg \underline{a}_j \& \underline{b}_j; \;\; \nu^j_{10} = \underline{a}_j \& \neg \underline{b}_j; \;\; \nu^j_{11} = \underline{a}_j \& \underline{b}_j$$

From these four types of components, $u_{lm}$ with $l, m \in \{0, 1\}$ is computed by the formula $u_{lm} := \sigma(\nu^j_{lm}) = (1/n) \sum_{j=1}^{n} \nu^j_{lm}$.

Using the measures $m_1 - m_5$ of [6] leads to interval logics of various kinds. For each measure, a pair of connectives (conbot, contop) were derived (cf. [6], Theorem 6.3), giving the upper bound **top**($\underline{d}$) and the lower bound **bot**($\underline{d}$) of the value $\underline{d}$ for the composite structure.

In general, $m_i = m(F; G(u_{lm}))$ is a function of two parameters, $F$ and $G(u_{lm})$, where $u_{lm}$ is parameterized by $l, m \in \{0, 1\}$ as seen from the formulas above. Relation $G(u_{lm})$, the so called *constraint table* can be found, e.g., in [38] and is explained in detail in [6]. The constraint table has two distinct forms, one for the object language connectives, the other for the rules of inference that belong to the metalanguage [38].

The inequality restricting the possible values of measure $m(F; G(u_{lm}))$, expressing the logical values that fall within the interval, is written in its general form as:

$$\textbf{contop} \geq m(F; G(u_{lm})) \geq \textbf{conbot}.$$

## *Checklist Paradigm and other approaches*

As far as we are aware, the papers of Bandler and Kohout [6] and [9] appear to be the first in the literature on the topic of checklist paradigm.

So far in this paper we have employed $m(F) = m_1(F) = 1 - u_{10}$. With this measure, we have obtained for the **ply** (implication operator) the bounds [6]:

$$\min(1, 1 - a + b) \geq m_1(\mathbf{ply}) \geq \max(1 - a, b),$$

where the Lukasiewicz implication represents one bound, while the other bound (plybot) is the Kleene-Dienes implication operator, as seen above.

We have already noted that choosing for $F$ the connective type **and** (cf. Bandler and Kohout [9, 13]) and $m_1$, we obtain the bounds

$$\min(a, b) \geq m_1(\mathbf{and}) \geq \max(0, a + b - 1).$$

These bounds are formally identical with those of Schweitzer and Sklar [51] giving the bounds on *copulas* which play an important role in their theory of *norms* and *conorms*. Surprisingly, these checklist paradigm bounds also coincide with Novák's recent derivation of bounds on fuzzy sets approximating classes of Vopěnka's Alternative Set Theory [46]. Hisdal derives the same inequalities as the bounds on some connectives of her TEE model and comments on a possible link (cf. Appendix A2 "The TEE model and Bandler and Kohout's checklist paradigm" in [21]). In the context of modalities in fuzzy logics, checklist paradigm-like inequalities for $F \in \{\mathbf{and}, \mathbf{or}\}$ were recently also independently discovered by Resconi, Klir and St. Clair [49]. Yet all these models are neither formally nor epistemologically identical. This indicates the need for a more precise meta- and metametalogical formulation of many-valued based mathematical systems, that would include in their full definition a part formulating their "mathematical epistemology".

Measures other than $m_1$ yield other interesting results as demonstrated by Bandler and Kohout in their 1980 paper [6]. If for $F$ an implication operator type of connective is chosen again, but this time evaluation "by performance" using $m_2 = u_{11}/(u_{10} + u_{11})$ is performed, this yields the inequality

$$\min(1, b/a) \geq m_2(F) \geq \max(0, (a + b - 1)/a),$$

where plytop is in this instance the well-known G43 implication of Goguen-Gaines (cf. eg. [6]).

Still another contracting measure $m_3 = u_{11} \bigvee (u_{00} + u_{01})$ yields [6]:

$$\max[\min(a, b), 1 - a] \geq m_3(F) \geq \max(a + b - 1, 1 - a).$$

The lower contrapositivization of $m_3$ yields the measure $m_4 = (u_{11} \bigvee (u_{00} + u_{01})) \bigvee (u_{00} \bigvee (u_{01} + u_{11}))$ gives the following bounds [6]:

$$\min[\max(a+b-1, 1-a), \max(b, 1-a-b)] \le m_4 \le \min[\max(1-a, b), \kappa a, \kappa b],$$

where the cripsness $\kappa a$ is defined as $a \vee (1-a)$.

The measure $m_5 = m_2 \bigvee u_{00} + u_{11}$ yields:

$$\max[\min(1, b/a), 1-a] \ge m_5 \ge \max[(a+b-1)/a, 1-a].$$

For the proofs of the results presented in this section and for further explanation, see [6], Sections 5 and 6.

## *On the Role of Crispness and Fuzziness in Interval Approximate Reasoning*

Crispness and its dual fuzziness, two important characteristics of fuzzy sets, were introduced by Bandler and Kohout in their 1978 paper [3, 4]:

**Definition.**

1. *The **crispness** of a truth value $a \in V(=[0,1])$ is $\kappa a = a \vee (1-a)$.*
2. *The **crispness** of a fuzzy subset $B$ is $\kappa B = \bigvee \kappa \mu_B x$ (by harsh criterion).*
3. *The dual concept of **fuzziness** of a truth value and of fuzzy subset is defined as $\phi a = 1 - \kappa a$ and $\phi B = 1 - \kappa B$.*
4. ***Crispness by mean criterion** is defined as*

$$\kappa_m B = \frac{\sum_U \kappa \mu_B x}{\text{card}(\text{supp } B)}$$

5. *The **height** of a fuzzy set $B$ is defined as $hB = \bigvee_U \mu_B x$.*
6. *The **plinth** of a fuzzy set $B$ is defined as $pB = \bigwedge_U \mu_B x$.*

For these definitions, we have $pB = 1 - hB^c$, and

$$\kappa B = p(B \cup B^c) = 1 - h(B \cup B^c).$$

Later, independently, fuzziness was also introduced and investigated in detail by Yager [55] and further studied in a more abstract setting by Higashi and Klir [20].

In the context of interval logics, crispness, fuzziness, height and plinth provide important characteristics of fuzzy propositions, predicates and power sets. The higher the degree of fuzziness of a proposition is, the wider is the margin of its imprecision.

Interval connectives-pairs generated by the measure $m_1$ are of theoretical and practical interest, because of the theorem of Bandler and Kohout [13] concerning the fuzziness of the logics generated by $m_1$. This theorem links the ***width of the interval*** that the values of two many-valued (fuzzy) logic propositions composed by a connective of logical type $F$ attain, with the imprecision of that proposition, measured by the degree of unnormalized ***fuzziness***. The theorem holds for all non-trivially two-argument logic types of connectives.

If we define the ***unnormalized fuzziness of x*** (cf. Bandler and Kohout 1978 [3]) as $\phi x = \min(x, 1 - x)$ then for $x$ in the range $[0, 1]$, $\phi x$ is in the range $[0, 0.5]$, with value 0 iff and only iff $x$ is *crisp*, and value 0.5 iff $x$ is equal to 0.5. We have:

**Gap Theorem.** (Bandler and Kohout, 1986, [13]):

$$a \textbf{ andtop } b - a \textbf{ andbot } b = a \textbf{ ortop } b - a \textbf{ orbot } b =$$

$$a \textbf{ plytop } b - a \textbf{ plybot } b = \min(\phi a, \phi b);$$

$$a \textbf{ ifftop } b - a \textbf{ iffbot } b = a \textbf{ eortop } b - a \textbf{ eorbot } b = 2\min(\phi a, \phi b).$$

The width of the interval produced by an application of a pair of associated connectives, (i.e., **top** and **bot** connectives) characterizes the margins of imprecision of an interval logic expression. The interval between the **top** connective and the bottom connective is directly linked to the concept of fuzziness $\phi$. Hence the ***margins of imprecision*** can be directly measured by the degree of $\phi$.

# 3 GROUPS OF LOGIC TRANSFORMATIONS OF INTERVAL CONNECTIVES

Logic transformations are useful in investigating the mutual interrelationships of interval logic connectives. Let transformations on basic propositional functions $f(x, y)$ of 2 arguments be given as follows:

$$I(f) = f(x, y), \quad D(f) = \neg f(\neg x, \neg y),$$

$$C(f) = f(\neg x, \neg y), \quad N(f) = \neg f(x, y).$$

In the set of the above transformations $T_p = \{I, D, C, N\}$, the individual transformations are called ***identity, dual, contradual, negation*** transformation, respectively. It is known that for the crisp (2-valued) logic these transformations determine the Piaget group [47]. This group of transformations is a realization of abstract Klein 4-element group. Adding new non-symmetrical transformations to those defined by Piaget enriches the algebraic structure of logical transformations [31, 35].

We have shown [9, 13] that for a given contraction/approximation measure, there are 16 inequalities linking the **top** and **bot** types of connectives. Important transformations linking **top** and **bot** logical types of connectives can be found and their global structure further investigated.

It is known that the global structure of systems of various logic connectives can be investigated by looking at abstract group properties of the group of their transformations. Let us recall that a realization of an abstract group is any group of concretely performable operations which has the same algebraic structure as the given abstract group. It is well known that any abstract group can be concretely realized by a family of permutations. The global structure imposed on many-valued connectives by certain type of checklist paradigm contracting measures is captured by the $S_{2\times2\times2}$ group [35] characterizing algebraically their transformations.

## 3.1 The Piaget Group of Transformations

### *Klein Abstract Group and Its Realization by the Piaget Group of Logic Transformations*

Let 4 transformations on basic propositional functions $f(x,y)$ of 2 arguments be given as follows:

$$I(f) = f(x,y), \quad D(f) = \neg f(\neg x, \neg y), \quad C(f) = f(\neg x, \neg y), \quad N(f) = \neg f(x,y).$$

In the set of the above transformations $T_p = \{I, D, C, N\}$ the individual transformations are called ***identity, dual, contradual, negation*** transformation, respectively. It is well known that for the crisp (2-valued) logic these transformations determine the Piaget group [47]. This group of transformations is a realization of abstract Klein 4-element group. It is also known that the Piaget group of transformation is satisfied by some many-valued logics (cf. Turksen [54, 53], Dubois and Prade [18], Kandel [22]). All the logics described in these just quoted references, link only separately within each group of transformations, the connectives of (what we call) the **conbot** or **contop** family. The structures described by these authors do not provide a crosslink, a ***mutual coupling*** of the **conbot** and **contop** connectives. The checklist paradigm, however, generates also these richer cases. Let us look at some useful examples of such crosslinks now.

### *A Mutual Link of TOP and BOT Connectives by the Piaget Group of Logic Transformations Generated by the Checklist Paradigm*

Given a logical connective (or a set of connectives) **con**, the set ***closed*** with respect to some family of logic transformations $T$ will be denoted $\mathcal{C}(T; \mathbf{con})$. This will also be written $\mathcal{C}_T(\mathbf{con})$ where appropriate.

Example: $\mathcal{C}(T_P; \{\equiv_{\mathbf{top}}, \bigoplus_{\mathbf{bot}}\} = \{\equiv_{\mathbf{top}}, \bigoplus_{\mathbf{bot}}, \equiv_{\mathbf{bot}}, \bigoplus_{\mathbf{top}}\}$.
Here (cf. Bandler and Kohout [9], [13]):

$$a \equiv_{\mathbf{top}} b = \min(1 - a + b, 1 - b + a),$$

$$a \bigoplus_{\mathbf{bot}} b = \max(a - b, b - a),$$

$$a \equiv_{\mathbf{bot}} b = \min(1 - a - b, a + b - 1),$$

$$a \bigoplus_{\mathbf{top}} b = \max(a + b, 2 - a - b)$$

(for $T_p$ as defined in [35]).

**Theorem 3.1.1.** *The system of connectives* $\{\equiv_{\mathbf{top}}, \bigoplus_{\mathbf{bot}}, \equiv_{\mathbf{bot}}, \bigoplus_{\mathbf{top}}\}$ *obeys the Piaget group of transformations. Hence it possesses the abstract structure of the Klein 4-element group.*

## 3.2 An 8-Element Group of Logic Transformations

Kohout and Bandler [31, 35] added non-symmetrical transformations to those defined by Piaget. This addition enriches the algebraic structure of logical transformations.

Adding

$$LC(f) = f(\neg x, y), \quad RC(f) = f(x, \neg y),$$
$$LD(f) = \neg f(\neg x, y), \quad RD(f) = \neg f(x, \neg y)$$

to the above defined four symmetrical transformations we obtain a new 8-element group of transformations $T = \{I, D, C, N, LC, RD, LC, RD\}$.

The corresponding abstract group can be characterized as follows.

**Theorem 3.2.1.** *The 8-element group* $\{T, *\}$ *that captures the structure of of the above defined logic transformations is commutative and its elements satisfy the following equations:*

$$LC = LD * N = RC * C = RD * N,$$

$$RC = LD * D = LC * C = RD * N,$$

$$LC * RC = C, \quad LC * LD = N, \quad LC * RD = D,$$

$$LD = LC * N = LD * C = RC * D,$$

$$RD = RC * N = LC * D = LD * C;$$

$$N^2 = C^2 = D^2 = LC^2 = LD^2 = RC^2 = RD^2 = I.$$

The above 8-element group is called $\mathcal{S}_{2\times2\times2}$ group in the standard terminology of group theory.

For $m_1$ generated connectives we have the following two theorems.

**Theorem 3.2.2.** *The closed set* $C\{T; \rightarrow_6\}$ *generated by the connective*

$$a \rightarrow_6 b = \max(1 - a, b)$$

*represents a concrete realization of the abstract group* $\mathcal{S}_{2\times2\times2}$ *under the family of transformations* $T$.

**Theorem 3.2.3.** *The closed set* $C\{T; \rightarrow_5\}$ *generated by the connective*

$$a \rightarrow_5 b = \min(1, 1 - a + b)$$

*represents a concrete realization of the abstract group* $\mathcal{S}_{2\times2\times2}$ *under the family of transformations* $T$.

These theorems demonstrate that $\equiv$ and $\subseteq$ are assuming independent epistemological features and there is scope for $\equiv$ independent from that customarily defined by $\wedge$ and $\rightarrow$ (i.e., as $a \equiv b$ iff $(a \rightarrow b) \wedge (b \rightarrow a)$).

The interval logic system based on $m_1$ can be characterized by such groups of transformations. If we are interested in extending the interval logics into the domain of *paraconsistency*, we must base our systems on different measures, such as $m_2$ (cf. Section 2 above).

## 3.3 On Some Checklist Paradigm Generated Realizations of the $\mathcal{S}_{2\times2\times2}$ group

One area where conflicts may arise is when we separate assertions and rejections of propositions. Assertion and rejection of a proposition belong properly to the domain of meta-operations. In those kinds of logic, for which the mutual duality of assertion and rejection are their meta-properties the object language defined **ply** operators will be *contrapositive*, i.e., if $a \rightarrow b$ is equivalent to $\neg b \rightarrow \neg a$. The duality of inference rules (such as modus ponens and tollens) is assured by the contrapositive property of the ply operator [3, 6, 4] that enters the rules.

One such system that is not contrapositive is that determined by the measure $m_2$ (cf. Section 2.2 above). The $\mathcal{S}_{2\times2\times2}$ group also brings some order to the connectives generated by this measure, as Theorem 3.2.2 above clearly indicates. For the proof of this theorem see Kohout and Bandler [35].

**Theorem 3.3.1.** *The closed set of connectives generated from the ply-top implication operator* $a \rightarrow_4 b = \min(1, b/a)$ *by the transformation* $T$ *is listed below. This set of connectives together with* $T$ *is a realization of the abstract group* $S_{2\times2\times2}$:

$$g_1 = I(\rightarrow_4) = \min(1, b/a) \quad g_2 = C(\rightarrow_4) = \min(1, 1 - b/1 - a)$$

$$g_3 = D(\rightarrow_4) = \max(0, b - a/1 - a) \quad g_4 = N(\rightarrow_4) = \max(0, a - b/a)$$

$$g_5 = LC(\rightarrow_4) = \min(1, b/1 - a) \quad g_6 = LD(\rightarrow_4) = \max(0, 1 - a - b/1 - a)$$

$$g_7 = RC(\rightarrow_4) = \min(1, 1 - b/a) \quad g_8 = RD(\rightarrow_4) = \max(0, a + b - 1/a).$$

The corresponding implication operator operating on negative propositions given by the formula $\max(0, (a + b - 1)/a)$ also generates a similar group of transformations.

We have seen that the pair of implication operators $(\rightarrow_6, \rightarrow_5)$ represents the bounds of interval-valued inference system produced by $m_1$ (cf. Section 2 above and [6]). The pair of implication operators $(g_1, g_8)$ of Theorem 3.3.1, on the other hand, represents the bounds of interval-valued inference system produced by $m_2$.

# 4 SPECIAL TYPES OF COMPOSITIONS OF RELATIONS

## 4.1 Triangle and Square Relational Products

On the abstract level of computations within a relational inference system, the knowledge is represented by means of fuzzy relations which are many-valued logic based extensions of crisp (i.e. non-fuzzy) mathematical relations. This condition of extendibility guarantees that any fuzzy computation will default into a standard crisp (i.e. non-fuzzy) computation with crisp (non-fuzzy) data. Relational representation of knowledge makes it possible to perform all the computations and decision making in a uniform relational way [34], by means of ***special relational compositions*** called triangle and square products. The triangle ***sub-product*** $R \triangleleft S$, the triangle ***super-product*** $R \triangleright S$, and ***square product*** $R \square S$ were first introduced by Bandler and Kohout in 1977, and are referred to as the BK-products in the literature. Their theory and applications have made substantial progress since then. See the survey in [34] with a list of 50 selected references on the mathematical theory and applications of BK-products in various fields of science and engineering.

Let us look now at the definitions and meaning of the products. Where $R$ is a relation from $X$ to $Y$, and $S$ a relation from $Y$ to $Z$, a ***product relation*** $R * S$ is a relation from $X$ to $Z$, determined by $R$ and $S$. There are several types of product used to produce product-relations [15], [34]. Each product type performs a ***different logical action*** on the intermediate sets, as ***each logical type*** of the product enforces a ***distinct specific meaning*** on the resulting product-relation $R * S$. When the relations are fuzzy, there is a further wide choice of realization for each of the four product kinds defined below, because several many-valued logics provide ***implication operator*** and other connectives suitable for interval computation. In order to explain clearly the need for, and the significance of, different logical types of relational products, we begin with crisp relations, and then extend these to fuzzy.

| BK-Product Type | Set-based Definition |
|---|---|
| Circle product: | $x(R \circ S)z \Leftrightarrow xR$ intersects $Sz$ |
| Triangle Subproduct: | $x(R \lhd S)z \Leftrightarrow xR \mathrel{\tilde{\subseteq}} Sz$ |
| Triangle Superproduct: | $x(R \rhd S)z \Leftrightarrow xR \mathrel{\tilde{\supseteq}} Sz$ |
| Square product: | $x(R \square S)z \Leftrightarrow xR \cong Sz$ |

Before the relational products can be of real service in interval computing, they must be fuzzified. This can be accomplished by replacing the standard (crisp) set-theoretical operations by their fuzzy versions: $\tilde{\subseteq}$, $\tilde{\supseteq}$ and $\cong$. Each of these symbols represents a family of fuzzy subset, superset operators, and fuzzy powerset set equality, respectively. The *degree* to which fuzzy set $A$ is contained in fuzzy set $B$ (in symbols $A \mathrel{\tilde{\subseteq}} B$) is determined by the formula

$$\pi(\mathbf{A} \subseteq \mathbf{B}) = \bigwedge_{x \in U} (\mu_{\mathbf{A}}x \rightarrow \mu_{\mathbf{B}}x).$$

Similarly,

$$\pi(\mathbf{A} \equiv \mathbf{B}) = \bigwedge_{x \in U} (\mu_{\mathbf{A}}x \leftrightarrow \mu_{\mathbf{B}}x)$$

defines a fuzzy set equality. The symbols $\rightarrow, \leftarrow$ represent many-valued logic implication operators, $\leftrightarrow$ is a many-valued logic equivalence operator. We have already encountered a variety of these in the section on the checklist paradigm above. A useful short introductory overview of implication operators is [14].

Because there are number of distinct many-valued logics, a particular choice of many-valued logic (MVL) connectives determines the specific properties of the ensuing fuzzy set system, or system of relations. Thus the specific choice of an MVL implication operator $\rightarrow$ or MVL equivalence operator $\leftrightarrow$ will determine the specific properties of the fuzzy set-inclusion and set-equality operators, respectively. For further details see [4].

The set inclusion is in a specific relationship with the system of power sets in each particular set theory. This relationship is given by the formula $\pi(\mathbf{A} \subseteq \mathbf{B}) = \mu_{\mathcal{P}(B)}A$. For this reason, the knowledge of the mathematical properties of many-valued logic based (fuzzy) *power set theories* is essential [4], [5] for proper understanding of the logical properties of various fuzzy set inclusions. However, once the concept of the fuzzy powerset is clearly understood and appropriately utilized, the way of fuzzifying formulas defining relational compositions becomes obvious. On what grounds these can be selected from a repertory of fuzzy *power set* theories can be seen from [4, 5].

We have looked so far at the definitions of BK-products in terms of operators acting on subsets. Crisp (Boolean) relations as well as their many-valued logic based extensions (both, conservative and non-conservative) can, however, be also expressed in terms of relational matrices. In order to switch over to the matrix notation that is often more convenient computationally because of its explicit handling of logic values, we shall express now the relational products in their pure logic form. In the logic formulas, $R_{ij}$ will represent the fuzzy degree to which the statement $x_i R y_j$ holds.

| BK-Product Type | Many-Valued Logic Formula |
|---|---|
| Circle product: | $(R \circ S)_{ik} = \bigvee_j (R_{ij} \wedge S_{jk})$ |
| Triangle Subproduct: | $(R \triangleleft S)_{ik} = \bigwedge_j (R_{ij} \rightarrow S_{jk})$ |
| Triangle Superproduct: | $(R \triangleright S)_{ik} = \bigwedge_j (R_{ij} \leftarrow S_{jk})$ |
| Square product: | $(R \square S)_{ik} = \bigwedge_j (R_{ij} \leftrightarrow S_{jk})$ |

The customary logical symbols for the logic connectives AND, OR, both *implications* and the *equivalence* in the above formulas represent the connectives of some many-valued logic, have to be *chosen* according to the properties of the products that are required. The details of choice of the appropriate many-valued connectives are discussed for example in [6, 12, 14].

The two different forms of relational compositions listed above, namely, the

- subset-based and
- logical-matrix-based

are algebraically equivalent, thus producing the same mathematical results. It should be, however, noted that these product forms may be different from the point of view of the computational efficiency [32]. As we are concerned in this paper with explaining what types of many-valued logics to use for defining relational structures in fuzzy interval inference, we shall not discuss the questions of computational performance of specific relational computing architectures here. because it depends of the complexity of specific data. The Activity Structures methodology [25, 23, 27], on which our design of relational architectures is based, can, however, deal also with these questions, cf. for example [41, 42].

In relational computations, it is important to distinguish what we call ***harsh*** fuzzy products (defined above) from a different family, the family of ***mean*** products. Given the general formula $(R@S)_{ik} ::= \#(R_{ij} * S_{jk})$, a mean product is obtained by replacing the outer connective $\#$ by $\sum$ and normalizing the resulting product appropriately. The mean products are very effective in medical applications, although their mathematical theory does not take such a neat form as that of harsh products.

The mean products provide an effective inference mechanism in medical CLINAID applications (cf. Section 5 below). Their superiority has been shown empirically, for most recent extensive experiments. In some non-medical applications, such as urban studies [52], harsh products, however, work equally well. See [34] for further references.

In representing clinical knowledge structures, not only quantitative but also qualitative notions are involved. Product-relations formed by the relational products represent new entities composed from the original data. Their specific semantics defines the conceptual meaning which is partially dependent on the conceptual meaning of the original data-relations. If $R$ is the relation between ***patients*** and ***individual symptoms***, and $S$ a relation between ***symptoms*** and ***diseases***, $R * S$ will be a relation between ***patients and diseases***. The diagnostic clinical interpretation of each distinct logical type (e.g. the triangular square product types) of these product-relations has a ***distinct clinical meaning***:

$x(R \circ S)z$: $x$ has at least one symptom of illness $z$.
$x(R \lhd S)z$: $x$'s symptoms are among those which characterize $z$.
$x(R \rhd S)z$: $x$'s symptoms include all those which characterize $z$.
$x(R \square S)z$: $x$'s symptoms are exactly those of illness $z$.

In the above example, the semantic descriptors of the medical domain, namely "symptom" and "illness" are mutually linked in semantic relationships describing how these logically interact:

- ___*has-at-least-one*___,
- ___*include-all-those-which*___,
- ___*are-exactly-those-of*___,
- etc.

It can be easily seen, that these have their semantic counterparts in terms of relational compositions:

- _$(R \circ S)$__,
- _$R \rhd S$__,
- _$R \square S$_

respectively. The latter, in turn, have a set-based logic semantics:

- _*R intersects S*_,
- _$R \supseteq S$_,
- _$R \cong S$_,

respectively. Thus the structure of semantic descriptors specific to a knowledge domain must be *coherent* with the logical semantics of specific types of products.

## 4.2 Use of Triangle and Square Relational Products in Interval Reasoning

### *Families of Fuzzy Products of the Same Logical Type that have to be Distinguished*

We have seen that the relational products can be expressed in two different notational forms: set-theoretical and logic-based. The table of definitions given above contains two different notational forms:

1. fuzzy set based notation using the concept of set inclusion;
2. many-valued logic based notation, which uses the logical connectives $\wedge$ and $\leftarrow$.

In both forms the right choice of many-valued logic connectives is important. So, let us look at some criteria for selecting these. The customary logical symbols for the logic connectives **and**, **or**, both *implications* and the *equivalence* in the above given definitions of the relational products represent the connectives of some many-valued logic. We have seen that when the relations are fuzzy, there is a wide choice of realization for each of the four product kinds, because a number of many-valued logic *implication operators* and other connectives can be used in interval computations, each giving it different inferential properties.

Let us call the set of MVL-connectives chosen to define a specific kind of a product *the MVL-base of the product*. There are two different facets of the problem of selecting the appropriate MVL-base:

- The first is specifying relational inference structures in the form where connectives are specified only as generic logic types: **and**, **or**, **ply**, etc. The choice of generic logic connectives determines whether the product a triangle sub-product, triangle superproduct, or a square product.
- The second, more particular problem is the choice of specific valuation (many-valued logic algebra operations) for the generic logic connective-types. For example, for the **and** connective:

- one can chose the usual fuzzy **and**, namely, val(A **and** B)=min($a, b$);
- another possible choice is what is called Bold intersection, namely, val(A **and** B)=max($0, a + b - 1$),
- and there are many other choices.

So, one can say that the *generic* relational inferential and knowledge representation structures (with generic logic types of connectives) are ***instantiated*** by the choice of a specific many-valued system of connectives. By this instantiation, the system becomes a specific relational inference system. In practice, this instantiation might be changed in order to adapt to different circumstances that will arise during the run time of the computation. The details of choice of the appropriate many-valued connectives are discussed in [6, 12, 14, 34].

## 4.3 Further Work on Relational Triangle and Square BK-products

Triangular and square products play a pivotal role in relational knowledge representation and reasoning (see [34] for a recent survey and selected bibliography).

- Crisp *circle* product is well known function and relational composition. It has been first extended to fuzzy by Zadeh.
- The triangle ***subproduct*** $R \triangleleft S$, the triangle ***superproduct*** $R \triangleright S$, and ***square product*** $R \square S$ were introduced in their general form by Bandler and Kohout in 1977.
  - The square product, however, stems from Riguet (1948) [50], needing only to be made explicit [15].
  - Independently of Bandler and Kohout, Sanchez defined a an "$\alpha$-composition" which is a special form of the triangle product using implication operator of Breuwerian logic, and is used in the theory of relational equations.

Since then other developments have occurred. The products in their non-fuzzy form were more recently rediscovered and described in 1986 by Diognon, J. P. Monjardet, B. Roubens, and P. Vincke, [17, 19]. Aubin and Frankowska in their book of 1990 [2] uses a crisp triangle product, which they, however call, "square". Extensive theoretical work on Bandler and Kohout's products as well as on their several modifications has also been conducted by DeBeats and and Kerre. For recent survey of our work on the triangle and square products see [34].

# 5 AN APPLICATION: THE BASIC KNOWLEDGE HANDLING MECHANISMS OF CLINAID BY MEANS OF RELATIONAL INFERENCE

## 5.1 Special Features of Medical Reasoning

### *The Need to Process Information that may be Incomplete, Imprecise, Unreliable, Contradictory or Otherwise Imperfect*

CLINAID is a knowledge-based system that ***extends*** classical relational inference to fuzzy interval logics in order to best utilize information that may be incomplete, imprecise, unreliable, contradictory, or otherwise imperfect. Opponents of such extensions might argue that they question the quality of the knowledge generated by systems which utilize incomplete, imprecise or contradictory information, no matter what methodology is applied. But such criticism is based on a serious misunderstanding of the potential of interval computing. Indeed, the basic premise of our approach can be put in even stronger terms: ***One must question the quality of knowledge generated by any computing system, no matter what methodology is applied.*** That is exactly what we do, by introducing the ***interval computation methods*** which define the ***bands of precision***, reliability and consistency of the results.

Information that is available in real life, in medical applications and elsewhere, is always incomplete, imprecise or otherwise imperfect. The quality of any knowledge that is generated by the computing systems, be they based on precise logical methods, numerical computations, or AI heuristics ***ought to be questioned!*** No point inference or computation is acceptable to us with imperfect data. Intervals, the upper and lower bounds of reliability of the results have to be introduced. One must specify the margins of imprecisions and reliability of computations and check repeatedly if they are satisfied. Of course, when data are completely precise the approach must default into the classical precise inference, and the upper and lower bounds ought to collapse into one precise point. If, on the other hand, the data are too incomplete, imprecise or unreliable, it might make the interval of possible values too wide. Such results will not be accepted, but tagged as unusable by the Knowledge Based System.

To achieve this goal [30] in the field of medical computing Clinaid, the CLINAID system has to be equipped to deal with:

- *The Problem of INCOMPLETENESS:* Not all signs and symptoms characterizing the disease are always observable or present.
- *The Problem of LOCALITY OF INFERENCE:* Not all signs and symptoms are relevant in a particular given diagnostic context.
- *The Problem of COMPLEXITY OF INFERENCE:* The complexity of exhaustive matching of all disease descriptors and indicators (e.g. signs, symptoms, tests, etc.) may be excessively high.

Incompleteness leads to interval inference, the technical details of which are discussed below. Locality and complexity are dealt with by introducing semantic hierarchies of clinical or architectural knowledge that we shall explain now.

## *Dynamics of Medical Diagnostic Process*

To see how the proposed work fits into the global framework of previous work in the field, further essentials of the diagnostic process and of its organization will be briefly described here. Let us start with the experts' activities and their mapping into the CLINAID conceptual [1, 39] and formal relational schemes [29, 44, 45, 43].

When a physician sees a patient, his activity consists of the following ***processes*** [33]:

1. Observation and acquisition of relevant patient data.
2. Conceptual classification and filtration of relevant observational data for the next process 3.
3. Clinical decision and patient management.

All these components interact with one another. This interaction of clinical activities manifests itself in the properties of the dynamic process of clinical inference. ***Incompleteness, locality of inference*** and ***directed complexity reduction*** play a substantial role here [40]. In order to deal successfully with these factors, triangle and square relational products of relations [15, 12] were used to represent inference structures in *Clinaid* [28].

All the above activities have to be represented in CLINAID and this diversity is reflected in the virtual architecture of the system, which consists of the following co-operating units (basic shell substratum) [30]:

1. Diagnostic Unit (comprised of several parallel co-operating centers).
2. Treatment Recommendation Unit.
3. Patient Clinical Record Unit.
4. Co-ordination and Planning Unit.

The strategy of clinical inference used over this diagnostic hierarchy which includes syndromes, is realized in *Clinaid* by means of fuzzy triangle products [15] that were explained above, in Section 4. Within each specific body system of CLINAID, (e.g. cardiac or respiratory, etc.) the inferential levels have to be related appropriately. Each body system has specific conceptual semantics that assign a clinical meaning to the relational compositions.

The inferential process is performed in a hierarchical or heterarchical manner, involving the following levels:

1. Symptoms and Sign Level
2. Risk Factors Level
3. Body System Level
4. Syndrome Level
5. General Disease Level
6. Specific Disease Level
7. Etiological Level

There are 11 body systems generally accepted in general medicine, with 200–400 signs and symptoms and 100–400 diseases per body system, approximately.

A fully comprehensive medical system must have the capability of dealing with a number of diverse body systems. The multiplicity of body systems in which the given signs and symptoms of a particular patient are interpreted, defines logically a multiplicity of contexts. It requires ***special precautions*** for the inferential system to deal with this multiplicity of contexts ***correctly***, as will be seen later. Formally, the inference is dealt with by means of triangle and square relational products.

## 5.2 Relationships over the Basic Relational Structures of CLINAID

Relational knowledge structures that enter the inferential process are divided into a number of semantic types each of which has a distinct meaning. Let us list now, the medical meaning (semiotic descriptors) of the sets entering into the relations of individual levels used in the sequel.

| Name: | A set of: |
|---|---|
| $B$ | body systems. |
| $D$ | specific diseases. |
| $V$ | variability index of observation events. |
| $P$ | patients. |
| $S$ | symptoms and physical signs. |

These types of entities enter into mutual relationships defining types of fuzzy relations. In the list of these relations is given below. The expression

$$\mathcal{R}_F(X \rightsquigarrow Y)$$

denotes, as before, a lattice of all fuzzy relation from the set $X$ to the set $Y$.

| Name: | Definition: | Relation |
|---|---|---|
| $PVS$ | $\mathcal{R}_F(P \times V \times S)$ | between ***patients, variability index of observations*** and ***signs / symptoms*** |
| $VBS$ | $\mathcal{R}_F(V \times B \times S)$ | between ***variability index of observations*** and ***signs / symptoms*** |
| $PS$ | $\mathcal{R}_F(P \rightsquigarrow S)$ | from ***patients*** to ***symptoms and physical signs.*** |
| $PD$ | $\mathcal{R}_F(P \rightsquigarrow D)$ | from ***patients*** to ***specific diseases.*** |
| $SB$ | $\mathcal{R}_F(S \rightsquigarrow B)$ | from ***symptoms and physical signs*** to ***body systems.*** |
| $SD$ | $\mathcal{R}_F(S \rightsquigarrow D)$ | from ***symptoms and physical signs*** to ***specific diseases.*** |

Now, we shall briefly explain the meaning and purpose of the relations employed in the inference. Given are two ternary relations, $PVS$ and $VBS$, from which the composed relations $PBS$ and $PB$ are computed. The medical domain semantics (i.e. semiotic descriptors) is outlined below:

- $PVS$ is a ternary (3-place) relation relating the set of patients $P$ with the set of all time instances $V$ at which the observations of relevant signs and symptoms $S$ are made. This relation belongs to the family of relations capturing the relevant empirical findings.
- $VBS$ is a ternary relation relating the variability index of observational events $V$ with the set of body systems $B$ and the set of signs and symptoms $S$. This relation belongs to the family of relations capturing medical theories and normative facts. The $VBS$ relation captures the relationship between signs and symptoms and body systems. In the example discussed here, the relationship of $S$ and $B$ does not depend on the variability index, as the relationship is stable in time.

The actual observations of patients captured by $PVS$ relation have to be matched with the knowledge contained in the relation $VBS$ over the set $V$. This yields the computation

$$PBS_{imk} = (PVS \lhd VBS)_{imk} = \bigoplus_{j=l} (PVS_{ijk} \rightarrow VBS_{lmk}),$$

where card $V$ and card $S$ are the cardinalities (sizes) of the sets of relevant elements, over which the relational composition (matching) is performed.

From these two relation, we compute the others:

- First, the relation $PB$ computes the degree of involvement of each body system relative to the composition over available symptoms and signs (and generally also over other observables) characterizing individual patients.

$$PB_{im} = (PBS\Box^{*}PBS)_{im} = \bigoplus_{k}(PBS_{imk} \wedge (PBS_{imk} \equiv PBS_{imk}))$$

- The relevant ***syndromes*** in a specific body system are computed by the fuzzy triangle product $PY = PS \triangleleft SY$ from patients to syndromes given by the formula

$$PS_{ij} = (PS \triangleleft SY)_{ik} = \bigoplus_{j}(PS_{ij} \rightarrow SY_{jk})$$

- The working ***diagnoses*** of a patient is computed by the fuzzy triangle product $PDY = PS \triangleleft SDY$. This relational composition, the result of which ranges from patients to diseases is given by the formula

$$PDY_{ikl} = (PS \triangleleft SDY)_{ikl} = \bigoplus_{j}(PS_{ij} \rightarrow SDY_{jkl})$$

The connectives used by this method are divided into two classes: **contop** and **conbot**, computing the top and bottom point of the interval of the possible values of the products.

## *An Example: Identification of Relevant Body Systems by* CLINAID

Table 1 presents a small fragment of the knowledge structure of CLINAID, capturing the relationship of signs/symptoms to several body systems.

We have used the following ***connectives in the examples*** of this section:

**contop**: $a \rightarrow_5 b = \min(1, 1 - a + b)$;
$a \equiv b = \min(1 - a + b, 1 - b + a)$.

**conbot**: $a \rightarrow_6 b = \max(1 - a, b)$;
$a \equiv b = \max(a + b - 1, 1 - (a + b))$.

$\bigoplus$: $\bigoplus(\ldots) = (1/n)\sum(\ldots)$.

Table 1: Relation of Signs/symptoms to Body systems

| Signs/symptoms | *Body systems* | | | |
|---|---|---|---|---|
| | RS | CVS | MS | HEM |
| SOB | 0.9 | 0.9 | 0.3 | 0.6 |
| PND | 0.1 | 0.9 | 0.1 | 0.6 |
| Cyanosis | 0.8 | 0.8 | 0.3 | 0.3 |
| Tachycardia | 0.5 | 0.9 | 0.2 | 0.6 |
| Fine basal creps | 0.2 | 0.9 | 0.2 | 0.6 |
| Ankle swelling | 0.3 | 0.9 | 0.2 | 0.6 |
| Hepatomegaly | 0.3 | 0.7 | 0.1 | 0.6 |

In the example shown in Table 1 the patient exhibits the following set S of signs and symptoms:

$s_1 =$ SOB/0.8 (Shortness of breath);

$s_2 =$ AS/0.6 (Ankle swelling);

$s_3 =$ HM/0.7 (Hepatomegaly);

$s_4 =$ ANX/0.6 (Anxiety);

$s_5 =$ JVPH/0.8 (Jugular venous pressure high);

$s_6 =$ CMRM/0.7 (Cardiac murmurs);

$s_7 =$ PLP/0.7 (Palpitations); and

$s_8 =$ RVEP/0.75 (Right ventricular enlargement).

An experienced clinician would easily recognize these as representing heart failure. From this fuzzy set, and using the clinical knowledge structure of Table 1 CLINAID will compute the relation $PB$ that suggests the involvement of individual body systems. This relation is computed by the products

$$PVS \triangleleft VBS$$

and $PBS \square^* PBS$ given above. The results of this computation are listed in Table 2.

Table 2: Interval-Based Diagnosis: Identification of Possible Body Systems of Patient $p_1$

| Body Systems | $p_1PB$ | | $1 - p_1PB$ | |
|---|---|---|---|---|
| RS | 0.5 | 0.43 | 0.57 | 0.5 |
| CVS | 1.0 | 0.67 | 0.33 | 0.0 |
| MS | 0.5 | 0.33 | 0.67 | 0.5 |
| HEM | 0.9 | 0.2 | 0.8 | 0.1 |
| value | max-positive | min-positive | max-negative | min-negative |
| Grounds for: | *Acceptance* | | *Rejection* | |

Table 2 shows that the system computed:

- *high* positive evidence for the involvement of cardiovascular system CVS,
- *low*, negligible positive and negative evidence (around 0.5) for respiratory system RS and
- *substantial* negative evidence for the involvement of the muscular system MS.

There is, however, great uncertainty in indicating the possibility of the involvement of the hematological system HEM. From the clinical point of view, the wide interval $[0.2, 0.9]$ indicates that the given signs and symptoms are not as specific for HEM as they are for the cardiovascular system CVS. Hence, the CVS is the winning candidate body system. It also shows that the presented signs and symptoms are not reliable enough to decide about the accepting or rejecting HEM body system.

# 6 TOWARD SUCCESSFUL UTILIZATION OF INTERVAL METHODS IN SOFT COMPUTING

Further extension of this approach requires to pay attention to those systems that are not contrapositive, and can also deal with conflicts and contradiction. Hence our aim is to analyze the epistemological and ontological status of some basic notions of checklist paradigm based interval logic systems that are needed if we wish to analyze not only the valid inference but also semantic conflicts in fuzzy logics. This requires also a careful foundational analysis of the systems involved. In such a foundational analysis we ought to distinguish sharply [24]:

(1) mathematical questions,

(2) logical questions,

(3) ontological, epistemological and metaphysical questions,

and carefully look at their interrelationship, of the pairs of checklist paradigm generated pairs of connectives that form the semantic space of interval truth values.

In order to bring out the characteristic features of the metatheory of paraconsistency in approximate reasoning systems, we need to employ an adequate method of conceptual analysis.

Interplay of formal techniques and conceptual issues has to be examined carefully, if one attempts to extend the interval-based inference to include essential paraconsistent features [26]. The word "paraconsistent" (meaning "beyond the consistent") was coined by Miró Quesada [48] to apply to the studies that are inconsistent but not trivial. In interval fuzzy logic, paraconsistency "to a degree" may appear for example, if $\mu(S) > 1 - \mu(S)$ and at the same time $\mu(\neg S) > 1 - \mu(S)$. This may appear for logics in which the implication operator is ***not contrapositive***, so that $(a \rightarrow b) \neq (\neg b \rightarrow \neg a)$.

# REFERENCES

[1] J. Anderson, L. J. Kohout, W. Bandler, and C. Trayner, "A knowledge-based clinical decision support system using new techniques", In: A. H. Levy and B. T. Williams, editors, *Proc. of AAMSI Congress 1985*, American Association for Medical Systems and Informatics, Washington, DC, May 1985, pp. 187–192.

[2] J. P. Aubin and H. Frankowska, *Set-valued analysis*, Birkhauser, Boston, 1990.

[3] W. Bandler and L. J. Kohout, "Fuzzy relational products and fuzzy implication operators", In: *International Workshop on Fuzzy Reasoning Theory*

*and Applications*, Queen Mary College, University of London, London, September 1978.

[4] W. Bandler and L. J. Kohout, "Fuzzy power sets and fuzzy implication operators", *Fuzzy Sets and Systems*, 1980, Vol. 4, pp. 13–30. Reprinted in: *Readings In Fuzzy Sets for Intelligent Systems*, D. Dubois, H. Prade and R. Yager (eds.), Morgan Kaufmann Publishers, San Mateo, Calif., 1993.

[5] W. Bandler and L. J. Kohout, "Fuzzy relational products as a tool for analysis and synthesis of the behaviour of complex natural and artificial systems", In: P. P. Wang and S. K. Chang, editors, *Fuzzy Sets: Theory and Applications to Policy Analysis and Information Systems*, Plenum Press, New York and London, 1980, pp. 341–367.

[6] W. Bandler and L. J. Kohout, "Semantics of implication operators and fuzzy relational products", *Internat. Journal of Man-Machine Studies*, 1980, Vol. 12, pp. 89–116. Reprinted in: E. H. Mamdani and B. R. Gaines, eds., *Fuzzy Reasoning and its Applications*, Academic Press, London, 1981, pp. 219–246.

[7] W. Bandler and L. J. Kohout, "Fast fuzzy relational algorithms", In: A. Ballester, D. Cardús, and E. Trillas, editors, *Proc. of the Second Internat. Conference on Mathematics at the Service of Man*, (Las Palmas, Canary Islands, Spain, 28 June–3 July), Universidad Politechnica de las Palmas, Las Palmas, 1982, pp. 123–131.

[8] W. Bandler and L. J. Kohout, "The four modes of inference in fuzzy expert systems", In: R. Trappl, editor, *Cybernetics and Systems Research 2*, North-Holland, Amsterdam, 1984, pp. 581–586.

[9] W. Bandler and L. J. Kohout, "Unified theory of multiple-valued logical operators in the light of the checklist paradigm", In: *Proc. of the 1984 IEEE Conference on Systems, Man and Cybernetics*, IEEE, New York, 1984, pp. 356–364.

[10] W. Bandler and L. J. Kohout, "The interrelations of the principal fuzzy logical operators", In: M. M. Gupta, A. Kandel, W. Bandler, and J. B. Kiszka, editors, *Approximate Reasoning In Expert Systems*, North-Holland, Amsterdam, 1985, pp. 767–780.

[11] W. Bandler and L. J. Kohout, "Probabilistic vs. fuzzy production rules in expert systems", *Internat. Journal of Man-Machine Studies*, 1985, Vol. 22, pp. 347–353.

[12] W. Bandler and L. J. Kohout, "A survey of fuzzy relational products in their applicability to medicine and clinical psychology", In: L. J. Kohout and W. Bandler, editors, *Knowledge Representation In Medicine and Clinical Behavioural Science*, an Abacus Book, Gordon and Breach Publ., London and New York, 1986, pp. 107–118.

[13] W. Bandler and L. J. Kohout, "The use of checklist paradigm in inference systems", In: C. V. Negoita and H. Prade, editors, *Fuzzy Logic In Knowledge Engineering*, Verlag TÜV Rheinland, Köln, 1986, Chapter 7, pp. 95–111.

[14] W. Bandler and L. J. Kohout, "Fuzzy implication operators", In: M. G. Singh, editor, *Systems and Control Encyclopedia*, Pergamon Press, Oxford, 1987, pp. 1806–1810.

[15] W. Bandler and L. J. Kohout, "Relations, mathematical", In: M. G. Singh, editor, *Systems and Control Encyclopedia*, Pergamon Press, Oxford, 1987, pp. 4000 - 4008.

[16] W. Bandler and L. J. Kohout, "Special properties, closures and interiors of crisp and fuzzy relations", *Fuzzy Sets and Systems*, 1988, Vol. 26, No. 3, pp. 317–332.

[17] J. P. Diognon, M. Monjardet, B. Roubens, and P. Vincke, "Biorders families, valued relations and preference modelling", *J. Math. Psychol.*, 1986, Vol. 30, pp. 435–480.

[18] D. Dubois and H. Prade, *Fuzzy Sets and Systems: Theory and Applications*, Academic Press, New York, 1980.

[19] J. C. Fodor, "Traces of fuzzy binary relations", *Fuzzy Sets and Systems*, 1992, Vol. 50, No. 3, pp. 331–341.

[20] M. Higashi and G. J. Klir, "On measures of fuzziness and fuzzy complements", *International Journal of General Systems*, 1982, Vol. 8, No. 3, pp. 169–180.

[21] E. Hisdal, *Infinite-valued logic based of two-valued logic and probability. Pt. 1.4. The TEE model for grades of membership*, ISBN 82-7368-054-1 140, University of Oslo, Inst. of Informatics, Box 1080, Blindern Oslo 3, Norway, October 1990.

[22] A. Kandel, *Fuzzy mathematical techniques with applications*, Addison Wesley, Reading, Mass., 1986.

[23] L. J. Kohout, "Activity structures as a tool for design of technological artifacts", *Systems and Cybernetics: An International Journal*, 1987, Vol. 18, No. 1, pp. 27–34.

[24] L. J. Kohout, "Theories of possibility: Meta-axiomatics and semantics", *Fuzzy Sets and Systems*, 1988, Vol. 25, pp. 357–367.

[25] L. J. Kohout, *A Perspective on Intelligent Systems: A Framework for Analysis and Design*, Chapman and Hall & Van Nostrand, London & New York, 1990. In 1991, received an international prize from The International Institute for Advanced Studies In Systems Research: *"The best book of the year in the area of AI Systems"*.

[26] L. J. Kohout, "Paraconsistency in activity structures", In: G. E. Lasker, editor, *3rd Internatinal Symposium on Systems Research, Informatics and Cybernetics*, Germany, August 12–18, 1991.

[27] L. J. Kohout, "Activity Structures: A methodology for design of multi-environment and multi-context knowledge-based systems", In: L. J. Kohout, J. Anderson, and W. Bandler, editors, *Knowledge-Based Systems for Multiple Environments*, Ashgate Publ. (Gower), Aldershot, U.K., 1992, Chapter 5.

[28] L. J. Kohout, J. Anderson, and W. Bandler, *Knowledge-Based Systems for Multiple Environments*. Ashgate Publ. (Gower), Aldershot, U.K., 1992. Awarded *"Outstanding Scholarly Contribution Award"* by the Systems Research Foundation In 1993.

[29] L. J. Kohout, J. Anderson, W. Bandler, A. Behrooz, S. Gao, and C. Trayner, "Activity structures based architectures for knowledge-based systems, part I: Dynamics of localised fuzzy inference and its interaction with planning", *Fuzzy Sets and Systems*, 1991, Vol. 44, No. 4, pp. 405–420.

[30] L. J. Kohout, J. Anderson, W. Bandler, S. Gao, and C. Trayner, "CLINAID: A knowledge-based system for support of decisions in the conditions of risk and uncertainty", In: L. J. Kohout, J. Anderson, and W. Bandler, editors, *Knowledge-Based Systems for Multiple Environments*, Ashgate Publ. (Gower), Aldershot, U.K., 1992, Chapter 10.

[31] L. J. Kohout and W. Bandler, *Checklist paradigm and group transformations*, Technical Note EES-MMS-ckl91.2, Dept. of Electrical Engineering, University of Essex. U.K, 1979.

[32] L. J. Kohout and W. Bandler, "Relational-product architectures for information processing", *Information Science*, 1985, Vol. 37, pp. 25–37.

[33] L. J. Kohout and W. Bandler, "Knowledge representation, clinical action and expert systems", In: L. J. Kohout and W. Bandler, editors, *Knowledge Representation In: Medicine and Clinical Behavioural Science*, an Abacus Book, Gordon and Breach Publ., London and New York, 1986, pp. 1–8.

[34] L. J. Kohout and W. Bandler, "Fuzzy relational products in knowledge engineering," In: V. Novák et al., editor, *Fuzzy Approach to Reasoning and Decision Making*, Academia and Kluwer, Prague and Dordrecht, 1992, pp. 51–66.

[35] L. J. Kohout and W. Bandler. "How the checklist paradigm elucidates the semantics of fuzzy inference," In: *Proc. of the IEEE Internat. Conference on Fuzzy Systems 1992*, IEEE, New York, 1992, pp. 571–578.

[36] L. J. Kohout and W. Bandler, "Use of fuzzy relations in knowledge representation, acquisition and processing", In: L.A. Zadeh and J. Kacprzyk, editors, *Fuzzy Logic for the Management of Uncertainty*, John Wiley, New York, 1992, pp. 415–435.

[37] L. J. Kohout and W. Bandler, "Interval-valued systems for approximate reasoning based on the checklist paradigm", In: P. Wang, Paul, editor, *Advances In Fuzzy Theory and Technology, Vol. 1*, Bookwrights Press, Durham, N.C., 1993, pp. 167–193.

[38] L. J. Kohout and W. Bandler, "Modes of interval-based plausible reasoning vieved via the checklist paradigm", In: B. Bouchon-Meunier, L. Valverde, and R. R. Yager, editors, *IPMU'92 Advanced Methods In Artificial Intelligence (Lecture Notes in Computer Science Vol. 682)*, Springer, Berlin, 1993, pp. 256–264.

[39] L. J. Kohout, W. Bandler, J. Anderson, and C. Trayner, "Knowledge-based decision support system for use in medicine", In: G. Mitra, editor, *Computer Models for Decision Making*, North-Holland, Amsterdam, 1985, pp. 133–146.

[40] L. J. Kohout, A. Behrooz, J. Anderson, S. Gao, C. Trayner, and W. Bandler, "Dynamics of localised inference and its embedding in activity structures based IKBS architectures", In: *Proc. of Second IFSA Congress*, International Fuzzy Systems Association, July 1987, pp. 740–743.

[41] L. J. Kohout and S. M. A. Mohamad, "Development of support tools and methodology for design and validation of multi-environmental computer architectures", In: L. J. Kohout, J. Anderson, and W. Bandler, editors, *Knowledge-Based Systems for Multiple Environments*, Ashgate Publ. (Gower), Aldershot, U.K., 1992, Chapter 17.

[42] L. J. Kohout, S. M. A. Mohamad, and W. Bandler, "A support tool for evaluation of knowledge enginnering structures" In: *FLAIRS - 89 Proceedings*, The Florida Artificial Intelligence Research Society, 1989, pp. 113–117.

[43] L. J. Kohout and I. Stabile, "Interval-valued inference in medical knowledge-based system Clinaid", *Interval Computations*, 1993, No. 3, pp. 88–115.

[44] L. J. Kohout and I. Stabile, "Interval-valued information retrieval and inference in medical knowledge-based system Clinaid," In: *Conf. on Numerical Analysis with Automatic Result Verification*, Lafayette, Louisiana, February 25 - March 1, 1993.

[45] L. J. Kohout, I. Stabile, W. Bandler, and J. Anderson, "CLINAID: Medical knowledge-based system based on fuzzy relational structures", In: M. Cohen and D. Hudson, editors, *Comparative Approaches in Medical Reasoning*, World Scientific, 1995, (in press).

[46] V. Novák. On the position of fuzzy sets in modelling of vague phenomena. In: R. Lowen and M. Roubens, editors, *IFSA '91 Brussels, vol. Artificial Intelligence*, International Fuzzy Systems Association, 1991, pp. 165–167.

[47] V. Pinkava, *Introduction to Logic for System Modelling*, Gordon and Breach, London and New York, 1988.

[48] G. Priest, R. Routley, and J. Norman, (eds.), *Paraconsistent Logic: Essays on the Inconsistent*, Philosophia Verlag, München, Wien, 1989.

[49] G. Resconi, G. J. Klir, and U. St.Clair, "Hierarchical uncetrainty metatheory based upon modal logic", *Internat. J. of General Systems*, 1992.

[50] J. Riguet, "Relations binaires, fermetures, correspondences de Galois", *Bull. Soc. Math. France*, 1948, Vol. 76, pp. 114–155.

[51] B. Schweizer and A. Sklar, *Probabilistic metric spaces*, North Holland, New York, 1983.

[52] E. Stiller, S. Løvstad, W. Bandler, and V. a Mancini. "Representing the social dimension: Design and methodology for an urban planning system", In: M. B. Fishman, editor, *Proceedings of 2nd Florida Artificial Intelligence Research Symposium*, The Florida AI Research Symposium, P. O. Box 12560, St Petersburgh, Florida 33733, 1989, pp. 216–220.

[53] I.B. Turksen. "Containment and Klein groups of fuzzy propositions", *Working Paper 79-010, Dept. of Industrial Eng., University of Toronto, Canada*, 1979.

[54] I. B. Turksen. "Klein groups in fuzzy inference", In: *Proc. of the American Control Conference*, American Automatic Control Council, 1984, pp. 556–560.

[55] R. Yager, "On the measure of fuzziness and negation. Part I: Membership in the unit interval", *International Journal of General Systems*, 1979, Vol. 5, No. 4, pp. 221–229.

# 13

# COMPUTING UNCERTAINTY IN INTERVAL BASED SETS

**Luis Mateus Rocha*, Vladik Kreinovich**, and R. Baker Kearfott*****

**Department of Systems Science and Industrial Engineering, T.J. Watson School, State University of New York at Binghamton, Binghamton, NY 13902, e-mail ba05099@bingsuns.cc.binghamton.edu*

***Department of Computer Science, University of Texas at El Paso, El Paso, TX 79968, email vladik@cs.utep.edu*

****Department of Mathematics, University of Southwestern Louisiana, U.S.L. Box 4-1010, Lafayette, LA 70504-1010, USA, email: rbk@usl.edu*

## ABSTRACT

The kinds of uncertainty present in different interval based techniques of representing uncertainty in knowledge-based systems will be discussed. Interval-Valued Fuzzy Sets (IVFS) will be shown to describe both fuzziness and nonspecificity in their membership degrees, while a structure called an *evidence set* further introduces conflict. A more realistic model of uncertainty is described by $L-$fuzzy sets, interval-valued $L-$fuzzy sets, and $L-$evidence sets, where $L$ is a finite set of possible degrees of confidence. Measures of uncertainty of such structures are examined.

## 1 INTRODUCTION

The majority of existing knowledge-based systems use real numbers to describe experts' degrees of confidence in different statements. In many cases, algorithms for reasoning with uncertainty are based on the quantitative estimations of the current uncertainty of knowledge.

Recently, *interval-based* formalisms, such as interval-valued fuzzy sets (IVFS), and evidence sets (a generalization of IVFS proposed in Rocha [21]) have been proposed for more adequate description of experts' uncertainty. To efficiently



*R. B. Kearfott and V. Kreinovich (eds.), Applications of Interval Computations*, 337–380.

use these formalisms in reasoning, we thus need to design uncertainty measures for these interval-based generalizations of traditional uncertainty formalisms.

In the following, measures of uncertainty for interval based sets are derived from the information measures described by Ramer et al. [17, 18, 19]. It will be shown that the numerical characteristics of uncertainty present in interval based set structures may be better captured in a discrete formulation where the real unit interval of degrees of confidence is replaced by a linearly ordered finite set. This discrete formulation is more coherent with human cognitive abilities.

I will start with a review of relevant mathematical background in Section 2; the basic concepts of uncertainty representation, as well as evidence sets, will be explained and formally defined. In Section 3, measures of uncertainty will be developed, while in Section 4, the $L-$fuzzy set alternatives will be defined. Finally, in Section 5, the importance of evidence sets will be stressed as models capable of expressing all the recognized forms of uncertainty.

*Important Comment.* Some of the notions utilized in this paper, such as degree of membership in a fuzzy set as well as the construction of belief functions in the Demspter-Shafer theory of evidence, are introduced with interpretations designed to facilitate their understanding by the expected audience of this book. These interpretations do not necessarily follow accurately the points of view of all the authors involved. Further, particularly regarding the Dempster-Shafer theory of Evidence, the interpretations here pursued may go against the orthodox interpretations of this theory. For a more complete discussion of these topics, based on the traditional interpretations, refer to Rocha [23, 25].

# 2 EVIDENCE SETS AS A DESCRIPTION OF UNCERTAINTY

In this section, we will outline a new method of describing human reasoning called *evidence sets.* This method was originally proposed in Rocha [21, 23]. To describe this new notion, we will first outline the existing approaches, explain why they are sometimes inadequate in describing human reasoning, and how this new method can be applied in these situations.

## 2.1 It is Necessary to Represent Expert Knowledge in a Computer

In many fields (geology, medicine, etc), there are experts who can make outstanding decisions. Unfortunately, the number of these super-experts is usually small, and we cannot use them for every problem we encounter. We cannot show all the patients with heart disease to the greatest specialist.

The natural idea is therefore to design a computer-based system that would use the knowledge of these experts and make conclusions and decisions based on that knowledge. Such systems need not be completely autonomous. If we can design a system that will prompt one or several different answers, this system will also be of great value to the corresponding community. Such helper systems are called *expert systems*, or *knowledge-based systems*.

To make these systems successful, we must be able to incorporate the experts' knowledge inside the computer. First, we need to describe statements that comprise the experts' knowledge. This has been done starting from the pioneer papers on Artificial Intelligence. However, our knowledge is not simply the set of statements that we make. In some of these statements, we are more certain; in some, we are less certain. This different certainty is an important part of our knowledge; for example, if different statements lead to different conclusions, we need to compare their relative certainty to decide which of these statements we trust more. Therefore, in addition to simply describing the statements, we must also describe the degree of our certainty in these statements.

The notion of uncertainty, or the available formalization of uncertainty, is thus very relevant to any discussion of the modeling of linguistic/mental abilities. From Zadeh's (1971, 1978) approximate reasoning to probabilistic and so-called evidential reasoning (Schum [26]), uncertainty is more and more recognized as a very important issue in cognitive science and artificial intelligence regarding the problems of knowledge representation and the modeling of reasoning abilities. (A good collection of papers in this area is found in Shafer and Pearl [27].)

## 2.2 Fuzzy Logic: Historically First Description of Uncertainty

### *The Basic Idea of Fuzzy Logic: Using Numbers ("Fuzzy Values") From the Interval* [0, 1] *to Describe Degrees of Confidence*

In the ideal case, when there is no uncertainty, each statement has only two possible degrees of confidence: "true" and "false". In the majority of computers, "true" is represented by 1, and "false" is represented by 0. Therefore, if we want to represent the intermediate degrees of confidence (that correspond to partial certainty), it is natural to use real numbers from the interval $[0, 1]$. This idea was originally proposed by Lotfi Zadeh in [36]. (The corresponding idea was later, in (Zadeh [37]), called *fuzzy logic*.)

Numbers from the interval $[0, 1]$ are actually used to describe uncertainty in the majority of real-life expert systems.

### *Where Do Fuzzy Values Come From?*

How can we assign a number to the expert's uncertainty? The ideal situation is when an expert herself can directly provide us with this number. This is, alas, rarely possible; therefore, we need indirect methods of describing this degree of certainty:

- For example, we can ask an expert to indicate her degree of certainty in a certain statement $A$ by choosing a number on a scale, say, from 0 to 10. Then, if she, e.g., picks 6, we take 0.6 as the desired degree $m_A$. (In general, we take $i/S$, where $i$ is the picked number, and $S$ is the largest number on this scale.)
- Another possibility is to ask an expert to compare the hypothetical alternatives "you get \$100 if $A$ is true, and 0 otherwise" with lotteries of the type "you get \$100 with probability $p$." For $p$ close to 0, the lottery is clearly worse than the $A$-conditioned alternative; for $p$ close to 1, the lottery is better. Therefore, there exists a value of the probability $p$ for which the probabilistic lottery and an $A$-conditioned lottery are of the same value to the expert. This value $p$ is then used to characterize the expert's uncertainty (it is called the *subjective probability*).

These two methods are most frequently used. Several other methods of assigning uncertainty values exist.

## *Logical Operations with Fuzzy Values*

The main objective of the described formalism is to be used in knowledge-based systems; in particular, fuzzy values (i.e., values from the interval $[0, 1]$) are used to characterize the expert's degree of confidence in the statements $A_1, ..., A_n$ that form a knowledge base. The goal of the knowledge-based system is to answer queries. In particular, we must be able to answer queries $Q$ that are simply logical combinations of the basic statement: e.g., the query can take one of the forms $\neg A_i$ ("not $A_i$"), $A_i \& A_j$ ("$A_i$ and $A_j$"), $A_i \vee A_j$ ("$A_i$ or $A_j$"), etc. Since we are not 100% sure that all the statements $A_i$ that form our knowledge base are true, we cannot answer such queries by simply answering "yes" or "no". Since we are not certain whether $Q$ is true or not, a natural solution is to describe our degree of confidence in $Q$.

For the three simple queries described above, we thus have the following problem: given the degrees of confidence $m_A$ and $m_B$, compute the degrees of confidence of the queries $\neg A$, $A \& B$, and $A \vee B$. Since the values $m_A$ and $m_B$ are the only information that we can use, the resulting estimates must be functions of $m_A$ and $m_B$. These functions are called, correspondingly, $\neg-$, $\&-$, and $\vee-$*operations*. We will denote them by $f_\neg$, $f_\&$, and $f_\vee$. So, if we know the degrees of confidence $m_A$ and $m_B$ in two statements $A$ and $B$, we will have the following:

- as a degree of confidence $m_{\neg A}$ in the statement "not $A$", we will take $f_\neg(m_A)$;
- as a degree of confidence $m_{A \& B}$ in the statement "$A$ and $B$", we take the value $f_\&(m_A, m_B)$;
- as a degree of confidence $m(A \vee B)$ in the statement "$A$ or $B$", we take the value $f_\vee(m_A, m_B)$.

There are many choices of $\neg$, $\&-$, and $\vee-$operations. Computationally the simplest choice is $f_\neg(a) = 1 - a$, $f_\& = \min$, and $f_\vee = \max$. The use of this negation operation can be justified by the second method of determining $m_A$: if the subjective probability of $A$ is $m_A$, then the subjective probability $m_{\neg A}$ of $A$'s negation must be, of course, equal to $1 - m_A$.

## *Statements with Parameters; Membership Functions ("Fuzzy Sets") as a Description of their Uncertainty*

In some cases, we have a general statement $P(x)$ about the values from a given set $X$, and its degree of certainty depends on the precise value to which it is applied. For example, we may say that, under certain conditions, "the temperature $x$ is high." The degree of our certainty $m_{P(x)}$ in this statement may differ depending on the actual value $x$. To describe this varying degree of certainty, we must assign to every $x \in X$ our degree of confidence in a statement $P(x)$. As a result, we get a function $m_P : X \to [0,1]$. Such a function is called a *membership function* corresponding to the property $P$.

If for every $x \in X$, the statement $P(x)$ is either absolutely true or absolutely false (such statements are called *crisp*), this function takes the values 0 (if $P(x)$ is false) and 1 (if $P(x)$ is true). In this case, we can also characterize the property $P$ by the *set* of all $x \in X$ for which $P(x)$ is true. The function $m_P$ is then a *characteristic function* of this set. Due to this analogy, in the general case, the pair $(X, m_P)$ is called a *fuzzy set*.

*Important Comment.* The traditional definitions of fuzzy logic and fuzzy sets are based not on degree of confidence but on the notion of a continuous degree of truth, between true and false, defined on the real unit interval for the former, and a continuous degree of membership for the latter. For a good introduction to these notions refer to Klir et al [10]. Previously, evidence sets have been defied with these interpretations Rocha [21, 23, 25].

## *Operations with Fuzzy Sets*

Let us consider the situation when we have two properties $P(x)$ and $Q(x)$ defined on the same universal set $X$. These properties are characterized by the membership functions $m_P(x)$ and $m_Q(x)$. According to the general idea of fuzzy operations, the logical combinations $\neg P(x)$, $P(x)\&Q(x)$, and $P(x) \vee Q(x)$ are characterized by the membership functions $m_{\neg P}(x) = f_\neg(m_P(x))$, $m_{P\&Q}(x) = f_\&(m_P(x), m_Q(x))$, and $m_{P\vee Q}(x) = f_\vee(m_P(x), m_Q(x))$. In particular, for the choice of computational operations (described above), we get: $m_{\neg P}(x) = 1 - m_P(x)$, $m_{P\&Q}(x) = \min(m_P(x), m_Q(x))$, and $m_{P\vee Q}(x) = \max(m_P(x), m_Q(x))$.

If $P$ and $Q$ are crisp properties, then:

- the set of all $x$ for which $\neg P$ is true, is the *complement* to the set $\{x|P(x)\}$;
- the set of all $x$ for which $P\&Q$ is true, is the *intersection* of the sets $\{x|P(x)\}$ and $\{x|Q(x)\}$;
- the set of all $x$ for which $P \vee Q$ is true, is the *union* of the sets $\{x|P(x)\}$ and $\{x|Q(x)\}$.

In view of this analogy, similar terms are used to describe the corresponding fuzzy sets; namely:

- a pair $(X, m_{\neg P})$, where $m_{\neg P}(x) = f_{\neg}(m_P(x))$, is called the *complement* to the fuzzy set $(X, m_P)$;
- a pair $(X, m_{P\&Q})$ is called the *intersection* of the fuzzy sets $(X, m_P)$ and $(X, m_Q)$;
- a pair $(X, m_{P\vee Q})$ is called the *union* of the fuzzy sets $(X, m_P)$ and $(X, m_Q)$.

## 2.3 Interval-Valued Fuzzy Sets

### *Main Idea*

Traditional fuzzy logic is based on the assumption that we can characterize the expert's degree of confidence in a statement $A$ by a single number $m_A$. At first glance, the above-described methods of gauging uncertainty do exactly that: by using them, we can get better and better approximations to the actual degree of uncertainty. To do that, we must:

- In the first scheme (estimating on a scale), ask an expert to describe his degree of confidence on a less and less granular scale:
  - first, we can describe the degree of confidence on a scale from 0 to 10 and thus, get the desired values with an accuracy 10%;
  - then, we can describe the degree of confidence on a scale from 0 to 100 and thus, get the desired values with an accuracy 1%;
  - etc.

- In the second scheme (comparing with lotteries), if we have already located the subjective probability with, say, an accuracy 10%, as $\approx$ 60%, we can then compare the $A$–conditioned outcome with lotteries with probabilities 55%, 56%, ..., 65%, and thus get the subjective probability with an accuracy 1%.

In reality, however, these methods are not practical. Hardly anyone can place his degrees of confidence on a scale from 0 to 100 so that there is a clear and meaningful difference between, say, the values 61 and 62 on this scale. Similarly, we will not be able to meaningfully choose between an $A$–conditioned event with subjective probability approximately 60% and a lottery with a probability, say, 61%.

In all these cases, on each step of the uncertainty-gauging procedure, instead of the ideal real number (the degree of confidence $m_A$), we have an *interval* of possible values of $m_A$.

- For example, if we marked our degree of confidence by a value 6 on a scale from 0 to 10, this does not mean that our degree of confidence is exactly 0.6; it rather means that the actual (unknown) degree of confidence is closer to the point 6 than to the point 5 and 7; in other words, that the actual degree of confidence belongs to the interval [0.55,0.65].
- Similarly, if we have found that the $A$–conditioned event is better than a lottery in which we win with a probability 50%, but worse than a lottery with a 60% win, this means that the subjective probability of $A$ belongs to the interval $[0.5, 0.6]$.

In the idealized case, we will be able to repeatedly narrow this interval down to the unique numerical value. In practice, after a few narrowings, we get to the limit of the expert's ability to make distinctions. So, in real life, whatever methods we apply to gauge the expert's degree of confidence, we always end up with an *interval* $\mathbf{m}_A$, and not with a real number $m_A$.

It is therefore natural to correspondingly redefine all notions of fuzzy logic in interval terms.

## *Operations with Interval-Valued Degrees of Confidence*

In particular, we can use the logical operations described above with numerical degrees of confidence to define corresponding operations with interval degrees of confidence. Namely, the interval degree of confidence $\mathbf{m}_A$ means, crudely speaking, that the actual (unknown) degree of confidence $m_A$ can take any value from the interval $\mathbf{m}_A$. Therefore, possible values of degree of confidence $m_{\neg A}$ in $\neg A$ are values $f_{\&}(m(A))$ for $m_A \in \mathbf{m}_A$. Similar arguments can be repeated for &– and ∨–operations. As a result, we get the following formulas:

$$\mathbf{m}_{\neg A} = f_{\neg}(\mathbf{m}_A) = \{f_{\neg}(m_A) | m_A \in \mathbf{m}_A\}.$$

$$\mathbf{m}_{A\&B} = f_{\&}(\mathbf{m}_A, \mathbf{m}_B) =$$

$$\{f_{\&}(m_A, m_B) | m_A \in \mathbf{m}_A,\ m_B \in \mathbf{m}_B\}.$$

$$\mathbf{m}_{A\vee B} = f_{\vee}(\mathbf{m}_A, \mathbf{m}_B) =$$

$$\{f_{\vee}(m_A, m_B) | m_A \in \mathbf{m}_A,\ m_B \in \mathbf{m}_B\}.$$

These formulas are similar to the basic formulas of interval arithmetic, with the only difference that now, they are applied to logical rather than to arithmetic operations. In particular, for standard operations $f_{\neg}(a) = 1 - a$, $f_{\&} = \min$, and $f_{\vee} = \max$, we can get the explicit formulas:

$$[m^-_{\neg A}, m^+_{\neg A}] = [1 - m^+_A, 1 - m^-_A].$$

$$[m^-_{A\&B}, m^+_{A\&B}] = [\min(m^-_A, m^-_B), \min(m^+_A, m^+_B)].$$

$$[m^-_{A\vee B}, m^+_{A\vee B}] = [\max(m^-_A, m^-_B), \max(m^+_A, m^+_B)].$$

These operations are described, e.g., in Gorzalczany [5].

## *Interval-Valued Fuzzy Sets (IVFS) and Operations with Them*

To describe the uncertainty in a property $P(x)$ with a free variable $x$, we need an *interval-valued* membership function defined as $\mathbf{m}_P(x) = \mathbf{m}_{P(x)}$. The pair $(X, \mathbf{m}_P)$ is naturally called an *interval-valued fuzzy set* (*IVFS* for short). The above-described interval logical operations define complement, intersection, and union of interval-valued fuzzy sets (see, e.g., Gorzalczany [5]).

*Historical comment.* The notions of interval-valued fuzzy sets and operations with them were introduced by Turksen [29] and Gorzalczany [5]; and also by

Atanassov ([1]; see also Atanassov and Gargov [3]) under the name of ***intuitionistic fuzzy sets***. For the latest developments, see, e.g., Bo Yuan et al. [35] and Atanassov [2].

Interval-valued fuzzy sets (IVFS) can be also viewed as a particular case of ***type 2 fuzzy sets*** introduced by Zadeh in [37]. A type 2 fuzzy set $m_P$ assigns to each element $x \in X$ a *fuzzy set* $m_P(x) \subseteq [0,1]$, i.e., a function $m_P(x) : [0,1] \rightarrow [0,1]$ that assigns to each $a \in [0,1]$, the degree of confidence that $a$ is a possible degree of confidence of the statement $P(x)$. IVFS is a specific case of a type 2 fuzzy set for which for all $x \in X$, the set $m_P(x)$ is a crisp interval. The greatest advantage of IVFS over more complicated type 2 fuzzy sets is their comparative simplicity:

- Generic type 2 fuzzy sets need an extensive amount of information that is usually not obtainable in relevant applications.
- On the other hand, an IVFS does not require much more information than a fuzzy set, simply upper and a lower bounds to its interval of membership.

## 2.4 Evidence Sets

### *Problems with Fuzzy Sets and with Interval-Valued Fuzzy Sets*

If we have a single expert, then it is reasonable to describe this expert's degree of confidence in a given statement $P$ by an interval $\mathbf{m}_P$. Knowledge that forms an expert system is rarely taken from a single expert. Single experts can make mistakes. Hence, to increase the reliability of the knowledge base, we must interview several experts. Different experts may have different confidences. The confidence of each of the experts can be characterized by an interval. How should the resulting confidence be characterized? In traditional expert systems, a single number (or, at best, an interval) is used to describe the degree of confidence in a statement. Hence, a number $m_A$ (or an interval $\mathbf{m}_A$) that describes the system's confidence in a given statement $A$ is obtained by some combination of the values $m_A^{(1)}, \ldots, m_A^{(N)}$ (corr., $\mathbf{m}_A^{(1)}, \ldots, \mathbf{m}_A^{(N)}$) provided by $N$ experts.

For numbers, the simplest combination is an *arithmetic averaging* $m_A = (1/N)\sum m_A^{(i)}$. The problem with such averaging is that it assigns the same value of $m_A$ to different knowledge situations. Let us give a simple example:

- If an expert does not have any knowledge about whether $A$ is true or not, then it is natural to describe this expert's degree of confidence in $A$ by a value $a$ for which the degrees of confidence in $A$ and $\neg A$ coincide. For $f_\neg(a) = 1 - a$, this degree of confidence is 0.5. If two experts have no knowledge about $A$, then $m_A^{(1)} = m_A^{(2)} = 0.5$, and therefore, $m_A = 0.5$.
- Let us now consider a different knowledge situation: we have two experts, one of them believes $A$ to be true, another strongly believes that $A$ is false. In this case, $m_A^{(1)} = 1$, $m_A^{(2)} = 0$, and hence, $m_A = 0.5$.

In both cases, we have $m_A = 0.5$, but in the first case, we do not have any knowledge at all, while in the second case, we have two conflicting opinions, which probably means that one of these opinions is correct.

The simplest case if when we have a *consistent* set of intervals, i.e., when the intersection of the intervals $\mathbf{m}_A^{(i)}$ is not empty. Then, we can take this intersection as the desired combined degree of confidence $\mathbf{m}_A$. In general, however, these intervals are not necessarily consistent.

For general intervals, there are two natural combination rules: as $\mathbf{m}_A$, we can take either the arithmetic average of the intervals $\mathbf{m}_A^{(i)}$ (in the sense of interval arithmetic), or the (convex envelope of) their union, i.e., the interval $[\min m_A^{(i)-}, \max m_A^{(i)+}]$. In both cases, similar examples show that by averaging, we lose some important information about the experts confidences.

*Philosophical comment.* The information about conflicts between different experts (and even within the opinions of a single expert) is important not only for expert systems per se, but also in the understanding of human cognitive categorization processes (Lakoff [11]; Rocha [21, 23]), and of human reasoning processes. It is therefore necessary to abandon the existing "objectivist" theories and formalisms (objectivist in the sense that confidences are treated in these formalisms as consistent approximate description of objective reality), and embark in a belief-based "subjectivist" theory that describes not an ideal (difficult to generate) consistent set of beliefs, but a real-life (thus, constructive) possibly inconsistent set of beliefs. We believe that such a constructivist epistemology is indispensable in modern approaches to cognition, systems research, and cybernetics (Medina-Martins and Rocha [14]; Medina-Martins [13]; Rocha [22, 24]).

## *The Idea of a Body of Evidence*

In order not to lose this information, we would like to keep in the expert system not only the average opinion of the experts, but also individual opinions as well. Well, keeping all opinions may be an overshoot: Indeed, we consider all interviewed experts as experts of more or less equal standing. (Otherwise, if one of them is inferior to the others, why should we include him in this group?) We are not experts, so we have no way of telling which of these experts are more correct. Therefore, it makes no sense to describe which expert holds which opinions. However, it makes sense to describe different opinions, and to describe how many experts hold each of these opinions.

This description can also be further clarified. What matters is not how many experts hold a certain opinion: if ten experts do, then it is important to know whether this is ten out of ten, or ten out of a hundred. So, what really matters in describing an importance of an opinion is the *fraction* of experts who hold this opinion.

For the problem of ascribing degrees of confidences to statements from the knowledge base, *opinions* stand for intervals of degrees of confidence. Therefore, to describe the experts' confidence in a given statement $A$, we must keep all intervals $I$ assigned to $A$ by different experts, with fraction of experts who believes that $I$ is the right interval. These fractions are non-negative numbers that add up to 1. Therefore, we can say that we have a *probability measure* on the set of all the intervals (such that the probability is different from 0 only for finitely many intervals). This probability measure $\mathcal{M}_A$ describes all the experts' evidence that we have and is therefore called a *body of evidence*.

*Important Comment.* The orthodox definitions of Dempster-Shafer theory of evidence (see Shafer [28]) interpret the probability measure $\mathcal{M}_A$ – the basic probability assignment – not as a construction based on the opinions of *several* experts, but on the quantification of belief a *single* expert can offer based on available evidence. Each portion of belief refers to a particular subset of values of a proposition associated with some context; the sum of the basic probability assignment values for the whole set of such subsets – a body of evidence – must be equal to one. Evidence sets were elsewhere defined based on this traditional interpretation of evidence theory. For a more complete discussion of the philosophical implications of the different interpretations, particularly regarding the modeling of human cognitive categories see Rocha [21, 23, 25].

In our explanation of this idea, we assumed that all experts are equal. Experts may carry different weights: e.g., one expert may be a lonely researcher, while another "expert" may actually be a group. In this case, it is reasonable to assign weights to different experts. If we normalize these weights so that their sum is equal to 1, then, as a result, we still get a probability measure on the set of all intervals, i.e., a body of evidence.

The set of all possible bodies of evidence is denoted by $\mathcal{E}$. Intervals $I$ for which $\mathcal{M}(I) \neq 0$ are called *focal elements*. The set of all focal elements will be denoted by $\mathcal{F}$.

## *Probability Distribution: A Degenerate Case of a Body of Evidence*

An interesting case of a body of evidence is when each expert can definitely specify her degree of confidence. In this case, all intervals $I = [I^-, I^+]$ with $\mathcal{M}(I) \neq 0$ are *degenerate* (*"singletons"*), i.e., they consist of a single value: $I^- = I^+$. In this case, the probability distribution $\mathcal{M}$ is defined not on set of all intervals, but on the interval $[0, 1]$. Such distributions were first used for describing uncertainty by Hirota [7].

## *Consonant and Non Consonant Bodies of Evidence*

In some cases, the experts more or less agree with each other, and the difference in their opinions is caused by the fact that some of the experts are more cautions in their statements, and some are bolder. In this case, if we order the experts in the order of cautiousness, we conclude that the opinion of the first expert leads to the smallest intervals, the opinion of the second expert is described by a slightly larger interval that contains the first one, etc. In other words, in these cases, the corresponding probability measure $\mathcal{M}_A$ is concentrated on the *nested* sequence of intervals $I_1 \subset I_2 \subset \ldots \subset I_N$. Such bodies of evidence are called *consonant*.

In other cases, the expert opinions are inconsistent; in terms of intervals, this can mean, e.g., that some focal elements $I$ and $J$ are disjoint. In these *non consonant* cases, there is a *conflict* between the opinions of different experts.

## *If the Degree of Confidence is Represented by a Body of Evidence, How Should a Query be Answered?*

Let us assume that in the knowledge base, the information about the experts' confidence in a statement $A$ is described by a body of evidence $\mathcal{M}_A$. If a user asks exactly the query "$A$?", what do we answer?

When the degree of confidence was described by a number, we simply returned this number; when it was an interval, we returned the interval. So, when the degree of confidence is described by a body of evidence, we can, in principle, return this very body of evidence. This will be a very detailed description of the experts' opinions; however, it will be difficult to understand for a normal user. So, we need some simplified output. One possibility to simplify the output is as follows: if the user is already accustomed to intervals, then, we can let the user not only formulate the query, but also try to guess the interval; then, we will answer to what extent this guess is consistent with the given body of evidence.

Suppose that the user asks a query $A$, and also guesses an interval $I$ of possible values of degree of confidence. Since our confidences are described in probabilistic terms, we cannot definitely say whether this interval is a correct description of confidence or not. We can only give, crudely speaking, the ***probability*** of $I$ being the correct interval. Then, depending on what the user wants, we can give three different answers:

- First, we can interpret the question as the query about the probability that $I$ is *exactly* the interval specified by an expert. Then, the answer to this question will simply be $\mathcal{M}(I)$. There is, however, a problem with this interpretation. Since the probabilities are different from 0 only for finitely many intervals (actually, on those intervals that have actually been described by experts), and there are infinitely many possible intervals $I$, in almost all cases, the answer to the query in this interpretation will be the senseless 0. So, this first interpretation does not make much sense.

- Second, we can interpret the question as the query about the probability that only values from $I$ are possible, in other words, as the probability that the actual interval is either $I$, or a subset of $I$. This probability is called the ***belief*** in an interval $I$, and it is denoted by

$$\text{bel}(I) = \sum_{J \subseteq I} \mathcal{M}(J).$$

- Third, we can interpret the question as the the query about the probability that some values from $I$ are possible; in other words, as the probability that the actual interval $J$ has a non-empty intersection with $I$. This probability is called the ***plausibility*** of an interval $I$, and it is denoted by

$$\mathrm{pl}(I) = \sum_{J:\, J\cap I\neq\emptyset} \mathcal{M}(J).$$

If we are not sure whether the user means the second or the third interpretation, we can deliver her ***both*** probabilities, i.e., the ***interval*** $[\mathrm{bel}(I), \mathrm{pl}(I)]$.

*Historical comment.* The notions of belief and plausibility were introduced in the so-called Dempster-Shafer theory of evidence (see, e.g., Shafer [28]).

*Comment.* In addition to belief and plausibility, we can return other characteristics of the body of evidence. These characteristics describe our ***fuzzy*** (uncertain) knowledge and are therefore called ***fuzzy measures***. Fuzzy measures deal with degrees of evidence representing the degree to which each crisp subset of an universal set (in our case, of the interval $[0,1]$) satisfies a property in question. For general definitions of fuzzy measures, see, e.g., Klir et al. ([10, 8], especially [8], p. 266). An even more general definition was proposed in Wang and Klir [31].

## *Logical Operations with Bodies of Evidence*

Suppose that we know bodies of evidence $\mathcal{M}_A$ and $\mathcal{M}_B$ that describe the experts' degree of confidence in two statements $A$ and $B$. According to $\mathcal{M}_A$, the probability of an interval $I$ to be the actual interval degree of confidence of $A$ is $\mathcal{M}_A(I)$, and the probability for an interval $J \subseteq [0,1]$ to characterize the confidence in $B$ is equal to $\mathcal{M}_B(J)$. How should the degrees of confidence in $\neg A$, $A\&B$, and $A \vee B$ be characterized?

Since in the description of the body of evidence, we did not keep track of who believes in what, we cannot attach opinions of different experts. Since we thus do not know how the expert's opinions on $A$ and $B$ are related, it is natural to assume that these opinions are ***independent*** probability measures. In other words, for every two intervals $I$ and $J$, the probability that $I$ is an interval degree of confidence in $A$ and $J$ is a degree of confidence in $B$ is the product of the corresponding probabilities: $\mathcal{M}_A(I)\cdot\mathcal{M}_B(J)$. With this probability, the degree of confidence in $A\&B$ (corr., in $A \vee B$) is equal to $f_\&(I,J)$ (corr., to $f_\vee(I,J)$), where $f_\&$ and $f_\vee$ are the interval operations described above. If an interval $K$ has only one representation of this type (i.e., as $I\&J$ or as $I \vee J$ for some $I$ and $J$ with $\mathcal{M}_A(I) \neq 0$ and $\mathcal{M}_B(J) \neq 0$), then we can take this product as the desired probability $\mathcal{M}_{A\&B}(K)$. If an interval $K$ admits several such representations, then, to define $\mathcal{M}_{A\&B}(K)$, we must add up the product probabilities stemming from all of them. As a result, we get the following formulas:

$$\mathcal{M}_{A\&B}(K) = \sum_{I,J:I\&J=K} \mathcal{M}_A(I) \cdot \mathcal{M}_B(J),$$

$$\mathcal{M}_{A\vee B}(K) = \sum_{I,J:I\vee J=K} \mathcal{M}_A(I) \cdot \mathcal{M}_B(J).$$

This formulas were first described in Rocha [21, 23] (a similar idea is described for general interval probabilistic operations in Berleant [4]).

## *Combining Two Bodies of Evidence*

A body of evidence is the result of interviewing several experts. It sometimes happens that we have processed one group of experts and have come up with a body of evidence $\mathcal{M}$; independently, another group of knowledge engineers has interviewed another group of experts, and has come up with another body of evidence $\mathcal{N}$. It is then natural to combine the bodies $\mathcal{M}$ and $\mathcal{N}$ that express the experience of both groups of experts into a single body of evidence.

If the bodies of evidence describe the opinion of similar experts, then, according to the definition of the body of evidence, we can simply take the arithmetic average of the corresponding bodies.

It can also happen that these two bodies of evidence describe the opinions of *complementary* groups of experts, i.e., two groups who are experts in two different aspects of the phenomenon and who, therefore, respect each other's opinions. In this case, if inside each group, all experts have the same opinion, i.e., if each group is described by intervals $I$ and $J$ respectively, then, to combine $I$ and $J$, we can simply take the intersection $I \cap J$ of these intervals. (If the intersection is empty, then something is wrong, because we consider complementary groups of experts, and their opinions should not contradict each other.) In the general case, if $\mathcal{M}$ and $\mathcal{N}$ are the only information that we have, then it is natural to assume (just like we did for logical operations) that the corresponding probability distributions are independent. Thus, for each pair of intervals $I$ and $J$, we get the interval $I \cap J$ with the probability $\mathcal{M}(I) \cdot \mathcal{N}(J)$. The only difference between this and logical connectives is that, unlike $I\&J$ which is defined for each pair of intervals, the intersection makes sense (i.e., is non-empty) only for some pairs. So, in order to get the correct formulas for the resulting probabilities, we must discount these senseless cases, and consider only the cases for which $I \cap J \neq \emptyset$. As a result, we get the following formula for the combined body of evidence:

$$\mathcal{C}(K) = \frac{1}{D} \sum_{I,J:I \cap J = K} \mathcal{M}(I) \cdot \mathcal{N}(J),$$

where

$$D = 1 - \sum_{I,J:I \cap J = \emptyset} \mathcal{M}(I) \cdot \mathcal{N}(J).$$

This formula is known as the *Dempster's rule of combination*, and it was originally proposed in the *Dempster-Shafer theory of evidence.*

Several other combination rules have been proposed that do not simply discard inconsistent cases $I \cap J = \emptyset$. The application of these formulas to bodies of evidence is described in Rocha [21, 23].

## *Evidence Sets and Operations with Evidence Sets*

In view of the above discussion, it is natural to represent the notions $P(x)$ with a free variable $x$ by assigning to every $x \in X$ a body of evidence that describes the expert's confidence that $x$ satisfies the property $P$. Such description was proposed in Rocha [21] under the name of *evidence set.* An evidence set is thus defined as a pair $(X, \mathcal{M}_P)$, where $\mathcal{M}_P : X \to \mathcal{E}$ is a function from the *universal set* $X$ to the set $\mathcal{E}$ of all possible bodies of evidence; this function $\mathcal{M}_P$ is called a *membership function* of the evidence set. The notion of evidence set was first described in Rocha [21, 23]. (A mathematically equivalent formulation has also been recently proposed by Zhu and Lee [38] under the name of *interval-valued logic of belief.*)

The above-described operations $f_\&$ and $f_\vee$ with bodies of evidence enable us to describe union and intersection of evidence sets as

$$\mathcal{M}_{P \cap Q}(x) = f_\&(\mathcal{M}_P(x), \mathcal{M}_Q(x)) \text{ and}$$

$$\mathcal{M}_{P \cup Q}(x) = f_\vee(\mathcal{M}_P(x), \mathcal{M}_Q(x)).$$

*Historical Comment.* These formulas first appeared in Rocha [21, 23]. Zhu and Lee [38] view evidence sets as describing lower and upper probabilities (i.e., bounds on probabilities) rather than describing evidential strengths of different intervals. Because of this difference in interpretations, they end up with ***different*** operations with evidence sets. Both evidence sets (Rocha [21]) and interval based logics of belief (Zhu and Lee [38]) are very recent. To develop them further, we need to establish sound rules and operations leading to a strong theory of uncertain, belief-based, reasoning. Therefore, all of the proposed operations and connectives must be evaluated for their applicability in coherent computational schemes, as well as accepted theories of cognition and artificial intelligence. In other words, belief-based interval set structures (such as evidence sets) must be evaluated (as models of categorization processes and human reasoning) within some larger uncertain reasoning framework.

**Example.** Interval-valued fuzzy sets are a particular example of evidence sets in which for every $x$, the probability measure $\mathcal{M}_P(x)$ is concentrated on a single interval with probability 1. These are the simplest possible evidence sets.

In the next two subsections, we will describe other particular cases of evidence sets: probabilistic sets and consonant evidence sets.

## *Probabilistic Sets as a Particular Case of Evidence Sets*

If for each $x \in X$, the body of evidence is a probabilistic distribution, evidence sets become ***probabilistic sets*** in the sense of Hirota [7]. A probabilistic set is a pair $(X, \mathcal{M}_P)$, where the function $\mathcal{M}_P$ assigns to each element $x \in X$, a probability distribution defined on $[0, 1]$. (For a more complete definition of probabilistic sets see Hirota [7].)

## *Consonant Evidence Sets*

An interesting case occurs when we require every body of evidence $\mathcal{M}_P(x)$ to be consonant for each $x \in X$, i.e., when we require that the measure $\mathcal{M}_P(x)$ is concentrated on a nested structure of interval focal elements $I_1 \subset I_2 \subset \ldots \subset I_N$. The simplest case is when the number $N$ of these intervals does not depend on $x$. Such evidence sets are called ***consonant***.

A consonant evidence set, instead of using a ***single*** interval of possible degrees of confidence to formalize the nonspecificity of the degree of membership of element $x$ of $X$ in a set $P$ (as is the case of IVFS), uses ***several*** nested intervals with different evidential strength (subjective probability) for each interval. These

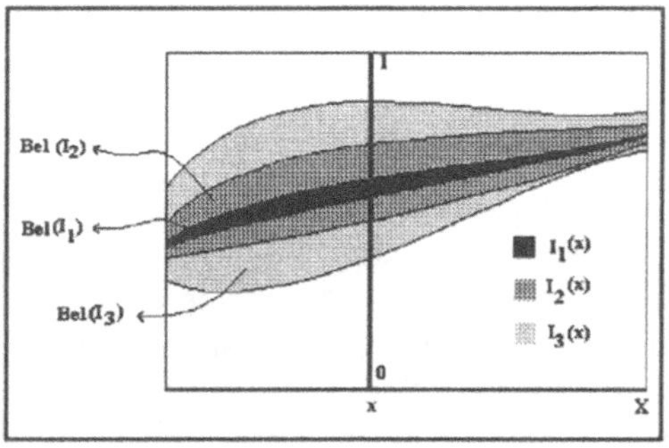


**Figure 1** Consonant evidence set with three focal elements

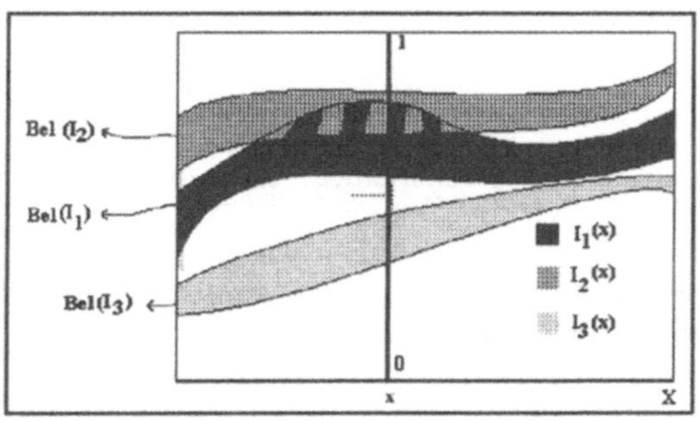


**Figure 2** Non-consonant evidence set with three focal elements

nested intervals state our graded evidence claims regarding the membership of element $x \in X$ in $P$.

# 3 DIFFERENT MEASURES OF UNCERTAINTY, AND HOW TO DESCRIBE THEM NUMERICALLY

For each description of the experts' uncertainty, it is desirable to gauge to what extent this knowledge is uncertain and incomplete. For example, we need numerical values to describe how much we can rely on the knowledge, and how much more data we need to add to the knowledge to get the complete knowledge base.

Also, we have talked about knowledge ***representation***, but an important part of a knowledge-based system is ***reasoning***, that enables the system to answer the queries. In many reasoning systems, the choice of the next reasoning step is based on the measurement of current uncertainty. This is true in systems ranging from information retrieval (Nakamura and Iwai [16] to artificial intelligence models (Rocha [20]; Medina-Martins and Rocha [14]; Medina-Martins, Rocha, et al [15]; Medina-Martins [13]). So, to improve the reasoning skills of such systems, we also need to gauge uncertainty in the knowledge base.

## 3.1 Three Characteristics of Uncertainty

Klir [8] classifies different aspects of uncertainty of our knowledge into three main forms:

- *conflict*, meaning that different pieces of knowledge disagree with each other;
- ***nonspecificity***, meaning that there are several different (well-defined) possibilities consistent with our knowledge.

These two forms of uncertainty correspond to the case when all the statements from the knowledge base are well-defined. Klir combines them under the name of ***ambiguity***.

- The third form of uncertainty is *fuzziness*, when the words that we use do not have well-defined meaning. (This is typical for the words from natural language.)

In the following subsections, formulas will be proposed for measuring the three different kinds of uncertainty for evidence sets. As a result, we will also get formulas for interval valued fuzzy sets, because they are a particular case of evidence sets.

First, we will describe the measure of uncertainty for the body of evidence $\mathcal{M}_P(x)$ that characterizes each element $x \in X$, and then combine the resulting

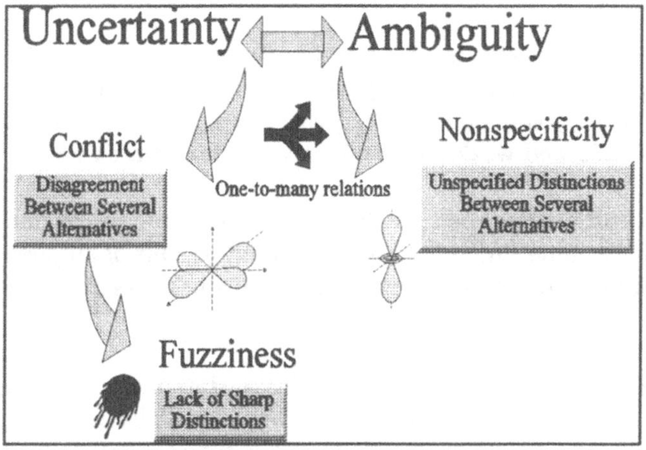


**Figure 3** Taxonomy of uncertainty

characteristics into a characteristic of the entire evidence set. In practically all our formulas and motivations, we will assume (for simplicity) that the universal set $X$ is finite.

## 3.2 Formalization of Fuzziness

### *Fuzziness: Informal Definition*

***Fuzziness*** is usually identified with lack of sharp distinctions. Other synonyms of the word fuzzy include: ***blurred, indistinct, unclear, vague, ill-defined, out of focus, not clear, indefinite; shadowy, dim, obscure; misty, hazy, murky, foggy; confused*** (Random House Webster's Dictionary [32]).

### *Fuzziness of a Degree of Confidence*

In a non-fuzzy (crisp) case, if an expert makes a statement $A$, and another expert makes an opposite statement $\neg A$, these two statements cannot be both true.

In a fuzzy case, two experts can make directly opposite statements about the same value: e.g., that a given body temperature is high, and that this same temperature is not high, and they can both be true (in the sense that in some cases, e.g., for a young and healthy person such a temperature is of no danger, while for an old and weak patient it may be life-threatening).

*Comment.* In the crisp case, the fact that we are able to conclude that both $A$ and $\neg A$ at the same time would mean a paradox. Out of all possible dictionary meaning of fuzziness, this "paradoxical" one is best described by the synonym *confused.*

In view of the possibility to have, for fuzzy statements $A$, $A$ and $\neg A$ at the same time, it is natural to take the degree of confidence $m_{A\&\neg A}$ that $A$ and $\neg A$ are both true as a measure $F(A)$ of the fuzziness of $A$. This definition of fuzziness was proposed by Yager [33, 34]. (Several other ways of describing fuzziness have been previously suggested; for a survey of these methods, see, e.g., Klir [8].).

In fuzzy logic, the degree of confidence $m_{A\&\neg A}$ that $A$ and $\neg A$ are both possible is described by a formula $F(A) = f_{\&}(m_A, f_{\neg}(m_A))$. In particular, for standard operations $f_{\&} = \min$ and $f_{\neg}(a) = 1 - a$, we get $F(A) = \min(m_A, 1 - m_A)$.

Several programming languages (e.g., C) do not have the function min, so we cannot directly use this formula. Instead, we can use the expression $\min(x, y) = (1/2)(x + y - |x - y|)$ that is evidently true for all $x$ and $y$. For our formula, the use of this expression leads to an equivalent formula

$$F(A) = (1/2)(1 - |2m_A - 1|).$$

The smallest value of $F(A)$ is 0; this value is attained only when $A$ is a crisp statement. The largest possible value of $F(A)$ is 1/2; it is attained when $m_A = 1/2$. So, the range of possible values of fuzziness is $[0, 1/2]$. It is often convenient to have a measure with a range from 0 to 1. To define such a measure, we can *re-scale* the range to $[0, 1]$ by multiplying $F$ by 2. As a result, we get an *index of fuzziness*, defined as $IF(A) = 2F(A) = 1 - |2m_A - 1|$.

## *Fuzziness of a Fuzzy Set*

A fuzzy set (i.e., a property $P(x)$ with a free variable $x \in X$) is crisp if for every $x$, the statement $P(x)$ is crisp. Every $x$ for which $P(x)$ is fuzzy adds additional fuzziness to the set $P$. Therefore, for finite universal sets $X$, it is natural to define fuzziness $F(m_P)$ of a fuzzy set $m_P$ as the sum of fuzzinesses of the degrees of confidence $m_P(x)$, i.e., as

$$F(m_P) = \sum_{x \in X} F(m_P(x)) = \frac{1}{2} \sum_{x \in X} (1 - |2m_P(x) - 1|).$$

The smallest possible value of $F(m_P)$ is 0; it is attained when the set $P$ is crisp. The largest possible value of $F(P)$ is attained when $m_P(x) = 1/2$ for all $x$; in this case, $F(m_P) = (1/2) \cdot |X|$. We can therefore re-scale fuzziness to a range $[0, 1]$ by defining the *index of fuzziness* of a fuzzy set $m_P$ as a ratio

$$IF(m_P) = \frac{F(m_P)}{(1/2) \cdot |X|} = \frac{1}{|X|} \sum_{x \in X} (1 - |2m_P(x) - 1|).$$

This ratio can also be expressed in terms of indices of fuzziness for $x \in X$ as

$$IF(m_P) = \frac{1}{|X|} \sum_{x \in X} IF(m_P(x)).$$

In other words, the index of fuzziness of a fuzzy set is equal to the arithmetic average of the indices of fuzziness for different $x \in X$.

For a crisp set, the index of fuzziness is a sum of 0's and is, therefore, equal to 0.

The definition of the index of fuzziness can be naturally generalized to continuous sets $X$: a natural limit transition leads to the necessity to consider integrals instead of the sums. For example, if $X = [a, b]$, we can define the index of fuzziness as

$$IF(m_P) = \frac{1}{b - a} \int_a^b (1 - |2m_P(x) - 1|) dx.$$

## *Fuzziness of an Interval Degree of Confidence and of an Interval-Valued Fuzzy Set*

If statement $A$ is characterized by an *interval* $\mathbf{m}_A = [m_A^-, m_A^+]$ of possible values of degrees of confidence, then the degree of confidence $\mathbf{m}_{A\&\neg A}$ is also an interval. For standard operations $f_\& = \min$ and $f_\neg(a) = 1 - a$, this interval is equal to

$$\mathbf{m}_{A\&\neg A} = [\min(m_A^-, 1 - m_A^+), \min(m_A^+, 1 - m_A^-)].$$

We want to characterize the degree of fuzziness by a single number $m_{A\&\neg A}$. The only thing that we know about this number is that it is somewhere on this interval. Therefore, it is natural to take the *midpoint* of this interval as the desired degree of fuzziness $F(\mathbf{m})$, i.e., define

$$F([m^-, m^+]) = \frac{1}{2}(\min(m^-, 1 - m^+) + \min(m^+, 1 - m^-)).$$

If we replace $\min(x, y)$ by $(1/2)(x + y - |x - y|)$, we will get an equivalent expression

$$F(\mathbf{m}) = \frac{1}{2}(1 - |m^- + m^+ - 1|).$$

The fuzziness so defined ranges from 0 to 1/2. The smallest possible fuzziness ($F(A) = 0$) is still attained only in two cases: when $A$ is absolutely true ($\mathbf{m}_A = [0, 0]$), and when $A$ is absolutely false (i.e., $\mathbf{m}_A = 1, 1]$). The fuzziness is the largest (i.e., equal to 1/2) not only when $\mathbf{m} = [1/2, 1/2]$, but also, e.g., when $\mathbf{m} = [0, 1]$ (i.e., when we know nothing about $A$).

By re-scaling, we can define the *index of fuzziness* $IF(\mathbf{m})$ with the range of values $[0, 1]$:

$$IF(\mathbf{m}) = 1 - |m^- + m^+ - 1|.$$

Similarly, we can define the *fuzziness* $F(\mathbf{m}_P)$ and the *index of fuzziness* $IF(\mathbf{m}_P)$ of an interval-valued fuzzy set $\mathbf{m}_P$ as

$$F(\mathbf{m}_P) = \frac{1}{2}\sum_{x \in X}(1 - |m_P^-(x) + m_P^+(x) - 1|)$$

and

$$IF(\mathbf{m}_P) = \frac{1}{(1/2) \cdot |X|} F(\mathbf{m}_P) = \frac{1}{|X|}\sum_{x \in X}(1 - |m_P^-(x) + m_P^+(x) - 1|).$$

Here also,

$$IF(\mathbf{m}_P) = \frac{1}{|X|}\sum_{x \in X} IF(\mathbf{m}_P(x)).$$

## *Fuzziness of Bodies of Evidence and Fuzziness of Evidence Sets*

When we describe the experts' degree of confidence in a statement $A$ by a body of evidence $\mathcal{M}$, we mean that this degree of confidence can take different interval values $I$ with probabilities $\mathcal{M}(I)$. Depending on which of the intervals $I$ turn out to be true, we will end up with different values of fuzziness. For each interval $I = [I^-, I^+]$, we already know how to describe the fuzziness: it is equal to $F(I) = (1/2) \cdot (1 - |I^- + I^+ - 1|)$. However, we do not know which of these intervals is the best representation of the current state of knowledge. Hence, the best we can do is to describe the *expected* value of fuzziness (in some reasonable sense).

The probability $\mathcal{M}(I)$ assigned to each interval means the ratio of experts that believe $I$ to be the correct description of our current confidence. Since we consider the experts to be equally good, we can say that the probability that an expert will turn out to be correct is the same for all the experts. Therefore, we can view this probability $\mathcal{M}(I)$ as the probability that $I$ will turn out to be the correct interval description of confidence according to the current state of knowledge. Hence, we can define the expected value of fuzziness as $F(\mathcal{M}) = \sum_I \mathcal{M}(I) \cdot F(I)$.

The value $F(I)$ is the difference between two expressions, one of them a constant. Therefore, the mathematical expectation of $F(I)$ can be represented as the difference between the two mathematical expectations, of which the first is simply a constant:

$$F(\mathcal{M}) = \frac{1}{2}(1 - \sum_I \mathcal{M}(I) \cdot |I^- + I^+ - 1|).$$

This value is the expectation of the fuzziness, and fuzziness ranges from 0 to 1/2. Therefore, this expected fuzziness also ranges from 0 to 1/2, so we can re-scale it to the index of fuzziness if we multiply it by 2:

$$IF(\mathcal{M}) = 1 - \sum_I \mathcal{M}(I) \cdot |I^- + I^+ - 1|).$$

Similarly, we can define the *fuzziness* $F(\mathcal{M}_P)$ and the *index of fuzziness* $IF(\mathcal{M}_P)$ of an evidence set $\mathcal{M}_P$ as

$$F(\mathcal{M}_P) = \sum_{x \in X} F(\mathcal{M}_P(x))$$

and

$$IF(\mathcal{M}_P) = \frac{F(\mathcal{M}_P)}{(1/2) \cdot |X|} = \frac{1}{|X|} IF(\mathcal{M}_P(x)).$$

## 3.3 Formalization of Nonspecificity

### *Nonspecificity: Informal Definition*

According to Random House Webster's Dictionary [32], the word "ambiguity" has two main meanings:

1. doubtfulness or uncertainty of meaning or intention: to speak with ambiguity;
2. the condition of admitting more than one meaning.

By ***nonspecificity***, we will understand the second meaning of ambiguity. Nonspecificity is thus usually identified with unspecified distinctions between several alternatives, that is, when we possess several alternatives that are equally possible, or probable. "[Nonspecificity] is connected with sizes (cardinalities) of relevant sets of alternatives." (Klir [8], p. 276).

### *Nonspecificity of a Finite Set*

To describe nonspecificity of a body of evidence, let us start with the known formulas for simpler cases. Let us first consider the situation when the only information that we have about an alternative $a$ is that it belongs to a certain finite set $B$. With expert systems in mind, a natural measure of non-specificity is the smallest number of binary ("yes"-"no") questions that we need to ask in order to find the exact value of $a$. In information theory, this number of binary questions is called the ***number of bits***. (For a detailed description of this concept in general uncertainty context, see, e.g., Klir and Folger [10], p. 149.)

*Comment.* If in addition to the set $B$, we know the probability $p(a)$ of each element $a \in B$, then the average number of "yes"-"no" questions needed to uniquely determine $a \in B$ is well known in information theory: it is equal to Shannon entropy of the given probability distribution:

$$N(p) = \sum_{a \in B} p(a) \log_2(p(a)).$$

We cannot apply Shannon's formula, however, because we do not know the probabilities $p(a)$.

If the set $B = \{a_1, a_2\}$ consists of two elements ($|B| = 2$), then a single question is sufficient (e.g., "is it $a_1$?"). If we ask $q$ questions, then there are $2^q$ possible combinations of answers. Therefore, after asking $q$ questions, we can uniquely select no more than $2^q$ different elements. On the other hand, if $B$ consists of exactly $2^q$ elements, then we can (in principle) enumerate these elements as $a_0, x_1, \ldots, a_{2^q-1}$ by using binary numbers with $q$ digits (from $0\ldots0_2 = 0_{10}$ to $1\ldots1_2 = (2^q - 1)_{10}$). Then, we can ask $q$ questions about these $q$ digits, and thus select $a \in A$ uniquely. So, when $|B| = 2^q$, then $q$ questions are necessary and $q$ questions are sufficient. Therefore, for such sets, we can define nonspecificity $N(X)$ as this number of questions $q = \log_2 |B|$.

When $2^{q-1} < |B| < 2^q$, then, to determine an element $a \in B$, we need at least $q$ questions. If we have several ($N$) such problems, however, then, to determine $a \in B$ in each of them, we need fewer than $Nq$ questions: namely, $\approx \log_2(|B|^N)$. The average number of questions turns out to be equal to $\log_2 |B|$. (For details, see, e.g., Ramer [18, 19].) Hence, for finite sets $B$, we can define $F(B) = \log_2 |B|$.

## *Nonspecificity of a Probability Distribution on the Family of Finite Sets*

Let us consider a slightly more complicated case, when we do not know exactly the finite set to which an alternative $B$ belongs, but we know that $a$ belongs to one of the finite sets $B$ from a given family $\mathcal{F}$, and we know the (subjective) probability $s(B)$ (evidential strength) of each of these sets $B$.

For each of these sets $B$, we know that on average, we will need $\log_2 |B|$ questions. We do not know, however, which of these sets $B$ is correct; instead, we know the probability $s(B)$ of each of these sets. So, we can define the nonspecificity as the *average* number of questions:

$$N(s) = \sum_{B \in \mathcal{F}} s(B) \log_2 |B|.$$

This formula was initially proposed by Hartley and called *Hartley measure of information* (or simply *Hartley measure*, for short). For a discussion of this formula in the knowledge representation context, see, e.g., Klir [8], p. 320.

## *Nonspecificity of an Interval*

The above formulas work well when the set of alternatives is finite. For interval-valued degrees of confidence, the set of possible alternatives $B$ is an interval

$\mathbf{m} = [m^-, m^+]$. An interval is infinite, and therefore, the above formulas lead to a meaningless infinite expression.

This infinite result is simply because we tried to find the number of binary questions that would determine $a \in B$ *uniquely*. To determine $a \in [m^-, m^+]$ uniquely, we need to know infinitely many binary digits of the binary expansion of $a$ (as opposed to finitely many binary digits in the finite case); to determine infinitely many digits, we need infinitely many binary questions.

In real life, however, we do not need to know the value $a \in [m^-, m^+]$ *exactly* (*uniquely*). For all practical purposes, it is sufficient to describe $a$ with a given accuracy $\delta$. In this case, to find $a$ with this accuracy, we can, e.g., divide the original interval $[m^-, m^+]$ into subintervals of size $2\delta$: $[m^-, m^- + 2\delta]$, $[m^- + 2\delta, m^- + 4\delta]$, ..., and determine (by asking binary questions) to which of these intervals does the actual value $a$ belongs. If it turns out that $a$ belongs to an interval

$$[m^- + 2k\delta, m^- + (2k+2)\delta],$$

then we can return the midpoint $\tilde{a} = m^- + (2k+1)\delta$ of this interval as the desired $\delta-$approximation of $a$.

An interval $\mathbf{m} = [m^-, m^+]$ of width $|\mathbf{m}| = m^+ - m^-$ can be divided into $\approx |\mathbf{m}|/(2\delta)$ subintervals of width $2\delta$. Therefore, we need $\approx \log_2(|\mathbf{m}|/(2\delta))$ binary questions to determine $a \in \mathbf{m}$ with a given accuracy. In other words, we can define *$\delta-$nonspecificity* $N_\delta$ as

$$N_\delta(\mathbf{m}) \approx \log_2(|\mathbf{m}|/(2\delta)).$$

If $\delta$ is sufficiently large (e.g., larger than the half-width of the interval $\mathbf{m}$), then we do not need any equations at all: the midpoint of the interval $\mathbf{m}$ is already the desired approximation. The above formula only makes sense for small $\delta$; the smaller $\delta$, the more accurate this formula is. (See [17, 18, 19].) For small $\delta$, we thus have

$$N_\delta \sim \log_2 |\mathbf{m}| - 1 - \log_2(\delta).$$

This expression consists of two terms: the second term depends only on the accuracy $\delta$, and the first term depends only on the interval $\mathbf{m}$. Therefore, if we are interested in designing a measure of nonspecificity of the interval $\mathbf{m}$ that will not depend on the accuracy, it is natural to take the first term of the above asymptotic expression as such a measure, i.e., define

$$N(\mathbf{m}) = \log_2 |\mathbf{m}|,$$

where $|\mathbf{m}|$ denotes the width of the interval $\mathbf{m}$:

$$|[m^-, m^+]| = m^+ - m^-.$$

## *Nonspecificity of an Interval-Valued Fuzzy Set*

The above formula can characterize nonspecificity of the interval valued degree of confidence. Let us show how this formula can be used to describe nonspecificity of interval-valued fuzzy sets.

Let $(X, \mathbf{m}_P)$ be an IVFS over a finite set $X$. To specify the exact values of $m_P(x)$, we must specify the values $m_P(x)$ for all $x \in X$. Specifying the value $m_P(x)$ for one $x$ does not bring us any closer to specifying $m_P(x')$ for some other $x' \neq x$. Therefore, we have $|X|$ independent specification problems. For each of these problems we already know that we can describe the needed number of binary questions by the value $N(\mathbf{m}_P(x)) = \log_2 |\mathbf{m}_P(x)|$. Since these problems are independent, to specify all of them, we must ask all the questions that correspond to all $x \in X$. In other words, the smallest number of questions needed to specify all the values $m_P(x)$ is equal to the sum of the numbers of questions needed to specify for each $x$. Therefore, we can define the nonspecificity of an interval-valued fuzzy set as

$$N(\mathbf{m}_P) = \sum_{x \in X} \log_2 |\mathbf{m}_P(x)|.$$

What is the range of this function? The above expression is a monotonically increasing function of the lengths of the intervals. The largest value of nonspecificity is thus attained when all the intervals have the largest possible width, and the smallest when these intervals are the narrowest.

All the intervals $\mathbf{m}_P(x)$ are sub-intervals of the interval $[0, 1]$. Therefore, the largest possible width of such an interval is 1 (when the interval coincides with $[0, 1]$), and the smallest value is 0 (attained when an interval degenerates into a point). Hence, the largest value of non-specificity (0) is thus attained when all the intervals coincide with $[0, 1]$ (i.e., correspond to complete ignorance), and the smallest possible value $-\infty$ corresponds to the case when one of these intervals is a point.

Since the non-specificity thus defined is always non-positive, it is convenient to define the *specificity* as a negative non-specificity $S(\mathbf{m}_P) = -N(\mathbf{m}_P)$. (This idea was pointed out to us by George Klir).

## *Nonspecificity of a Body of Evidence*

If our degree of confidence is characterized by the non-trivial body of evidence $\mathcal{M}$ (i.e., a body of evidence that is not described by a single interval), this means that we do not know exactly the interval $I$ of possible degrees of confidence, but we know that this interval $I$ belongs to a certain finite family $\mathcal{F}$, and we know the (subjective) probability $\mathcal{M}(I)$ (evidential strength) of each of these intervals $I \in \mathcal{F}$.

For each of these intervals $I$, we know that on average, we will need $\log_2 |I|$ (+ terms depending on $\delta$ only) questions. We do not know, however, which of these intervals $I$ is correct; instead, we know the probability $\mathcal{M}(I)$ of each of these sets. So, we can define the nonspecificity as the *average* number of questions:

$$N(\mathcal{M}) = \sum_{I \in \mathcal{F}} \mathcal{M}(I) \log_2 |I|.$$

## *Nonspecificity of an Evidence Set*

Similarly to the case of an IVFS, the number of binary questions necessary to specify a membership function is a sum of the numbers of binary questions necessary to specify the body of evidence $\mathcal{M}_P(x)$ for each $x \in X$. Therefore, for a finite universal set $X$, we can define the nonspecificity of an evidence set $(X, \mathcal{M}_P)$ as the sum of nonspecificities of these bodies of evidence:

$$N(\mathcal{M}_P) = \sum_{x \in X} N(\mathcal{M}_P(x)).$$

Thus defined nonspecificity ranges from $-\infty$ to 0; therefore, it is natural to use specificity (negative nonspecificity) $S(\mathcal{M}_P) = -N(\mathcal{M}_P)$.

# 3.4 Formalization of Conflict

## *Conflict: Informal Definition*

Conflict means disagreement between several alternatives.

## *Strife: A Measure of Conflict in a Body of Evidence*

A measure of conflict called *strife* was proposed by Klir and Ramer in [9]. The idea behind this measure is as follows. (This measure was formulated for a slightly different situation; we will reformulate the arguments so that they will be directly applicable to bodies of evidence.) A body of evidence $\mathcal{M}$ means that we have a finite family $\mathcal{F}$ of intervals, and for each interval $I \in \mathcal{F}$, we have a probability (evidential support) $\mathcal{M}(I)$. The interpretation given above is that one of these intervals (we do not know exactly which one) is the *actual* interval of degrees of confidence, and the probability that $I$ is that actual interval is exactly $\mathcal{M}(I)$.

One of the major goals of an expert system is to describe, for each query $Q$, our degree of confidence in $Q$. Ideally, we should be able to describe *all values* of degree of confidence that are consistent with out current knowledge about $Q$. At least, we should get *a value* of degree of confidence that is consistent with the current knowledge.

If our knowledge is described by interval-valued fuzzy sets, then for statements from the knowledge base we can simply return the corresponding intervals, or any values from these intervals.

If uncertainty is described by *consonant* bodies of knowledge, then it is not so easy to describe *all* possible values of degree of confidence (because we do not know which of the intervals is a correct one), but it is still easy to generate a value that is definitely a possible value of the degree of confidence: namely, we can take any value from the intersection of the corresponding intervals. no matter which of these intervals is the actual interval of degree of confidence, this point is possible.

For non-consonant bodies of evidence, however, this problem is no longer easy. An expert who believes that $I$ is an interval of possible values will return a value from $I$. Because of the conflict, however, this value may not belong to the *actual* (unknown) interval $J$. Strife, crudely speaking, is the measure of conflict-related uncertainty, i.e., the average number of binary questions that we have to ask to be able to generate a value from the actual interval.

A natural way to generate the value is as follows:

- First, since we have assumed that all experts are equally good, we pick an expert at random (with equal probability of picking each expert).

- Second, from the interval corresponding to this expert, we pick a value at random.

Let us describe this procedure in mathematical terms and estimate the probability that this procedure will succeed.

Due to our motivation of a body of evidence, picking an expert at random corresponds to picking each interval $I$ with a probability $\mathcal{M}(I)$. When an interval $I$ is picked, we can pick any value from $I$. Since we have no reasons to prefer one of the values to another one, it is natural to assume that we are picking each value with an equal probability, i.e., in mathematical terms, that we are picking the value $m \in I$ according to the uniform distribution on the interval $I$.

What is the probability that this procedure will succeed, i.e., that it will lead to the degree of confidence that belongs to the (unknown) actual interval?

To compute this probability, let us assume that $J$ is the actual interval. Let us consider the case when we have picked $I \in \mathcal{F}$ as the interval from which we are picking $m$. If the resulting point $m$ belongs to $J$, then $m$ is a correct choice, otherwise it is not. So, from all the points $m \in I$, only points from the intersection $I \cap J$ correspond to the correct answer. Since we are picking $m$ according to a uniform distribution on $I$, the probability that the resulting answer will be correct is equal to probability of a random point from $I$ to be in $I \cap J$, i.e., to $|I \cap J|/|I|$. This is a conditional probability, given that we have picked $I$.

*Comment.* When the interval $I$ is degenerate (i.e., consists of a single point) a special consideration is required, because in this case, $|I| = 0$ and the above formula is not applicable. In this case, if $I \subseteq J$, then the desired probability is equal to 1, otherwise, it is equal to 0. To cover this case, we will *interpret* the expression $|I \cap J|/|I|$ as follows:

- If $|I| > 0$, then this expression is understood to be a ratio of two real numbers;
- If $|I| = 0$, then this expression literally becomes $0/0$ (undefined), so its value is defined as follows:
  - If $I \subseteq J$, then this expression is interpreted to be 1.
  - If $I \not\subseteq J$, then this expression is interpreted to be 0.

The probability of picking $I$ is $\mathcal{M}(I)$; therefore, by the formula of total probability, the probability of picking the right answer is equal to

$$p^*(J) = \sum_{I \in \mathcal{F}} \mathcal{M}(I) \frac{|I \cap J|}{|I|}.$$

If this probability is 1, then we do not need any extra binary questions: we are already guaranteed that we are OK. If this probability is 0.5, this means that out of two picks, on average, one is correct. In this case, we need on average one binary question to choose the correct pick. If the probability is 1/4, then one out of four picks is correct. So, to choose a correct pick, we need $\log_2(4) = 2$ binary questions. In general, we need $-\log_2(p(J))$ binary questions, i.e.,

$$C^*(J) = -\log_2 \left( \sum_{I \in \mathcal{F}} \mathcal{M}(I) \frac{|I \cap J|}{|I|} \right).$$

This number $C^*(J)$ characterizes strife in case we know which of the intervals $I \in \mathcal{F}$ is the actual one. In reality, we do not know, so, it is natural to measure the conflict in a given body of evidence as the *expected* number of binary questions that are needed to eliminate the conflict, i.e., as the value

$$C(\mathcal{M}) = \sum_{J \in \mathcal{F}} \mathcal{M}(J) \cdot C^*(J),$$

or, as

$$C(\mathcal{M}) = -\sum_{J \in \mathcal{F}} \mathcal{M}(J) \cdot \log_2 \left( \sum_{I \in \mathcal{F}} \mathcal{M}(I) \frac{|I \cap J|}{|I|} \right).$$

This is the formal definition of a strife.

The smallest value of strife is 0; it corresponds to the case when there is no conflict at all. If we have many non-intersecting intervals, we get larger and larger values of strife, so strife takes values from 0 to $+\infty$.

## *Strife: A Measure of Conflict in a Body of Evidence*

To eliminate a conflict for all $x \in X$, we must eliminate a conflict for each $x \in X$. Therefore, the total number of binary questions needed to eliminate all conflicts for all $x \in X$, can be defined as the sum of the numbers of binary questions needed to eliminate the conflicts for each $x \in X$. In other words, we can define the strife of an evidence set $\mathcal{M}_P$ as

$$C(\mathcal{M}_P) = \sum_{x \in X} C(\mathcal{M}_P(x)).$$

## 3.5 Other Measures of Uncertainty

In addition to the three measures described above, other measures of uncertainty have been proposed by Lamata and Moral [12]; Klir [8]; and Harmanec and Klir [6]. In our terms, these measures combine nonspecificity and conflict into a single measure.

# 4 $L$–FUZZY SETS, INTERVAL BASED $L$–FUZZY SETS, AND $L$–EVIDENCE SETS

## 4.1 Motivations

Why are continuous domains important as membership degree (or possibility) spaces? Are they coherent with human cognitive abilities? It is undeniable that the concept of continuity makes no sense in the establishment of membership or possibility degrees. No sane human being will attempt to quantify some degree of truth (possibility) or membership in a set, with, say, an irrational number! Hypothetic degrees of confidence like $1/\sqrt{2}$ are beyond the intuitive grasp of human cognition. All applied fuzzy systems will use at most a very small number of decimals in describing membership degrees in their interface with human users. In fact, I would argue that human reasoning is not able to intuitively grasp the difference between membership degrees much beyond a 10-valued logic. Further, if the ultimate medium of implementation of fuzzy (interval) sets is the digital computer, then the original membership unit interval will be discretized at some point anyway.

## 4.2 Definitions of $L$–Fuzzy Sets, Interval-Valued $L$–Fuzzy Sets, $L$–Probabilistic Sets, and $L$–Evidence Sets

For these reasons, valid at least in the application realm of fuzzy logic and interval-valued fuzzy logic, the membership degrees need only be defined in a linearly ordered set $L = \{e_1, ..., e_n\}$ of $n$ elements. In this set, $e_1 < e_2 <$

... $< e_n$, where $a < b$ means that confidence described by $b$ is stronger than confidence described by the degree of confidence $a$. Correspondingly, we are moving from fuzzy sets to ***L–fuzzy sets***, defined as pairs $(X, m_P)$, where $m_P : X \to L$.

For $L$–fuzzy logic, an ***interval*** $I \subseteq L$ can be defined as a pair $[e_i, e_j]$ with $i \le j$. The cardinality $|I|$ of an interval $I = [e_i, e_j]$ is equal to $j - i + 1$; in particular, the cardinality of $L = [e_1, e_n]$ is $n$. The set of all intervals of $L$ will be denoted by $\mathcal{I}$. Using these intervals, we can define the membership function of an interval-valued $L$–fuzzy set as a function $\mathbf{m}_P : X \to \mathcal{I}$.

An ***L–body of evidence*** is now defined as a probability measure $\mathcal{M}$ on the set $\mathcal{I}$ of all intervals of $L$. If we define the set of all $L$–bodies of evidence by $\mathcal{E}$, then we can define an ***L–evidence set*** as a pair $(X, \mathcal{M}_P)$, where $\mathcal{M}_P$ is a function from $X$ to $\mathcal{E}$.

When for each $x \in X$, each interval $I$ for which $\mathcal{M}_P(x)(I) \ne 0$ is degenerate (i.e., $I = [e_i, e_i]$ for some $i$), then such $L$–evidence set will be called an ***L–probabilistic set.***

Let us now show how the measures of uncertainty described above will work for $L$–evidence sets.

## 4.3 Fuzziness in $L$–Evidence Sets

To apply the above formulas for fuzziness, we must assign the values $\in [0, 1]$ to our abstract values $e_1, ..., e_n$. We have already mentioned that one of the major methods of extracting the degrees of confidence from an expert is by asking her to mark a point on a scale from 0 to $S$; in this case, we get $S + 1$ possible values, and to these values, we put into correspondence the degrees of confidence 0, $1/S$, $2/S$, ..., $(S - 1)/S$, and 1: the first value is $e_1 = 0$, the second is $e_2 = 1/S$, ..., and the $i$–th is $e_i = (i - 1)/S$. In our case, we have $n$ different degrees of confidence, so $S + 1 = n$, and hence, we have a scale from 0 to $S = n - 1$. Therefore, the $i$–th value $e_i$ can be described by a number $(i - 1)/(n - 1)$. With these numbers, we can apply the above formulas that define the fuzziness $F$ and the index of fuzziness $IF$.

## 4.4 Nonspecificity in $L$–Evidence Sets

For $L$–fuzzy logic, each interval is finite. Therefore, we can define a nonspecificity $N(I)$ of an interval $I$ as $\log_2 |I|$, where $|I|$ is the number of elements in an interval $I$. The advantage of this definition is that it is always non-negative, because it describe the exact number of binary questions that we need to specify an element of the interval $I$. The smallest value of this nonspecificity is 0; it is attained when the interval consists of only one value. The largest value of this nonspecificity is $\log_2(n)$; it is attained when $I$ coincides with the set $L$ (i.e., when we do not know anything).

Similarly, for an $L$–*body of evidence*, we can define nonspecificity as an average nonspecificity of the corresponding intervals, i.e., as

$$N(\mathcal{M}) = \sum_I \mathcal{M}(I) \cdot \log_2 |I|.$$

Since nonspecificity ranges from 0 to $\log_2(n)$, its expected value $N(\mathcal{M})$ also ranges from 0 to $\log_2(n)$. The value 0 is attained if all the intervals are degenerate (i.e., consist of only one point); the value $\log_2(n)$ is attained if there is only one interval $I = L$, with probability $\mathcal{M}(I) = 1$. Hence, similarly to how we defined an index of fuzziness, we can define the *index of nonspecificity* if we divide the nonspecificity by $\log_2(n)$ and this scale it to the range $[0, 1]$:

$$IN(\mathcal{M}) = \frac{N(\mathcal{M})}{\log_2(n)} = \frac{1}{\log_2(n)} \sum_I \mathcal{M}(I) \cdot \log_2 |I|.$$

Finally, nonspecificity of an $L$–evidence set $(X, \mathcal{M}_P)$ can be defined as the sum of nonspecificities of the bodies of evidence $\mathcal{M}_P(x)$ that correspond to different $x \in X$:

$$N(\mathcal{M}_P) = \sum_{x \in X} \sum_I [\mathcal{M}_P(x)](I) \cdot \log_2 |I|.$$

Since for each $x \in X$, nonspecificity ranges from 0 to $\log_2(n)$, the resulting total nonspecificity ranges from 0 to $|X| \cdot \log_2(n)$. Hence, we can define an *index of nonspecificity* of an $L$–evidence set as

$$IN(\mathcal{M}_P) = \frac{1}{|X| \cdot \log_2(n)} N(\mathcal{M}_P) = \frac{1}{|X| \cdot \log_2(n)} \sum_{x \in X} \sum_I [\mathcal{M}_P(x)](I) \cdot \log_2 |I|.$$

We can also describe this index of nonspecificity in terms of indices for $x \in X$:

$$IN(\mathcal{M}_P) = \frac{1}{|X|} \sum_{x \in X} IN(\mathcal{M}_P(x)).$$

In other words, an index of nonspecificity of an $L-$evidence set can be defined as an arithmetic average of indices corresponding to different $x \in X$.

### *Crisp, Fuzzy, and Probabilistic Sets Have No Nonspecificity*

For an $L-$fuzzy set, all the intervals $I$ of degrees of confidence are degenerate (consist of only one point). Hence, for these sets, $\log_2 |I| = 0$ and the nonspecificity is equal to 0. Similarly, $L-$probabilistic sets have no nonspecificity.

Since crisp sets are a particular case of $L-$fuzzy sets, $IN = 0$ for crisp sets as well.

## 4.5 Conflict in $L-$Evidence Sets

For $L-$bodies of evidence, we can define strife similarly to the definition of strife for the usual evidence sets. The only difference is that when we have already picked an interval $I$, and we need to pick a value from this interval, we can now (since an interval is finite) do it simply by assigning equal probabilities to all elements of $I$. As a result, we get a similar formula for strife:

$$C(\mathcal{M}) = -\sum_{J \in \mathcal{F}} \mathcal{M}(J) \cdot \log_2 \Big( \sum_{I \in \mathcal{F}} \mathcal{M}(I) \frac{|I \cap J|}{|I|} \Big).$$

(The only difference is that now, the width $|I|$ of an interval means simply the number of elements in it.)

By definition, strife is a mathematical expectation of the number of binary questions that are necessary to eliminate the conflict. Since we have only $n$ possible degree of confidence, then by asking $\log_2(n)$ questions, we not only eliminate the conflict, but we determine the degree of confidence uniquely. Therefore, the strife is always $\leq \log_2(n)$; hence, we can defined an ***index of conflict*** $IC$ if we dividing $C$ by $\log_2(n)$ and thus, re-scale the range of strife to the interval $[0, 1]$:

$$IC(\mathcal{M}) = \frac{1}{\log_2(n)} C(\mathcal{M}) = -\frac{1}{\log_2(n)} \sum_{J \in \mathcal{F}} \mathcal{M}(J) \cdot \log_2 \Big( \sum_{I \in \mathcal{F}} \mathcal{M}(I) \frac{|I \cap J|}{|I|} \Big).$$

The conflict for the entire $L-$evidence set $(X, \mathcal{M}_P)$ can be defined as the sum of the conflicts in the bodies of evidence $\mathcal{M}_P(x)$ for different $x \in X$:

$$C(\mathcal{M}_P) = \sum_{x \in X} C(\mathcal{M}_P(x)) =$$

$$-\sum_{x \in X} \sum_{J \in \mathcal{F}} [\mathcal{M}_P(x)](J) \cdot \log_2 \left( \sum_{I \in \mathcal{F}} [\mathcal{M}_P(x)](I) \frac{|I \cap J|}{|I|} \right).$$

The range of this sum is from 0 to $|X| \cdot \log_2(n)$, so we can re-scale this value and define an *index of conflict* for an $L-$evidence set:

$$IC(\mathcal{M}_P) = \frac{1}{|X| \cdot \log_2(n)} C(\mathcal{M}_P) =$$

$$-\frac{1}{|X| \cdot \log_2(n)} \sum_{x \in X} \sum_{J \in \mathcal{F}} [\mathcal{M}_P(x)](J) \cdot \log_2 \left( \sum_{I \in \mathcal{F}} [\mathcal{M}_P(x)](I) \frac{|I \cap J|}{|I|} \right).$$

We can also describe this index of conflict in terms of indices for $x \in X$:

$$IC(\mathcal{M}_P) = \frac{1}{|X|} \sum_{x \in X} IC(\mathcal{M}_P(x)).$$

In other words, an index of conflict of an $L-$evidence set can be defined as an arithmetic average of indices corresponding to different $x \in X$.

### *Crisp, Fuzzy, and Interval-Valued Fuzzy Sets Have No Conflict*

Crisp, fuzzy, and interval-valued sets can be viewed as particular cases of $L-$evidence sets, in which for each $x \in X$, the measure $\mathcal{M}_P(x)$ is concentrated on only one interval with probability 1. From the above formulas, one can easily see that in this case, the measure of conflict defined above is indeed equal to 0.

## 5 3-D UNCERTAINTY UNIT CUBE

In the previous sections, we showed that the most realistic description of uncertainty is represented:

- *locally* (i.e., for each $x \in X$) by an $L-$body of evidence; and

- globally, by $L-$evidence sets.

We have also introduced the idea of describing uncertainty by three different characteristics: fuzziness, nonspecificity, and conflict. If we use the indices described above to gauge these three characteristics, we thus characterize each $L-$body of evidence $\mathcal{M}$ by a triple

$$\vec{U}(\mathcal{M}) = (IF(\mathcal{M}), IN(\mathcal{M}), IC(\mathcal{M})),$$

and each $L-$evidence set $\mathcal{M}_P$ by a triple

$$\vec{U}(\mathcal{M}_P) = (IF(\mathcal{M}_P), IN(\mathcal{M}_P), IC(\mathcal{M}_P)).$$

Since each index takes the values from the interval $[0,1]$, the resulting vector characteristics take the values from the ***unit cube*** $[0,1] \times [0,1] \times [0,1]$. Also, due to the above-mentioned relationship between the index of an $L-$evidence set and indices that correspond to different $x \in X$, we conclude that

$$\vec{U}(\mathcal{M}_P) = \frac{1}{|X|} \sum_{x \in X} \vec{U}(\mathcal{M}_P(x)).$$

In other words, if we calculate the uncertainties of an $L-$evidence set $\mathcal{M}_P$ for each element $x$ of $X$ (***local uncertainties***), and plot each element in this cube, then the point $\vec{U}(\mathcal{M}_P)$ that corresponds to ***global*** uncertainty will be the center of mass of the cloud of local uncertainties.

In this paper, we have described several different models of uncertainty. We can summarize our results about the ranges of uncertainty indices for different description of uncertainty in Table 1.

| | $IF$ | $IN$ | $IC$ |
|---|---|---|---|
| Crisp Sets | 0 | 0 | 0 |
| Fuzzy Sets | [0,1] | 0 | 0 |
| IVFS | [0,1] | [0,1] | 0 |
| Probabilistic Sets | [0,1] | 0 | [0,1] |
| Evidence Sets | [0,1] | [0,1] | [0,1] |

Table 1. Ranges of uncertainty indices for different descriptions of uncertainty

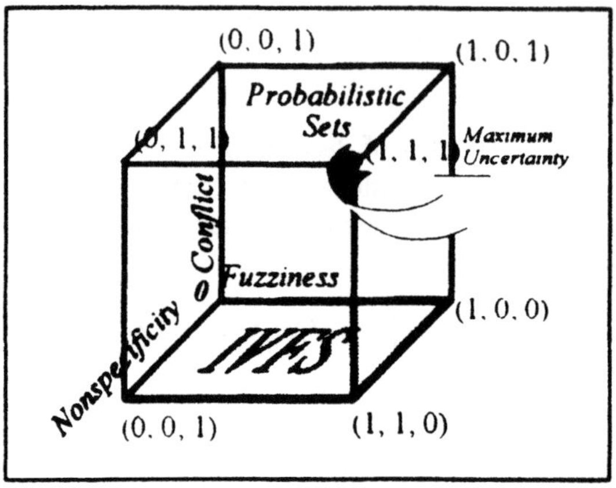


**Figure 4** 3-D Uncertainty Unit Cube

According to this table:

- A ***crisp set*** entails no uncertainty whatsoever. Thus, crisp sets occupy the ***origin*** $(0, 0, 0)$ of the coordinate system.
- ***Fuzzy sets*** introduce fuzziness, but they have no nonspecificity or conflict. Thus, fuzzy sets occupy the ***fuzziness axis***.
- ***Interval-valued fuzzy sets*** introduce nonspecificity. Thus, they capture two forms of uncertainty: fuzziness (as in the case of normal fuzzy set) and nonspecificity. They have no conflict, and therefore, they occupy the ***fuzziness-nonspecificity plane*** in our 3-D representation.
- ***Evidence sets*** introduce conflict. These sets occupy ***the 3-D part of the cube***.
- ***Probabilistic sets*** are a particular case of evidence sets that have only conflict, but no nonspecificity. They thus occupy the ***conflict-fuzziness plane***.

Since IVFS and probabilistic sets occupy only 2-D subsets of a 3-D space of uncertainty models, they are not sufficient to describe a generic uncertainty. For that, we need evidence sets.

*Comment.* It is worth noticing that not all points from the cube correspond to $L-$evidence sets: For example, we cannot have nonspecificity without fuzziness, since if we define an interval of membership, we will necessarily have non-crisp membership degrees.

## Conclusions

In this paper we have proposed measures of uncertainty for evidence sets and interval based set structures in general, which are indispensable in the development of practical models of artificial, uncertain, reasoning. Since evidence sets seem to capture all the relevant forms of uncertainty, and we have a way of measuring it, they offer a sufficient framework for the complete modeling of uncertainty in human reasoning.

## Acknowledgements

This work was supported in part by the NSF grants CCR-9203730, CDA-9015006, and EEC-9322370, and by the NASA grant No. NAG 9-757. The authors are most thankful to George Klir, Hung T. Nguyen, and to anonymous referees for valuable discussions.

## REFERENCES

[1] K. T. Atanassov, "Intuitionistic fuzzy sets," *Fuzzy Sets and Systems*, 1986, Vol. 20, pp. 87–96.

[2] K. T. Atanassov, "Operators over interval-valued intuitionistic fuzzy sets," *Fuzzy Sets and Systems*, 1994, Vol. 64, pp. 159–174.

[3] K. T. Atanassov and G. Gargov, "Interval valued intuitionistic fuzzy sets, " *Fuzzy Sets and Systems*, 1989, Vol. 31, pp. 343–349.

[4] D. Berleant, "Automatically Verified Arithmetic on Probability Distributions and Intervals", *This Volume*.

[5] M. B. Gorzalczany, "A method of inference in approximate reasoning based on interval-valued fuzzy sets", *Fuzzy Sets and Systems*, 1987, Vol. 21, pp. 1–17.

[6] D. Harmanec and G. J. Klir, "Measuring total uncertainty in Dempster–Shafer Theory: a novel approach," *International Journal of General Systems*, 1994, Vol. 22, pp. 405–419.

[7] K. Hirota, "Concepts of probabilistic sets," *Fuzzy Sets and Systems*, 1981, Vol. 5, pp. 31–46.

[8] G. J. Klir, "Developments in uncertainty-based information," In: M. Yovits (ed.), *Advances in Computers*, 1993, Vol. 36, pp. 255–332.

[9] G. J. Klir and A. Ramer, "Uncertainty in the Dempster–Shafer theory: a critical re–examination," *International Journal of General Systems*, 1990, Vol. 18, pp. 155–166.

[10] G. J. Klir and T. Folger, *Fuzzy Sets, Uncertainty, and Information*, Prentice Hall, 1988.

[11] G. Lakoff, *Women, Fire, and Dangerous Things: What Categories Reveal About the Mind*, University of Chicago Press, 1987.

[12] M. T. Lamata and S. Moral, "Measures of entropy in the theory of evidence," *International Journal of General Systems*, 1988, Vol. 14, No. 4, pp. 183–196.

[13] P. R. Medina–Martins, "Metalogues: an abridge of a genetic psychology of non-natural systems", *Communication and Cognition – Artificial Intelligence*, 1995, Vol. 12, Nos. 1–2 (Special Issue *Self–Reference in Biological and Cognitive Systems*), in press.

[14] P. R. Medina–Martins and L. M. Rocha, "The in and the out: an evolutionary approach," In: R. Trappl (ed.), *Cybernetics and Systems Research '92*, World Scientific Press, 1992, pp. 681–689.

[15] P. R. Medina–Martins, L. M. Rocha, et al., *CYBORGS: A Fuzzy Conversational System. Final Report for the NATO International Program on Learning Systems*, 1993.

[16] K. Nakamura and S. Iwai, "A representation of analogical inference by fuzzy sets and its application to information retrieval systems," In: M. M. Gupta and E. Sanchez (eds.), *Fuzzy Information and Decision Processes*, 1982, North–Holland, pp. 373–386.

[17] A. Ramer, "Concepts of fuzzy information on continuous domains", *International Journal of General Systems*, 1990, Vol. 17, pp. 241–248.

[18] A. Ramer and V. Kreinovich, "Maximum entropy approach to fuzzy control", *Information Sciences*, 1994, Vol. 81, No. 3–4, pp. 235–260.

[19] A. Ramer and V. Kreinovich, "Information complexity and fuzzy control", Chapter 4 in: A. Kandel and G. Langholtz (Eds.), *Fuzzy Control Systems*, CRC Press, Boca Raton, FL, 1994, pp. 75–97.

[20] L. M. Rocha, "Fuzzification of Conversation Theory," In: *Principia Cybernetica Conference*, Free University of Brussels, 1991.

[21] L. M. Rocha, "Cognitive Categorization revisited: extending interval valued fuzzy sets as simulation tools for concept combination," In: *Proceedings of the 1994 International Conference of NAFIPS/IFIS/NASA*, 1994, IEEE, pp 400–404.

[22] L. M. Rocha, "Von Foerster's cognitive tiles: semantically closed building blocks for AI and Alife." In: R. Trappl (ed.), *Cybernetics and Systems'94*, World Scientific Press, 1994, Vol. 1, pp. 621–628.

[23] L. M. Rocha, "Interval Based Evidence Sets", In: *Proceedings of the 1995 NAFIPS/IFIS Joint Conference*, IEEE Press, 1995 (in press).

[24] L. M. Rocha, "Artificial semantically closed objects," In: L. Rocha (ed.), *Special issue in Self-Reference in Biological and Cognitive Systems, Communication and Cognition – AI*, 1995, Vol. 12, No. 1–2.

[25] L. M. Rocha, "Evidence (interval based) Sets: Modelling Subjective Categories," *International Journal of General Systems*, 1995 (in press).

[26] D. A. Schum, *Evidential Foundations of Probabilistic Reasoning*, John Wiley and Sons, 1994.

[27] G. Shafer and J. Pearl (editors), *Readings in Uncertain Reasoning*, Morgan Kauffman, San Mateo, CA, 1990.

[28] G. Shafer, *A Mathematical Theory of Evidence*, Princeton University Press, Princeton, NJ, 1976.

[29] I. B. Turksen, "Interval valued fuzzy sets based on normal forms," *Fuzzy Sets and Systems*, 1986, Vol. 20, pp. 191–210.

[30] J. Vejnarova and G. J. Klir, "Measure of strife in Dempster–Shafer theory", *International Journal of General Systems*, 1993, Vol. 22, pp. 25–42.

[31] Z. Wang and G. Klir, *Fuzzy Measure Theory*, Plenum Press, N.Y., 1992.

[32] *Webster's Dictionary*, Random House, 1991.

[33] R. R. Yager, "On the measure of fuzziness and negation. part I: membership in the unit interval," *International Journal of Man-Machine Studies*, 1979, Vol. 11, pp. 189–200.

[34] R. R. Yager, "On the measure of fuzziness and negation. Part II: Lattices," *Information and Control*, 1980, Vol. 44, pp. 236–260.

[35] Bo Yuan, W. Wu, and Y. Pan, "On normal form based interval valued fuzzy sets and its application to approximate reasoning," *International Journal of General Systems*, 1995 (in press).

[36] L. A. Zadeh, "Fuzzy Sets." *Information and Control*, 1965, Vol. 8, pp. 338–353.

[37] L. A. Zadeh, "The concept of a linguistic variable and its application to approximate reasoning," *Information Science*, 1975, Vol. 8, pp. 199–249.

[38] Q. Zhu and E. S. Lee, "Evidence theory in multivalued logic systems", *International Journal of Intelligent Systems*, 1995, Vol. 10, pp. 185–199.

# 14

# SOFTWARE AND HARDWARE TECHNIQUES FOR ACCURATE, SELF-VALIDATING ARITHMETIC

**Michael J. Schulte and Earl E. Swartzlander, Jr.**

*Department of Electrical and Computer Engineering,*
*University of Texas at Austin, Austin, TX, 78712, USA*

## ABSTRACT

The need for accurate and reliable numerical applications has led to the development of several software tools and hardware designs for accurate, self-validating arithmetic. Software tools include variable-precision software packages, interval arithmetic libraries, scientific programming languages, computer algebra systems, and numerical problem solving environments. Hardware designs include coprocessors that support the directed rounding modes and exact dot products, variable-precision integer and floating point processors, and coprocessors for variable-precision interval arithmetic. In this survey, we examine various software and hardware techniques for accurate, self-validating arithmetic and discuss their strengths and limitations. We also discuss numerical applications that employ these tools to produce accurate and reliable results.

## 1 INTRODUCTION

Advances in VLSI technology, parallel processing, and computer architecture have led to increasingly faster digital computers. During each of the last three decades, the computational speeds of the fastest computers increased by a factor of roughly 100 and an increase by a factor of 1,000 is expected for the current decade [3]. This increase in computing power has lead to the development of computer systems that perform billions of arithmetic operations per second. Although these systems give researchers the ability to solve previously intractable problems, the large number of arithmetic operations has made it extremely important to monitor and control errors in numerical applications.



*R. B. Kearfott and V. Kreinovich (eds.), Applications of Interval Computations,* 381–404.

As noted in [57], [11], roundoff error and catastrophic cancelation can quickly lead to results which are completely inaccurate. Most computer systems, however, do not provide an efficient mechanism to determine the accuracy of the computation or to recompute inaccurate results with higher precision.

To improve the accuracy and reliability of numerical computations, several software tools for accurate, self-validating arithmetic have been developed. Accurate arithmetic is achieved by increasing the precision of the computation or using exact calculations to eliminate roundoff error. Self-validating arithmetic produces two values for each result, such that the mathematically correct result is guaranteed to lie between these two values [48]. This provides a rigorous bound for each result. Self-validating arithmetic is often achieved through interval arithmetic [52]. Techniques for accurate, self-validating arithmetic have been successfully applied to several numerical applications including global optimization [30], [35], function evaluation [13], finding roots of polynomials [58], solving the initial value problem for ordinary differential equations [66], enclosing solutions of integral equations [17], and solving systems of equations [55]. Other applications that have benefited from accurate, self-validating arithmetic are discussed in [46], [53], [4], [2].

Several software tools have been developed to provide accurate, self-validating arithmetic. Variable-precision arithmetic packages, such as [10], [73], [14], give the programmer the ability to specify the precision of the computation based on the problem to be solved and the desired accuracy of the results. Libraries for interval arithmetic provide self-validating arithmetic by producing two values for each result [44], [36], [33]. The two values correspond to the endpoints of an interval such that true result is guaranteed to lie on the interval. Scientific programming languages [68] support special instructions and data types for interval arithmetic and exact dot products [49]. Techniques for variable-precision arithmetic and interval arithmetic have been combined in the languages PASCAL-XSC [3], C-XSC [13], ACRITH-XSC [17], and VPI [23]. These extended scientific languages provide data types and special instructions for variable-precision numbers, intervals, complex numbers, vectors, and matrices. Computer algebra systems provide symbolic or exact solutions when traditional numerical tools are inadequate [31]. Problem solving environments incorporate a variety of numerical software tools into a single system [51].

The main disadvantage of software tools for accurate, self-validating arithmetic is their speed. Since the arithmetic operations are simulated in software, tremendous overhead occurs due to function calls, memory management, error and range checking, expression manipulation, changing rounding modes, and exception handling. The interval arithmetic routines discussed in [59] are approximately 40 times slower than their single-precision floating point equivalents. Routines which support variable-precision interval arithmetic (up to 56 decimal digits) are more than 1,200 times slower than the corresponding single-precision routines [59]. Arithmetic operations in the arbitrary precision library discussed in [64] are 50 to 100 times slower than floating point operations, even when no additional precision is required. Certain application programs that use computer algebra systems are 1,000 to 3,000 times slower than equivalent programs written in C [16].

To overcome the speed limitation of existing software tools, direct hardware support is required. To improve the accuracy and performance of vector and matrix operations, coprocessors which support exact dot products have been designed [43], [29], [39]. To facilitate the use of interval arithmetic, these coprocessors provide the four rounding modes specified by the IEEE 754 floating point standard [1]: round-to-nearest, round-toward-zero, round-toward-positive-infinity and round-toward-negative-infinity. However, to produce accurate results and tight intervals when multiple operations are performed, exact dot products should be combined with variable-precision arithmetic [34]. Other processors including CADAC [21], DRAFT [20], MGAP-2 [7], TRAC [5] and Cascade [18] have been designed to provide hardware support for variable-precision arithmetic. Coprocessors for variable-precision interval arithmetic and staggered interval arithmetic are discussed in [63] and [62], respectively. These coprocessors provide a variety of instructions and data types to efficiently support accurate, self-validating arithmetic.

In this survey, software tools and hardware designs for accurate, self-validating arithmetic are presented, along with a discussion of their strengths and limitations. Software tools are examined in Section 2, followed by a discussion of hardware designs in Section 3. Section 4 presents the hardware design and software interface of a variable-precision interval arithmetic coprocessor. Conclusions and areas for future research are presented in Section 5.

# 2 SOFTWARE TOOLS

Several software tools have been developed for accurate, self-validating arithmetic. These include variable-precision software packages, interval arithmetic libraries, scientific programming languages, computer algebra systems, and numerical problem solving environments.

## 2.1 Variable-Precision Software Packages

Since the early 1960's, a number of software packages have been developed for variable-precision floating point arithmetic [10], [73], [14], [65]. Variable-precision arithmetic gives the programmer the ability to specify the precision (i.e., the number of digits in the significand) of the computation based on the problem to be solved and the desired accuracy of the results. These software packages use subroutine calls to perform variable-precision arithmetic operations. For example, in MP [14], the variable-precision operation $C = A + B$ is executed by the subroutine

$$MPADD(A, B, C)$$

where $A$, $B$, and $C$ are variable-precision floating point numbers. Variable-precision software packages have been used in a variety of scientific applications including calculating numerical constants, finding roots of polynomials, and evaluating elementary functions [15].

Variable-precision floating point packages simulate the arithmetic operations with the built-in fixed point instructions of the computer. Since these routines are designed to be portable to a number of different architectures, they are unable to take full advantage of the arithmetic hardware. They also require time-consuming subroutine calls for each arithmetic operation, which results in long execution times. For example, hardware addition is roughly 1,000 times faster than variable-precision addition to 60 decimal digits using the software package discussed in [73]. Another disadvantage of early variable-precision software packages is they did not support the directed rounding modes, which are essential for interval arithmetic.

## 2.2 Interval Arithmetic Libraries

Interval arithmeticinterval arithmetic was originally proposed as a tool for bounding roundoff errors in numerical computations [52]. It is also used to determine the effects of approximation errors and errors that occur due to inexact inputs. Interval arithmetic produces two values for each result. The two values correspond to the lower and upper endpoints of an interval, such that the true result is guaranteed to lie on this interval. The width of the interval indicates the accuracy of the result. Interval arithmetic libraries have been successfully applied to several applications including global optimization [36], function evaluation [13], enclosing solutions of integral equations [17], and solving systems of equations [55].

INTLIB is a fixed-precision interval arithmetic library for FORTRAN 77 [36], [37]. INTLIB gives the programmer the ability to perform arithmetic operations on intervals, compute accurate standard functions for interval and point data values, and read, write, and access intervals. is portable to several computer systems and does not require hardware support for IEEE 754 arithmetic [1]. Interval arithmetic is performed by subroutine calls, which utilize the machine's intrinsic double precision floating point arithmetic and then simulate directed rounding. A package called INTBIS is used with INTLIB to find interval roots of nonlinear systems of equations [38].

Another library for fixed-precision interval arithmetic is the Programmer's Runtime Optimized Fast Interval Library (PROFIL) [44]. PROFIL is a C++ class library that uses a set of Basic Interval Arithmetic Subroutines (BIAS) to provide data types and arithmetic operations for booleans, integers, reals, intervals, vectors, matrices, and complex numbers. Operations on various combinations of these data types are also provided. Memory management routines and operator overloading are provided to simplify the task of writing extensive interval programs. An extension to PROFIL has been developed which contains test matrices, support for linked lists, and subroutines for local minimization and automatic differentiation. PROFIL and BIAS have been used in different research areas including global optimization and solving nonlinear systems [44]. Efficient coding of interval computations in PROFIL produces execution times that are roughly twice as long as equivalent floating point computations. For certain iterative algorithms, programs that employ PROFIL have much shorter execution times than similar algorithms that use standard floating point arithmetic [44].

InC++ is a set of C++ libraries for interval computations [33]. Unlike previous interval libraries, InC++ supports operations on open-ended, infinite, and discontinuous intervals. It also allows complements of intervals to be computed and manipulated. InC++ uses a combination of algebraic and numerical interval arithmetic techniques to determine ranges of functions and solutions to interval constraint satisfaction problems. This provides tighter bounds than techniques that only use numerical interval arithmetic, but requires longer execution times.

## 2.3 Scientific Programming Languages

In the 1980's, several scientific programming languages including ACRITH-XSC [8], PASCAL-SC [69], FORTRAN-SC [9], and MODULA-SC [26] were developed. These languages are extensions to existing programming languages, which allow the programmer to define abstract data types, overload operators, create dynamic arrays, and access subarrays (i.e., rows, columns, and blocks of data). Special data types are provided for vectors, matrices, and complex numbers. To support accurate, self-validating computation, these languages provide the following features:

1. Real and complex arithmetic operations with guaranteed accuracy
2. Directed rounding for interval arithmetic
3. Accurate dot products for vector and matrix computations
4. Elementary functions with high accuracy

Like the variable-precision packages and the interval arithmetic libraries, the scientific programming languages are designed to be portable to a large number of computer systems. Since the arithmetic operations are simulated in software, they have long execution times for numerically intensive problems. Although these languages support accurate dot products, they do not support variable-precision arithmetic, which is needed when successive dot product computations are performed [34]. A more detailed survey of scientific programming languages is given in [68]. In [2], the implementation of several applications including solving the least squares problem, inverting and multiplying matrices, performing the conjugate gradient algorithm, and solving a system of nonlinear equations is discussed. Techniques for developing efficient interval algorithms and using the scientific programming languages to implement these algorithms are presented.

Extensions to the scientific programming languages have recently been developed. These include PASCAL-XSC [3], C-XSC [13], ACRITH-XSC [17] , and VPI [23]. In addition to the features provided by the scientific programming languages, these languages support variable-precision data types for intervals, vectors, matrices, and complex numbers. They also provide several routines for numerical applications including solving linear and nonlinear systems of equations, evaluating polynomials, finding zeros of polynomials, and determining eigenvalues and eigenvectors of matrices. They provide modular programming to simplify the task of developing numerical applications [13]. Since variable-precision interval arithmetic allows values to be computed to arbitrary accuracy, it is useful for solving a wide variety of ill-conditioned problems [54]. Sample programs that use the extended scientific programming languages are presented in [3], [13], [45].

## 2.4 Computer Algebra Systems

For applications in which traditional numerical solutions are not sufficient, computer algebra systems are often used to provide symbolic or exact calculations. Commercially available computer algebra systems include Maple [19], MACSYMA [56], Mathematica [71], REDUCE [28], SCRATHPAD II [67], and muMath [72]. Computer algebra systems typically provide the following features:

1. Arbitrary precision integer and rational arithmetic
2. Arbitrary precision floating point computations
3. Numerical evaluation of special functions
4. Matrix and vector computations
5. Algebraic and functional equation solving
6. Symbolic expression manipulation
7. Automatic code generation
8. A graphical interface

Computer algebra systems are advantageous because the algebraic solutions are exact and numerical results can be computed to arbitrary accuracy. They can also provide a better understanding of the relationship between the input parameters and the resulting output. When numerical results are required or the computer algebra system is unable to compute an exact solution, a combination of symbolic computing and variable-precision computations may be used to produce highly accurate solutions [47], [58]. With this technique, symbolic computations are used first to obtain a formal expression that is very close to the exact result. This eliminates roundoff error in the early computations. After this, the formal expression is evaluated using variable-precision arithmetic. Computer algebra systems are useful for several applications including integration, differentiation, finding roots of polynomials, solving systems of equations, and performing vector and matrix operations [31].

Although computer algebra systems provide exact or extremely accurate solutions, they are often too slow for numerically intensive computations. According to a study done in [16],Maple programs are more than 1,000 times slower than equivalent programs written in C, and Mathematica programs are more than 3,000 times slower. For several applications, the computation time required by computer algebra systems is prohibitive.

## 2.5 Problem Solving Environments

Problem solving environments incorporate a variety of numerical software tools into a single system [51]. For example, AQUARELS [24], [25] is a problem solving environment that provides variable-precision interval arithmetic, perturbation techniques, software validation, symbolic computations, numerical libraries, and automatic code generation. AQUARELS is implemented as a FORTRAN extension that uses special data types and instructions for numerical computations. A preprocessor translates the extended FORTRAN into FORTRAN-77 by creating calls to the appropriate routines. Combining a variety of software tools, numerical problem solving environments give the programmer increased power and flexibility in solving numerical problems. However, like other software tools for accurate, self-validating arithmetic, numerical problem solving environments often have long execution times.

# 3 HARDWARE DESIGNS

To overcome the speed limitation of software tools, several hardware designs for accurate, self-validating arithmetic have been developed. Coprocessors that support accurate dot products and the directed rounding modes are designed for applications that use extensive vector and matrix computations. Other processors that have direct hardware support for variable-precision integer or floating point arithmetic improve the speed and accuracy of numerical computations. Coprocessors for variable-precision interval arithmetic and staggered interval arithmetic improve the speed, accuracy, and reliability of scientific computations. These coprocessors also provide special instructions and hardware support for vectors, matrices, and complex numbers.

## 3.1 Dot Product Coprocessors

Several hardware designs for accurate, self-validating arithmetic support exact dot products [49], in which all calculations are mathematically exact and only a single rounding is performed at the very end. The exact dot product of two vectors $\vec{X}$ and $\vec{Y}$ with $n$ floating point elements is defined as

$$\vec{X} \cdot \vec{Y} = \Box(\sum_{i=1}^{n} x_i y_i)$$

where □ is a specified rounding mode. Exact dot products greatly reduce numerical errors due to roundoff and catastrophic cancelation. To facilitate the use of interval arithmetic, dot product coprocessors typically provide the four rounding modes specified in the IEEE 754 floating point standard [1]: round-to-nearest, round-toward-zero, round-toward-positive-infinity and round-toward-negative-infinity.

Dot product coprocessors are presented in [43], [29], [39], [27], [42] and [12]. These coprocessors typically consist of operand memory, a floating point multiplier and adder, a barrel shifter, and a long accumulator. Long accumulators allow floating point numbers and products of floating point numbers to be accumulated without roundoff error [49]. Special logic is used to reduce carry propagation and simplify normalization and rounding of the final result [42]. A survey of software and hardware designs for long accumulators is presented in [11]. Dot product coprocessors are often used in conjunction with the scientific programming languages presented in Section 2.3 to provide accurate, high-performance arithmetic. Although dot product coprocessors greatly improve the accuracy and execution time of vector and matrix operations, they do not provide a direct method for determining the accuracy of the results or recomputing inaccurate results with higher precision.

## 3.2 Variable-Precision Processors

To improve the speed and accuracy of high-precision integer computations, several variable-precision integer processors have been designed. The Cascade processor [18] uses a radix-16 redundant decimal signed-digit representation to avoid carry-propagation and reduce roundoff error [6]. In addition to the basic arithmetic operations, Cascade provides hardware support for computing the square root and greatest common denominator. Several memory management functions are provided for allocating and deallocating variable-precision integers. Cascade is designed for applications such as encryption and decryption, Gröbner basis calculations, and geometric modeling [18]. The TRAC computer chains together 8-bit microprocessors to perform variable-precision integer arithmetic [5]. A generate-propagate-carry bus is used to connect processors working on the same variable-precision numbers. The DRAFT computer consists of eight 32-bit arithmetic units. Each ALU can be combined or isolated from its neighbors on either side. This provides a maximum precision of 256 bits. The DRAFT computer is used for solving problems in computational and experimental number theory. The MGAP-2 computer uses an array of redundant radix-4 processors to perform variable-precision integer arithmetic to a maximum precision of 2,048 bits [7]. MGAP-2 is designed primarily for signal processing applications. Although these processors improve the execution time for variable-precision integer applications, they do not provide variable-precision floating point computations, which are essential for many scientific applications.

The Clean Arithmetic with Decimal Base and Controlled Precision (CADAC) processor performs the basic arithmetic operations on variable-precision floating point numbers [21]. The arithmetic specification for CADAC is presented in [32]. A maximum significand length of 32 decimal digits is supported. CADAC uses a pipelined accumulator and table-lookups to perform decimal arithmetic. CADAC is programmed in the Numerical Turing Language, which supports variable-precision decimal arithmetic, directed rounding modes, and exception handling. Although CADAC has five rounding modes, it does not provide special instructions for intervals, vectors, matrices, or complex numbers.

## 3.3 Variable-Precision, Interval Arithmetic Coprocessors

In [61], a coprocessor is described that provides direct hardware support for variable-precision interval arithmetic. The coprocessor gives the programmer the ability to specify the precision of the computation, determine the accuracy of the results, and recompute inaccurate results with higher precision. A variable-precision number format is used that has an extended range and a precision of up to 1,024 bits. Variable-precision intervals are represented by two variable-precision numbers, which correspond to the interval endpoints. The coprocessor features built-in instructions for intervals, matrices, vectors, and complex numbers. A long accumulator is used to compute variable-precision numbers and accurate dot products. Further details on the coprocessor's software interface and hardware design are given in Section 4.

Another method for obtaining variable-precision interval arithmetic is through the use of staggered interval arithmetic [50]. With this technique, a staggered interval $x$ is represented as the sum of $n$ floating point numbers plus a standard interval $X$ consisting of two floating point bounds, such that

$$x = \sum_{i=1}^{n} x_i + X$$

The floating point components $x_i$ are generated such that they do not overlap (i.e., their exponents differ by at least the number of bits in the significand). Increasing the value of $n$ increases the maximum precision of the staggered interval. Staggered interval arithmetic has been applied to matrix factorization, polynomial evaluation, and eigenvalue computations [50]. Techniques for using staggered interval arithmetic to produce accurate, self-validating programs are given in [13], [50].

A processor that supports staggered arithmetic is presented in [62]. The processor employs two long accumulators and a pipelined array of arithmetic units to perform staggered interval arithmetic. The upper and lower interval bounds of each operation are stored in the two long accumulators. A conversion algorithm is used to convert the upper and lower bounds to a staggered interval, which is then stored as $n+2$ IEEE double-precision floating point numbers. Staggered interval arithmetic requires less storage space and fewer arithmetic operations than variable-precision interval arithmetic. It also allows traditional floating point hardware to be used. However, staggered interval arithmetic does not increase the range of the numbers, and the maximum precision of staggered intervals is limited.

# 4 A VARIABLE-PRECISION, INTERVAL ARITHMETIC COPROCESSOR

This section gives an example of software tools and hardware designs for accurate, self-validating arithmetic, by describing the software interface and hardware design of a variable-precision interval arithmetic coprocessor. The coprocessor gives the programmer the ability to specify the precision of the computation, determine the accuracy of the results, and recompute inaccurate results with higher precision. The variable-precision interval arithmetic coprocessor uses a combination of hardware and software to provide an efficient, flexible, and numerically reliable system.

## 4.1 Software Interface

The software interface to the coprocessor is an extension of the C++ programming language. To support numerical computations, additional data types are defined. These include combinations of variable-precision floating point numbers, intervals, complex numbers, vectors, and matrices. The numerical data types supported by the software interface are shown in Figure 1. Operations on these data types are handled by the coprocessor, while operations on standard data types (e.g., integers, floats, etc.) are handled by the main processor. The coprocessor provides special instructions and hardware support for:

1. Arithmetic operations (e.g., add, subtract, multiply, and divide)
2. Interval operations (e.g., midpoint, intersection, width, and hull)
3. Elementary function evaluations (e.g., sine, cosine, logarithm)
4. Dot product computations

| | | |
|---|---|---|
| `vp_float` | `vp_vector` | `vp_matrix` |
| `vp_interval` | `vp_ivector` | `vp_imatrix` |
| `vp_complex` | `vp_cvector` | `vp_cmatrix` |
| `vp_cinterval` | `vp_civector` | `vp_cimatrix` |

**Figure 1** Data Types Supported by the Software Interface

Figure 2 shows code which uses some of the numerical data types. In this example, $A$, $B$, and $C$ are variable-precision interval matrices and $X$ is a variable-precision interval vector. $A$, $B$, and $C$ have sizes $m$ by $k$, $k$ by $n$, and $m$ by $n$, respectively. Initially, the entries of $A$ and $B$ are read in, and each element in $X$ is set to zero. Next, $C$ is set to the interval matrix product of $A$ and $B$. After this, a loop is entered that sums the rows of $C$ into $X$. In this program, all variable-precision data types have a precision of 4 words.

Figure 3 gives a sample program which illustrates how variable-precision interval arithmetic is used to recompute inaccurate results. The program performs the computation

$$y = \frac{1 - \cos(x)}{x^2},$$

where precision(*prec*) sets the precision of the computation and width($y$) returns difference between the upper and lower interval endpoints of $y$. Initially, the user enters the floating point value $x\_float$, which is then converted to the interval $x$. After this, a loop is entered in which $y$ is computed using variable-precision interval arithmetic. If the width of $y$ is not less than the error tolerance, the precision is increased by one word (i.e., 32 bits) and $y$ is recomputed. This continues until the width of $y$ is below the error tolerance

```
main() {
  precision(4);                   /* set precision to 4 words */
  int m, n, k;                    /* matrix, vector sizes */
  cin >> m >> n >> k;             /* read in matrix sizes */
  vp_imatrix A(m,k), B(k,n);      /* input interval matrices */
  vp_imatrix C(m,n);              /* output interval matrix */
  vp_ivector X(n) = 0;            /* interval vector */
  cin >> A >> B;                  /* read in vectors A and B */
  C = A*B;                        /* matrix multiplication */
  for (i = 0; i < m; i++)         /* loop to sum rows of C */
    X = X + C[i];                 /* add row i of C to X */
  cout << X;                      /* output X */
}
```

**Figure 2** Code for Variable-Precision, Interval Data Types

or the maximum precision is reached. The results of running the program are shown in Figure 4 for $x = 10^{-10}$. When the precision reaches 4 words (i.e., 128 bits), the error tolerance of $10^{-10}$ is achieved. If $x = 10^{-10}$ and $y$ is computed using IEEE double-precision arithmetic, then catastrophic cancelation in $1 - \cos(x)$ causes a result of 0 to be produced, even though the correct result is around 1/2.

## 4.2 Hardware Design

The coprocessor hardware supports the variable-precision data types and special instructions described in the the previous subsection. The format for variable-precision numbers is shown in Figure 5. Intervals are represented by two variable-precision numbers, which correspond to the interval endpoints. Each variable-precision number consists of a 16-bit exponent field ($E$), a sign bit ($S$), a 2-bit type field ($T$), a 5-bit significand length field ($L$), and a significand ($F$) which consists of $L+1$ significand words ($F[0]$ to $F[L]$). The exponent is represented with a bias of $2^{15} = 32,768$. The sign bit is zero if the number is positive and one if it is negative. The type field indicates if a number is infinite, zero, or not-a-number. The length field specifies the number of 32-bit words in the significand. The words of the significand are stored from most significant $F[0]$ to least significant $F[L]$. The significand is normalized between 1 and 2.

```
main() {
  int prec = 1, LMAX = 32;      /* precision is 1 word */
  vp_interval x, y;
  vp_float x_float, tol = 1e-10, error = 1000;
  precision(prec);              /* get the input value */
  cout << "Enter input value:  ";
  cin >> x_float;
  x = x_fl;                    /* convert to an interval */
  while (error > tol && prec <= LMAX) {
    y = (1 - cos(x))/square(x);
    error = width(y);          /* test width of y for error */
    cout << "y = " << y;
    cout << "error = " << error
         << ", prec = "<< 32*prec << " bits";
    prec++;                    /* increment the precision */
    precision(prec);
  }
}
```

**Figure 3** Program for Recomputing Inaccurate Results

```
Enter input value:  1e-10
y = [+0.000000000000e+0, +0.691507017239e+13]
error = +0.691507016681e+13, prec = 32 bits
y = [+0.000000000000000000000e+0, +0.165340832625378147406e+4]
error = +0.165340832625378147350e+4, prec = 64 bits
y = [+0.4999995791830674316396876862240e+0,
     +0.5000003869766408334622712293O2e+0]
error = +0.807793573401822583543044780840e-6, prec = 96 bits
y = [+0.499999999999999991007468144344732776280 8e+0,
     +0.500000000000000000998558839261460291 50333e+0]
error = +0.18978120248269870138751734097715303647 5e-15,
prec = 128 bits
```

**Figure 4** Results with Recomputing

The value of a variable-precision floating point number $VP$ is

$$VP = (-1)^S \times F \times 2^{E-32,768}$$

Variable-precision floating point numbers have a maximum precision of 1,024 bits and their range is approximately

$$[2^{-32,768}, 2^{32,769}] \approx [10^{-9,864}, 10^{9,864}]$$

A block diagram of the hardware unit which performs variable-precision interval arithmetic is shown in Figure 6. Control signals are shown as dashed lines. The significand and exponent data paths are depicted as bold and plain lines, respectively. The main components of the hardware unit are the register file, a 32-bit by 32-bit multiplier, a 64-bit adder, a 4-word by 64-bit selector, a long accumulator consisting of 64 64-bit segments, a 64-bit shifter, and a 16-bit exponent adder. The register file consists of two memory units: a 64-word by 32-bit header memory, and a 256-word by 32-bit significand memory. Each header word contains the exponent, sign, type, and length of the variable-precision number, along with an 8-bit index which points to the most significant word of the corresponding significand.

Significand words which are read from the register file go into the multiplier, the selector, or the long accumulator. The selector performs comparison operations and determines which values go into the adder and the shifter. The multiplier takes two 32-bit significand words as inputs and computes a 64-bit product. The adder takes two 64-bit inputs and produces a 64-bit sum and a 1-bit carry. The shifter takes a 64-bit number and shifts it by up to 64 bits. The long accumulator stores intermediate results of variable-precision arithmetic operations.

Area and cycle time estimates based on data from a 1.0 micron CMOS standard cell library [2] indicate that the variable-precision interval arithmetic processor has a total chip area of 102 $mm^2$ and a worst case delay of 16 ns [61]. In comparison, an IEEE double precision coprocessor has a estimated chip area of 89 $mm^2$ and an estimated cycle time of 20 ns [61]. Execution time estimates given in [61] for variable-precision interval arithmetic operations indicate that the coprocessor is 2 to 3 orders of magnitude faster than an existing software package for variable-precision interval arithmetic [23]. In [60], estimated execution times for the coprocessor are reported for dot product computation, polynomial evaluation, and interval Newton Method. Direct hardware support for variable-precision interval arithmetic gives the coprocessor a tremendous speed advantage over existing software packages.

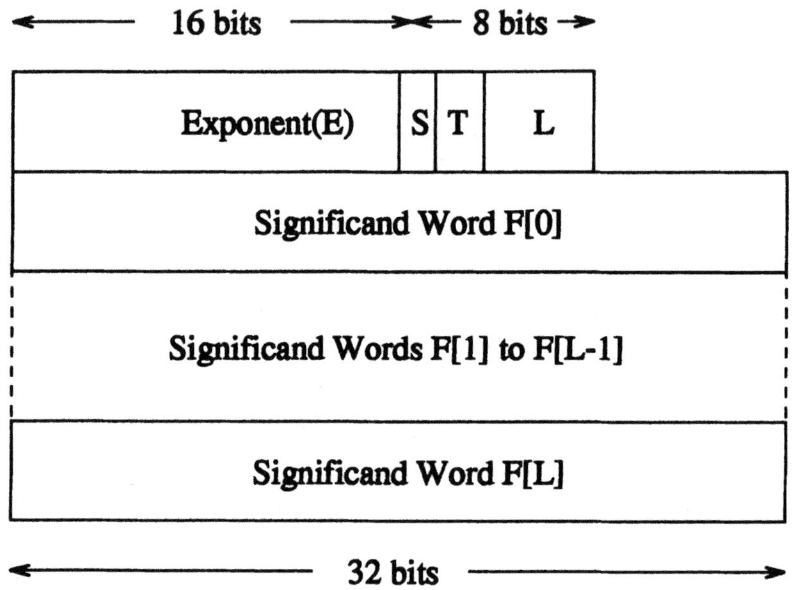


**Figure 5** Variable-Precision Floating Point Format.

# 5 CONCLUSIONS AND AREAS FOR FUTURE RESEARCH

A wide variety of software tools and hardware designs has been developed to assist programmers in developing, writing, and executing programs for accurate, self-validating arithmetic. The software tools range from fairly simple variable-precision arithmetic packages to complex problem solving environments. Hardware designs for accurate, self-validating arithmetic include dot product coprocessors, variable-precision integer and floating point processors, and coprocessors that support variable-precision interval arithmetic. Software tools typically provide more control and programming flexibility than hardware designs, but have much larger execution times for numerically intensive applications.

To further improve the speed and programmability of existing numerical applications, software tools and hardware designs for accurate, self-validating arithmetic should be combined into a single system. With this approach, specially designed numerical coprocessors provide high-speed support for the arithmetic operations, and software packages utilize this support to efficiently implement

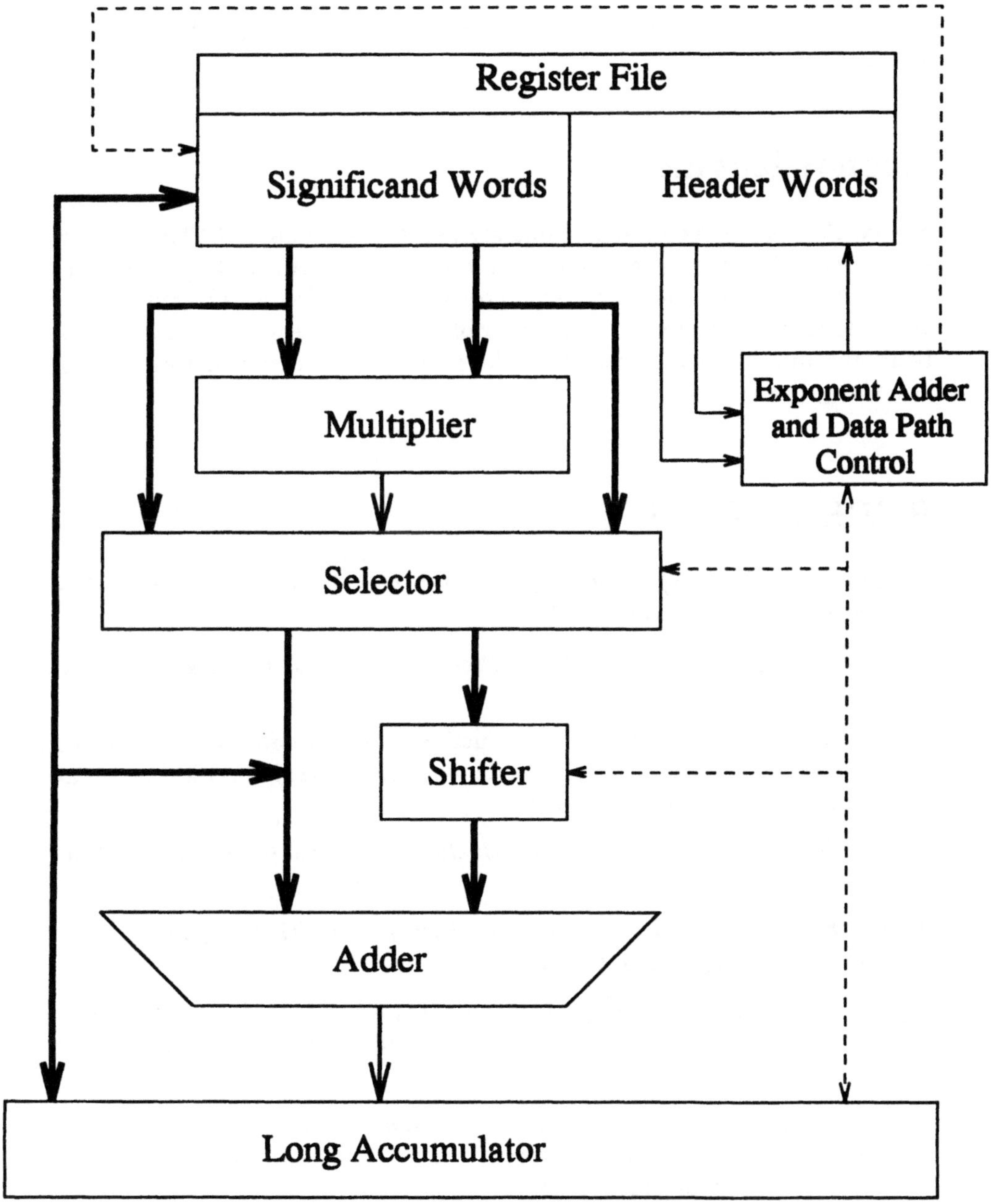


**Figure 6** Arithmetic Processor Hardware Design.

accurate, self-validating arithmetic. This gives the user the speed of special purpose hardware, along with the ease and flexibility provided by numerical software tools.

## Acknowledgments

Special thanks are extended to Professor Jeff Ely at Louis and Clark University for providing the variable-precision interval arithmetic package which was modified to simulate the coprocessor presented in Section 4. The authors are also grateful to Professor R. Baker Kearfott at the University of Southwestern Louisiana and the anonymous referees for providing comments that were useful in revising this survey.

## REFERENCES

[1] *IEEE Standard 754 for Binary Floating Point Arithmetic*, American National Standards Institute, 1985.

[2] *LSI Logic 1.0 Micron Cell-Based Products Databook*, LSI Logic Corporation, 1991.

[3] E. Adams and U. Kulisch, "Introduction", In E. Adams and U. Kulisch (editors), *Scientific Computing with Automatic Result Verification*, Academic Press, Inc., 1993, pp. 1–12.

[4] G. Alefeld and J. Herzberger, *Introduction to Interval Computations.* Academic Press, 1983.

[5] G. S. Almasi and A. Gottlieb, *Highly Parallel Computing*, Benjamin/Cummings, 1989, p. 487.

[6] A. Avizienis, "Signed-Digit Number Representations for Fast Parallel Arithmetic", *IRE Transactions on Electronic Computers*, 1961, Vol. 10, pp. 389–400.

[7] R. S. Bajwa, R. M. Owens, and M. J. Irwin, "Area-Time Tradeoffs in Micro-Grain VLSI Array Processors", *IEEE Transactions on Computers*, 1994, Vol. C-43, pp. 1121–1128.

[8] J. H. Bleher, A. E. Röder, and S. Rump, "ACRITH: High Accuracy Arithmetic - An Advanced Tool for Numerical Computation", In *Proceedings of the 7th Symposium on Computer Arithmetic*, 1985, pp. 318–321.

[9] J. H. Bleher, S. M. Rump, U. Kulisch, and J. Wolff von Gudenberg, "FORTRAN-SC: A Study of a FORTRAN Extension for Engineering Scientific Computation with Access to ARITH", ***Computing***, 39:93–110, 1987.

[10] B. I. Blum, "An Extended Arithmetic Package", ***Communications of the ACM***, 1965, Vol. 8. pp. 318–320.

[11] G. Bohlender, "What Do We Need Beyond IEEE Arithmetic?" In C. Ullrich, editor, ***Computer Arithmetic and Self-Validating Numerical Methods***, Academic Press, 1990, pp. 1–32.

[12] G. Bohlender and T. Teufel, "BAP-SC: A Decimal Floating-Point Processors for Optimal Arithmetic", In E. Kaucher, U. Kulisch, and Ch. Ullrich, editors, ***Computerarithmetic: Scientific Computation and Programming Languages***, B.G Teubner Verlag, 1987, pp. 31–58.

[13] K. Braune, "Standard Functions for Real and Complex Point and Interval Arguments with Dynamic Accuracy", In U. Kulisch and H. J. Stetter, editors, ***Scientific Computing with Automatic Result Verification***, Springer-Verlag, 1989, pp. 159–184.

[14] R. P. Brent, "A FORTRAN Multiprecision Arithmetic Package", ***ACM Transactions on Mathematical Software***, 1978, Vol. 4, pp. 57–70.

[15] R. P. Brent, "Multiple-Precision Zero Finding Methods and the Complexity of Elementary Function Evaluation", In J. F. Traub, editor, ***Analytic Computational Complexity***, Academic Press, Inc., 1976, pp. 151–176.

[16] B. Buchberger, "Gröbner Bases in MATHEMATICA: Enthusiasm and Frustration", In ***Programming Environments for High-level Scientific Problem Solving***, 1991, pp. 80–91.

[17] O. Caprani, K. Madsen, and O. Stauning. "Enclosing Solutions of Integral Equations", ***Reliable Computing***, 1995, Supplement (Extended Abstracts of APIC'95: International Workshop on Applications of Interval Computations, El Paso, TX, Febr. 23–25, 1995), pp. 45–51.

[18] T. M. Carter, "Cascade: Hardware for High/Variable Precision Arithmetic", In ***Proceedings of the 9th Symposium on Computer Arithmetic***, 1989, pp. 184–191.

[19] B. W. Char et al, "A Tutorial Introduction to Maple", ***Journal of Symbolic Computation***, 1986, Vol. 2, pp. 179–200.

[20] D. M. Chiarulli, W. G. Rudd, and D. A. Buell, "DRAFT: A Dynamically Reconfigurable Processor for Integer Arithmetic", In *Proceedings of the 7th Symposium on Computer Arithmetic*, 1985, pp. 309–318.

[21] M. S. Cohen, T. E. Hull, and V. C. Hamacher, "CADAC: A Controlled-Precision Decimal Arithmetic Unit", *IEEE Transactions on Computers*, 1983, Vol. C-32, pp. 370–377.

[22] G. F. Corliss, "Industrial Applications of Interval Techniques", In Ch. Ullrich, editor, *Computer Arithmetic and Self-Validating Numerical Methods*, Academic Press, 1990, pp. 91–113.

[23] J. S. Ely, "The VPI Software Package for Variable Precision Interval Arithmetic", *Interval Computations*, 1993, No. 2, pp. 135–153.

[24] J. Erhel and B. Philippe, "AQUARELS: A Problem-Solving Environment for Numerical Quality", In C. Brezinski and U. Kulisch, editors, *Computational and Applied Mathematics I: Algorithms and Theory*, North-Holland, 1992, pp. 127–136.

[25] J. Erhel and B. Philippe, "Design of a Toolbox to Control Arithmetic Reliability", In L. Atanassova and J. Herzberger, editors, *Computer Arithmetic and Enclosure Methods*, North-Holland, 1992, pp. 99–108.

[26] C. Falco-Korn, S. König, and S. Gutzwiller, "MODULA-SC: A Precompiler to Modula-2", In C. P. Ullrich, editor, *Contributions to Computer Arithmetic and Self-Validating Numerical Methods*, J. C. Baltzer, 1991, pp. 371–384.

[27] H. Fleurkens and R. Tangelder, "The High Level Design of a Chip for Scientific Computation", In *Proceedings of the CompEuro Conference*, 1991, pp. 811–815.

[28] R. Gerhard, *REDUCE: Software for Algebraic Computation*. Springer-Verlag, 1987.

[29] K. Hafner, "Chips for High Precision Arithmetic", In Ch. Ullrich, editor, *Computer Arithmetic and Self-Validating Numerical Methods*, Academic Press, 1990, pp. 33–54.

[30] E. Hansen, *Global Optimization Using Interval Analysis*. Marcel Dekker, 1992.

[31] D. Harper, C. D. Wooff, and D. E. Hodgkinson, *A Guide to Computer Algebra Systems*, John Wiley and Sons, 1991.

[32] T. E. Hull, M. S. Cohen, and C. B. Hull, "Specification for a Variable-Precision Arithmetic Coprocessor", In *Proceedings of the 10th Symposium on Computer Arithmetic*, 1991, pp. 127–131.

[33] E. Hyvönen and S. De Pascale, "InC++ Library Family for Interval Computations", *Reliable Computing*, 1995, Supplement (Extended Abstracts of APIC'95: International Workshop on Applications of Interval Computations, El Paso, TX, Febr. 23–25, 1995), pp. 85–90.

[34] W. M. Kahan and E. LeBlanc, "Anomalies in the IBM ACRITH Package", In *Proceedings of the 7th Symposium on Computer Arithmetic*, 1985, pp. 322–331.

[35] R. B. Kearfott, "A Review of Techniques in the Verified Solution of Constrained Global Optimization", *Reliable Computing*, 1995, Supplement (Extended Abstracts of APIC'95: International Workshop on Applications of Interval Computations, El Paso, TX, Febr. 23–25, 1995), pp. 105–106.

[36] R. B. Kearfott, M. Dawande, K. Du, and C. Hu, "A Portable FORTRAN 77 Elementary Function Library", *Interval Computations*, 1992, No. 3, pp. 96–105.

[37] R. B. Kearfott, M. Dawande, K. Du, and C. Hu, "Algorithm 737: INTLIB: A Portable FORTRAN 77 Interval Standard Function Library", *ACM Transactions on Mathematical Software*, 1994, Vol. 20, pp. 447–459.

[38] R. B. Kearfott and M. Novoa, "INTBIS, A Portable Interval Newton Bisection Package", *ACM Transactions on Mathematical Software*, 1990, Vol. 16, pp. 152–157.

[39] J. Kernhof et al., "A CMOS Floating-Point Processing Chip for Verified Exact Vector Arithmetic", In *ESSIRC 94*, 1994.

[40] R. Klatte, U. Kulisch, M. Neaga, D. Ratz, and Ch. Ullrich, *PASCAL-XSC: Language Reference with Examples*. Springer-Verlag, 1991.

[41] R. Klatte, U. Kulisch, A. Wiethoff, C. Lawo, and M. Rauch, *C-XSC: A C++ Class Library for Extended Scientific Computing*. Springer-Verlag, 1993.

[42] A. Knöfel, "Fast Hardware Units for the Computation of Accurate Dot Products", In *Proceedings of the 10th Symposium on Computer Arithmetic*, 1991, pp. 70–75.

[43] A. Knöfel, "Hardware Kernel for Scientific/Engineering Computations", In E. Adams and U. Kulisch, editors, *Scientific Computing with Automatic Result Verification*, Academic Press, Inc., 1993, pp. 549–570.

[44] O. Knüppel, "PROFIL/BIAS - A Fast Interval Library", *Computing*, 1994, Vol. 53, pp. 277–288.

[45] W. Krämer, "Multiple-Precision Computations with Result Verification", In E. Adams and U. Kulisch, editors, *Scientific Computing with Automatic Result Verification*, Academic Press, Inc., 1993, pp. 325–356.

[46] V. Kreinovich, "Data Processing Beyond Traditional Statistics: Applications of Interval Computations. A Brief Introduction", *Reliable Computing*, 1995, Supplement (Extended Abstracts of APIC'95: International Workshop on Applications of Interval Computations, El Paso, TX, Febr. 23–25, 1995), pp. 13–21.

[47] F. Krückeberg, "Computer Algebra and Variable Precision Computing", In *Proceedings of the 13th IMACS World Congress on Computation and Applied Mathematics*, 1991, pp. 65–66.

[48] U. Kulisch and H. J. Stetter, "Automatic Result Verification", In U. Kulisch and H. J. Stetter, editors, *Scientific Computing with Automatic Result Verification*, Springer-Verlag, 1989, pp. 1–6.

[49] U. W. Kulisch and W. L. Miranker, *Computer Arithmetic in Theory and in Practice*, Academic Press, Inc., 1981.

[50] R. J. Lohner, "Interval Arithmetic in Staggered Correction Format", In E. Adams and U. Kulisch, editors, *Scientific Computing with Automatic Result Verification*, Academic Press, 1993, pp. 301–321.

[51] J. C. Mason and I. Reid, "Numerical Problem-Solving Environments - Current and Future Trends", In J. C. Mason and M. G. Cox, editors, *Scientific Software Systems*, Chapman and Hall, 1988, pp. 223–237.

[52] R. E. Moore, *Interval Analysis*, Prentice Hall, 1966.

[53] R. E. Moore, *Reliability in Computing: The Role of Interval Methods in Scientific Computations*, Academic Press, 1988.

[54] R. E. Moore, "Computing to Arbitrary Accuracy", In C. Brezinski and U. Kulisch, editors, *Computational and Applied Mathematics I: Algorithms and Theory*, North-Holland, 1992, pp. 327–336.

[55] A. Neumaier, *Interval Methods for Systems of Equations*, Cambridge University Press, 1990.

[56] R. Pavelle and P. S. Wang, "MACSYMA from F to G", *Journal of Symbolic Computation*, 1985, Vol. 1, pp. 69–100.

[57] D. Ratz, "The Effects of the Arithmetic of Vector Computers on Basic Numerical Methods", In Ch. Ullrich, editor, *Contributions to Computer Arithmetic and Self-Validating Numerical Methods*, J. C. Baltzer, 1990, pp. 499–514.

[58] J. Schröder, "Verification of Polynomial Roots by Closed Coupling of Computer Algebra and Self-Validating Numerical Methods", In Ch. Ullrich, editor, *Contributions to Computer Arithmetic and Self-Validating Numerical Methods*, J. C. Baltzer, 1990, pp. 259–269.

[59] E. K. Reuter et al., "Some Experiments Using Interval Arithmetic", In *Proceedings of the 4th Symposium on Computer Arithmetic*, 1978, pp. 75–81.

[60] M. J. Schulte and E. E. Swartzlander, Jr., "Designs and Applications for Variable-Precision, Interval Arithmetic Coprocessors", *Reliable Computing*, 1995, Supplement (Extended Abstracts of APIC'95: International Workshop on Applications of Interval Computations, El Paso, TX, Febr. 23–25, 1995), pp. 166–172.

[61] M. J. Schulte and E. E. Swartzlander, Jr., "A Hardware Design and Arithmetic Algorithms for a Variable-Precision, Interval Arithmetic Coprocessor", In *Proceedings of the 12th Symposium on Computer Arithmetic*, 1995 (in press).

[62] M. J. Schulte and E. E. Swartzlander, Jr., "A Processor for Staggered Interval Arithmetic", In *Proceedings of the 1995 International Conference on Application Specific Array Processors*, 1995 (in press).

[63] M. J. Schulte and E. E. Swartzlander, Jr., Software Interface and Hardware Design for Variable-Precision Interval Arithmetic, *Reliable Computing*, 1995 (in press).

[64] J. Schwartz, "Implementing Infinite Precision Arithmetic", In *Proceedings of the 9th Symposium on Computer Arithmetic*, 1989, pp. 10–17.

[65] D. M. Smith, "Algorithm 693: A FORTRAN Package For Floating-Point Multiple-Precision Arithmetic", *ACM Transactions on Mathematical Software*, 1991, Vol. 17, pp. 273–283.

[66] H. J. Stetter, "Validated Solution of Initial Value Problem for ODE", In C. Ullrich, editor, *Computer Arithmetic and Self-Validating Numerical Methods*, Academic Press, 1990, pp. 171–187.

[67] R. S. Sutor, "The Scratchpad II Computer Algebra System", In *Proceedings of EUROCAL 85*, 1985, pp. 32–33.

[68] Ch. Ullrich, "Programming Languages for Enclosure Methods", In Ch. Ullrich, editor, *Computer Arithmetic and Self-Validating Numerical Methods*, Academic Press, 1990, pp. 115–130.

[69] J. Wolff von Gudenberg, "PASCAL-SC: A PASCAL Extension for Scientific Computation", In *Proceedings of the 10th IMACS World Congress on System Simulation and Scientific Computation*, 1982, pp. 402–482.

[70] W. V. Walter, "ACRITH-XSC: A Fortran-like Language for Verified Scientific Computing", In E. Adams and U. Kulisch, editors, *Scientific Computing with Automatic Result Verification*, Academic Press, Inc., 1993, pp. 45–70.

[71] S. Wolfram, *Mathematica : A System for Doing Mathematics by Computer*, Addison-Wesley, 1988.

[72] C. D. Wooff and D. E. Hodgkinson, *muMath: A Microcomputer Algebra System*, Academic Press, 1987.

[73] W. T. Wyatt, D. W. Lozier, and D. J. Orser, "A Portable Extended Precision Arithmetic Package and Library with FORTRAN Precompiler", *ACM Transactions on Mathematical Software*, 1976, Vol. 2, pp. 209–231.

# 15

# STIMULATING HARDWARE AND SOFTWARE SUPPORT FOR INTERVAL ARITHMETIC

**G. William Walster**

*SunSoft, A Sun Microsystems, Inc. Business, 2550 Garcia Avenue, MS UMTV12-40, Mountain View, CA 94043-1100, email bill.walster@Eng.Sun.Com*

## ABSTRACT

Commercial hardware and software system suppliers do not yet include support for interval arithmetic in their products. In this paper, the reasons for this lack of support are described. Based on these descriptions, recommendations are made to help stimulate end-user demand and commercial support for interval arithmetic.

## Disclaimer

The opinions expressed in this paper are those of the author and do not necessarily reflect either the plans of Sun Microsystems Inc., or the opinions of other Sun employees.

## 1 INTRODUCTION

Much has been written about the properties and benefits of interval arithmetic[1], including the ability to: bound the accuracy of floating point computations;

[1] Computers normally only use the small set of machine representable numbers. An interval is a *continuum* of numbers bounded by its endpoints. Intervals with machine representable endpoints are a convenient way to represent numbers that are otherwise inaccessible. Machine computations can be performed using interval arithmetic and directed rounding. (See the first three chapters of [5] for a concise introduction to interval arithmetic, and [12] for additional introductory material and an extended bibliography.)



*R. B. Kearfott and V. Kreinovich (eds.), Applications of Interval Computations*, 405–416.

provide numerical proofs; and, solve difficult, if not otherwise intractable problems. These proceedings contain numerous examples of the benefits of interval arithmetic applied to practical problems.

The benefits afforded by interval arithmetic have motivated substantial effort on the part of many researchers and developers. Numerous interval implementations and practical applications have been developed. Nevertheless, major computer system manufacturers have not yet introduced support for interval arithmetic.

## 2 THE PARTICIPANTS

The struggle to gain commercial acceptance of interval arithmetic involves four groups:

- The "interval community" consists of a growing number of developers and users of interval algorithms. Computer scientists and numerical analysts are the primary interval algorithm developers. Increasingly, end-users in science and industry are beginning to apply interval methods to solve significant commercial, scientific, and engineering problems.
- "System suppliers" manufacture commercial hardware and system software, such as compilers and application software development tools.
- "Application developers" write commercial applications using commercial hardware and system software.
- Finally, "end-users" employ hardware systems and applications to solve problems. In many cases, end users also develop custom applications.

The dynamics of the interactions between these four groups are driven by the forces motivating them. The following descriptions of these dynamics and forces are based on the author's personal experiences during more than twenty years in both academia and industry, while working to further the acceptance of interval methods.

## 2.1 Interval Community

R. E. Moore began his work on interval analysis over thirty years ago, while employed by Lockheed Missiles and Space Co. Inc. in Sunnyvale, CA. Following the publication of his book ***Interval Analysis*** [14], the interval community has steadily grown to include well over 500 people throughout the world. Baker Kearfott has recently established an interval interest alias and Vladik Kreinovich has created an interval web page containing information about interval computing (See [8] and [12]).

Although the motivation for interval arithmetic has always been distinctly "applied" in character, significant theoretical work has been done to establish the mathematical foundation of intervals (See [12] for an extended bibliography). The volume of work in the continuing quest for "sharp"[2] interval algorithms has led to interval analysis becoming a recognized branch of numerical analysis.

Members of the interval community are driven by the requirement to perform machine computations having known accuracy. They view any problems for which "sharp" interval methods do not yet exist as opportunities to expand the interval algorithm domain.

In the past, developers of interval algorithms have had to provide their own tools to perform even the most fundamental interval operations. While commercially available packages do exist, none are provided by system suppliers (See, for example, [3, 10, 11, 9, 13, 15, 16, 17, 18]). Consequently, no existing interval implementations take advantage of performance opportunities that can easily be obtained through hardware and compiler support.

When system suppliers provide hardware and compiler support for interval arithmetic, two positive outcomes will result: first, the interval community will be able to concentrate on sharp interval algorithm development rather than interval arithmetic implementations; second, performance improvements will increase the size of problems that can be solved in a given amount of time.

## 2.2 System Suppliers

System suppliers must succeed as businesses. Consequently, they are conservative and risk averse. They must show relatively short term financial return

[2] The term "sharp" is used to describe an interval result containing a small amount of unnecessary (or excess) width. Sharp interval algorithms produce sharp interval results.

from most decisions, and they have more problems to solve and opportunities to improve existing products than available resources can support. There are almost always credible advocates for a variety of diverse and often conflicting courses of action. In such circumstances, new technology such as interval arithmetic is not quickly embraced.

The safe way to allocate resources is to respond only to vocal customer demands. Yet, customers are not always aware of the opportunities provided by new technology. No customers were demanding a GUI when Apple introduced the Macintosh. This kind of innovation takes foresight and commitment from system suppliers. Most of all, it takes a willingness to step outside the domain of safety from within which most system suppliers prefer to compete.

What are some of the characteristics of this domain of safety? It is in reality a paradigm or set of implicit rules within which system suppliers compete. In general, this paradigm is dominated by speed, speed, and more speed. New features are important. Change within the existing paradigm is good, provided that it is not too radical. Multi-processing, multi-threading, and clusters are fashionable ways to achieve inexpensive speed. Objects and their associated technology promise to help solve the "software development crisis". However, speed remains the single most important goal on which most system supplier development resources are spent.

Why so much focus on speed? The obvious answer is that speed sells. Speed has characteristics that help to explain why the quest for it is so universal. For example, the rules are clear when competing for speed. Speed is simple to measure and market. Demand for more speed is easy to create by inventing new tools and features that require additional system resources.

The utility value of speed is *not* linear. There *is* a point of diminishing returns, beyond which other considerations will begin to compete with speed for development resources. Indeed, cracks may be appearing in the existing speed paradigm. As it becomes increasingly possible to perform billions and even trillions of floating point computations, it also becomes increasingly difficult to guarantee the accuracy of final results. However, experience has shown that even when both the need to bound errors, and the bounds provided by interval arithmetic are accepted, resistance can still be strong. The existing floating point paradigm will still be preferred until the expected rewards from market entry overcome the fear of embracing new technology.

## 2.3 Application Developers

Application developers are ***customers*** of system suppliers and ***producers*** for end-users. As a consequence, they share characteristics of both customers and producers. Like system suppliers, they constantly introduce change within their own products to drive the sale of upgrades. And like end-users, application developers resist incompatible system changes that require support for multiple versions of their existing products.

As with system suppliers, application developers are driven by the quest for volume. And just as with system suppliers, the lure of the high volume platform creates a barrier to innovation. It is difficult to persuade either system suppliers or application developers to embrace fundamentally new technology before market volume has been established. But without hardware, system software, ***and*** applications, users will not have a complete solution to their requirements.

Interval arithmetic provides much more than a method to bound errors in floating point computations. As even the most cursory examination of the interval literature will show, interval arithmetic has provided the foundation on which qualitatively new numerical algorithms have been built. As soon as the market is recognized for new applications based on these algorithms, application developers will aggressively compete to supply them.

## 2.4 End-Users

Like system suppliers and application developers, many end-users are conservative, risk averse, and addicted to speed. Still, there are important differences. End users are typically more resistant to change than either system suppliers or application developers. Computers are the steam engines of the information revolution. Users want their engines to "***just work***". They tend to be substantively focused on the problem being solved. Computers are just a solution tool, not an end in themselves.

Most users demand environmental stability. Change, even within the existing paradigm is an anathema. In spite of upgrades and bug fixes to operating systems and compilers, old "dusty decks" must continue to work. Often, clever "hacks" are used to work around, or even take advantage of existing bugs. It is not surprising that users will persist in using the existing floating point paradigm until a practical solution is commercially available.

Users are increasingly aware of their growing dependence on computers. While computers provide the economical power to solve increasingly difficult problems, vulnerability is a hidden cost. This may partially explain why the Pentium chip incident[3] sent such shock waves through the user community in late 1994 and early 1995. However, vulnerability alone does not completely explain the user community's reaction to the Pentium hardware bug. In comparison, rounding errors in normalized floating point arithmetic are much more pervasive and serious. Rounding errors and cancelation can conspire to completely destroy the accuracy of machine computations. Why were users so quick to demand a fix for the relatively innocent Pentium bug when they have not demanded a fix for floating point arithmetic? For the Pentium chip, the solution was obvious: fix the hardware bug. For floating point arithmetic, the solution is neither as obvious, nor as easy to implement. Many users are ignorant of numerical accuracy issues. Even among the technically sophisticated, few users are aware of the numerical properties of the algorithms they use. Even fewer users are aware that interval arithmetic is a viable alternative. In circumstances where no obvious solution to a problem is perceived to exist, continuing to ignore the problem may appear to be the only alternative.

# 3 STABLE EQUILIBRIUM

The status quo is likely to continue, without a ground swell of demand for an alternative to the existing floating point paradigm. System suppliers, application developers, and end-users are not likely to demand a change. The stability of this equilibrium is enhanced by the dynamics of high volume systems and applications. Users demand both applications and speed. The fastest hardware is of little value without applications. The system with the highest volume is preferred by application developers. In the beginning, systems supporting interval arithmetic will have low volume. Therefore, they will not attract application developers.

The dynamics of the interrelated needs of system suppliers, application developers, and end-users conspire to make it difficult to introduce the fundamental changes required by interval arithmetic. If these changes are to take place, they must begin with the interval community. System suppliers and application developers will not implement interval methods without end-user demand. And

[3] The initial release of Intel's Pentium chip contained a small obscure hardware bug. Intel initially refused to provide replacement chips unless users could demonstrate they had encountered the bug in their work. The user community protested so vocally that Intel was forced to reverse itself and agree to provide replacements to anybody who asked for them.

unless interval methods exist, end-users cannot demand them. The remaining sections of this paper describe steps that can be taken to destabilize the existing equilibrium and thereby begin a shift to the interval paradigm.

# 4 THE INTERVAL PARADIGM SHIFT

A number of conditions must be satisfied before a shift to the interval paradigm can take place:

- The technical foundation for intervals must exist, including the mathematics, hardware, and software. The mathematical foundation exists. As a consequence of the IEEE 754 floating point standard, some hardware support exists for directed rounding. Much more can be done to improve both hardware and software performance. Many man years have been spent to write a number of important commercial interval implementations (See [3, 10, 11, 9, 13, 15, 16, 17, 18]). Nevertheless, interval arithmetic will remain unnecessarily slow until system providers integrate interval support into compilers and hardware.

- Interval methods must have universal applicability, at least in principle. Skeptics will demand that interval methods produce sharp answers for every conceivable difficult problem. The interval community will need to accept these challenges and develop sharp interval methods wherever they are required. Many sharp interval methods continue to be developed for problems that were thought to be difficult or even impossible to solve. There is no reason why this trend will not continue until a critical mass of sharp interval algorithms has been accumulated.

- A number of interval conventions and standards will need to be established: First, interval I/O needs to be made as simple and intuitive as possible. It is difficult for users to count digits between lower and upper bounds to determine the accuracy of interval results. Second, standard notation needs to be devised for new interval operators such as the different types of interval relational operators. Third, the syntax and semantics of interval intrinsic functions need to be standardized.

  To implement some of these conventions, language extensions and language-independent interfaces will be needed. To promote portability, it is important that these extensions and interfaces become standards. To facilitate adoption, the interval community will need to unite behind any proposed standards. If the interval community cannot agree, any proposals will lack credibility.

# 5 SYSTEM SUPPLIER DEMAND

The possibility exists to create demand for interval arithmetic among system hardware and software suppliers. If system suppliers need interval arithmetic, they will be more inclined to add system support. The Pentium case demonstrated that system suppliers *must* find and fix floating point hardware and software bugs. Interval arithmetic can provide direct help to system suppliers in this effort. For example, interval software developers routinely use different, but mathematically equivalent, algorithms to compute the same quantity. Intersecting the resulting intervals can produce a sharper answer than any single algorithm. If the intersection is empty a bug *must* exist. Of course, the bug may be in the test. But, if the test can be excluded, the bug must be in the system. It is possible to exploit this idea to develop interval programs designed to detect floating point hardware and software problems. If successful, these algorithms may convert system hardware and software developers into consumers of interval arithmetic.

# 6 END-USER DEMAND

What developments need to be made before the interval community unleashes a media campaign to expose the floating point crisis and promote interval arithmetic as the solution? Before beginning any organized campaign, interval arithmetic must be easy to find and use. Even if the media campaign succeeds in sparking end-user interest, without software that is easy to access and run, interest will die and the campaign will fail. To solve this problem, the interval community must provide software for easy end-user access to interval arithmetic. Ideally, this software should be freely available through netlib, the Free Software Foundation, or any easily accessible and economical medium. The goal must be to create the widest possible distribution.

Three pieces of related software to provide easy end-user access are: Fortran 90 and C++ interval arithmetic implementations; an interval calculator program; and, an interval spreadsheet program.

- Fortran 90 and C++ interval arithmetic implementations need to be provided for the most significant high volume platforms. To do a complete implementation, some language extensions will be required to support interval I/O, relational operators, and intrinsics. A number of such implementations have been written. However, they have either been written by their authors in the process of developing new interval algorithms, or they are relatively expensive commercial implementations that are unknown outside the interval community (See, [10, 9], and [3, 11, 13, 15, 16, 17, 18]).

- An interval calculator program[4] will be helpful to both interval algorithm developers and new interval arithmetic users. An interval calculator will serve as a tool to help in the debugging of small interval expressions and algorithms, and it will also provide new users with a simple way to learn how to use interval arithmetic. An interval calculator program may also provide a testbed on which to refine alternative interval I/O conventions, and to implement generalized interval arithmetic to reduce the impact of correlated errors (see Hansen [4]).

- An interval spreadsheet program will provide users with additional opportunities to use interval arithmetic for both simple and complicated numerical computations. In particular, it will make possible routine computations in which errors or uncertainty in algorithm inputs can be represented by intervals. Again, Hansen's generalized interval arithmetic should be included to cope with correlated errors. It is worth noting that two interval spreadsheet programs already exist (See [1, 6, 7] and references therein). Both programs use interval global optimization [5] to produce the sharpest possible interval results.

Once these three pieces of software have been developed, tested, and are freely available, the media campaign has a chance to succeed. It will then be possible for end-users, computer scientists, and numerical analysts to engage in productive dialogue regarding the utility of interval methods. Tools will exist to test algorithms, answer questions, and expose areas where improvements remain to be made.

General availability of interval implementations and interval spreadsheets will make it practical for application developers to begin writing and distributing general purpose interval applications. These applications will enable scientists, engineers, and members of the business community to solve problems that are otherwise difficult or numerically intractable. Through this kind of exposure,

[4] The author is unaware of an interval calculator program.

end-user demand for interval arithmetic will grow and system suppliers will begin to pay attention.

To support the media campaign it will be important to collect and document actual instances in which floating point methods have produced inaccurate results. Of course, sharp interval alternatives will need to be presented as well. This information can be collected in the interval web page [12].

# 7 ACTION PLAN

There are a number of actions that can be taken to help speed the process of attracting the attention of system suppliers. These include:

- Support the interval interest group. Baker Kearfott maintains an email alias for this purpose. (See [8] for the email address.)
- Become vocal on the Internet. Communicating with interval colleagues, while important, is not sufficient. Outside visibility is critical. Every opportunity must be taken to show how interval methods can be applied to solve problems of interest in other disciplines. To this end, outstanding problems and success stories should be posted on the new interval web page (See [12] for the URL).
- Freely contribute interval tools and applications to help create broader interest and demand. This can be done in such a way that later commercialization is not compromised. For example, limits can be imposed on the size of problems that can be solved with free demonstration versions of software. One suggestion to encourage wide distribution is for editors and referees of interval papers to strongly encourage authors to make their software freely available [2].

# Conclusion

Once sufficient end-user demand for interval methods has been created, system suppliers and application developers will compete to respond. The competition will be to deliver new tools and applications to satisfy users who will then accept nothing less than the information provided by sharp interval methods. The challenge for the interval community is to create the end-user demand.

# REFERENCES

[1] A. B. Babichev, O. B. Kadyrova, T. P. Kashevarova, A. S. Leshchenko, and A. L. Semenov, "UniCalc, A novel approach to solving systems of algebraic equations", *Proceedings of the International Conference on Numerical Analysis with Automatic Result Verifications. Lafayette, Louisiana, USA, February - March, 1993, Interval Computations*, 1993, No. 2, pp. 29–47.

[2] G. F. Corliss, Personal communication.

[3] R. Hammer, M. Neaga, and D. Ratz, *PASCAL-XSC, New Concepts for Scientific Computation and Numerical Data Processing*, In: E. Adams and U. Kulisch, *Scientific computing with automatic result verification*, Academic Press, New York, 1993, pp. 15–44.

[4] E. R. Hansen, "A generalized interval arithmetic", In: K. Nickel (ed.), *Interval mathematics*, Springer-Verlag, Lecture Notes in Computer Science, 1975, Vol. 29, pp. 7–18.

[5] E. R. Hansen, *Global Optimization Using Interval Analysis*, Marcel Decker Inc., New York, 1992.

[6] E. Hyvönen and S. De Pascale, "InC++ library family for interval computations", *Reliable Computing*, 1995, Supplement (Extended Abstracts of APIC'95: International Workshop on Applications of Interval Computations, El Paso, TX, Feb. 23–25, 1995), pp. 85–90.

[7] E. Hyvonen and S. De Pascale, "Interval constraint Excel", *Reliable Computing*, 1995, Supplement (Extended Abstracts of APIC'95: International Workshop on Applications of Interval Computations, El Paso, TX, Feb. 23–25, 1995), pp. 91–101.

[8] R. B. Kearfott, Interval interest group alias: " `reliable_computing@interval.usl.edu`".

[9] R. B. Kearfott, M. Dawande, M., K.-S. Du, and C.-Y. Hu, "Algorithm 737: INTLIB: A Portable FORTRAN 77 Interval Standard Function Library", *ACM Trans. Math. Software*, 1994, Vol. 20, No. 4, pp. 447–459.

[10] R. B. Kearfott, "A Fortran 90 Environment for Research and Prototyping of Enclosure Algorithms for Constrained and Unconstrained Nonlinear Equations", *ACM Trans. Math. Software*, 1995, Vol. 21, No. 1, pp. 63–78.

[11] R. Klatte, U. Kulisch, A. Wiethoff, C. Lawo, and M. Rauch, *C-XSC A C++ Class Library for Extended Scientific Computing*, Springer-Verlag, New York, 1993.

[12] V. Kreinovich, Interval Computations URL:
"`http://cs.utep.edu/interval-comp/main.html`".

[13] C. Lawo, "C-XSC A Programming Environment for Verified Scientific Computing and Numerical Data Processing", In: E. Adams and U. Kulisch, *Scientific Computing with Automatic Result Verification*, Academic Press, New York, 1993, pp. 71–86.

[14] R. E. Moore, *Interval Analysis*, Prentice-Hall, Englewood Cliffs, New Jersey, 1966.

[15] S. M. Rump, "Accuracy Control and Estimation, Self-Validating Systems and Software Environments for Scientific Computation", *IFIP Trans. A, Comput. Sci. Technol. (Netherlands)*, 1992, Vol. A-2, pp. 49–56.

[16] W. V. Walter, "FORTRAN-XSC: A Portable Fortran 90 Module Library for Accurate and Reliable Scientific Computing", *Computing* (Suppl.), 1993, Vol. 9, pp. 265–286.

[17] W. V. Walter, "ACRITH-XSC: A Fortran-Like Language for Verified Scientific Computing", In: E. Adams and U. Kulisch, *Scientific Computing with Automatic Result Verification*, Academic Press, New York, 1993, pp. 45–70.

[18] J. Wolff von Gudenberg, "Programming Language Support for Scientific Computation", *Interval Computations*, 1992, Vol. 4, pp. 116–126.

# INDEX

# Applied Optimization

1. D.-Z. Du and D.F. Hsu (eds.): *Combinatorial Network Theory.* 1996
ISBN 0-7923-3777-8
2. M.J. Panik: *Linear Programming: Mathematics, Theory and Algorithms.* 1996
ISBN 0-7923-3782-4
3. R.B. Kearfott and V. Kreinovich (eds.): *Applications of Interval Computations.* 1996
ISBN 0-7923-3847-2

KLUWER ACADEMIC PUBLISHERS – DORDRECHT / BOSTON / LONDON

GPSR Compliance

*The European Union's (EU) General Product Safety Regulation (GPSR) is a set of rules that requires consumer products to be safe and our obligations to ensure this.*

*If you have any concerns about our products, you can contact us on ProductSafety@springernature.com*

In case Publisher is established outside the EU, the EU authorized representative is:

Springer Nature Customer Service Center GmbH
Europaplatz 3
69115 Heidelberg, Germany

**Batch number: 10585156**

Printed by Printforce, the Netherlands